Preface

The 15th International Workshop on Graph–Theoretic Concepts in Computer Science (WG '89) was held at Castle Rolduc, Kerkrade, the Netherlands, near Aachen, Federal Republic of Germany, from June 14 to 16, 1989. It was organized by Lehrstuhl für Informatik III of Aachen University of Technology.

The successful tradition of the WG workshops which take place in alternate places in Central Europe every year is reflected by the list of predecessor workshops and their organizers:

Berlin	1975	U. Pape
Göttingen	1976	H. Noltemeier
Linz	1977	J. Mühlbacher
Castle Feuerstein near Erlangen	1978	M. Nagl & H. J. Schneider
Berlin	1979	U. Pape
Bad Honnef	1980	H. Noltemeier
Linz	1981	J. Mühlbacher
Neunkirchen near Erlangen	1982	H. J. Schneider & H. Göttler
Haus Ohrbeck near Osnabrück	1983	M. Nagl & J. Perl
Berlin	1984	U. Pape
Castle Schwanenberg near Würzburg	1985	H. Noltemeier
Monastery Bernried near München	1986	G. Tinhofer & G. Schmidt
Castle Banz near Bamberg	1987	H. Göttler & H. J. Schneider
Amsterdam	1988	J. van Leeuwen

The program committee of the WG '89 Workshop consisted of:

H. Bunke,	University of Berne, Switzerland
B. Courcelle,	University of Bordeaux, France
J. van Leeuwen,	University of Utrecht, The Netherlands
M. Nagl,	Aachen University of Technology, West Germany
H. Noltemeier,	University of Würzburg, West Germany
G. Schmidt,	University of German Forces, West Germany
H. J. Schneider,	University of Erlangen, West Germany

In the call for papers contributions were solicited describing original results in the study and application of graph–theoretic concepts in various fields of computer science. From 48 submissions the program committee selected 27 for presentation. The selection reflects several current research directions that are representative for the topic of the workshop. However, certainly not all aspects could be covered by a three–day workshop.

The present volume contains the revised versions of nearly all the papers presented at the workshop. The revisions are based on the comments and suggestions received by the authors from referees and/or participants of the workshop. Several papers are in the form of preliminary reports on ongoing research. It is expected that more elaborate versions of these papers will eventually appear in scientific journals. The members of the program committee hope that this volume gives a good impression of activities in the topic of graph–theoretic concepts in computer science.

The workshop was attended by 50 participants from 10 countries (Canada, France, Israel, Italy, Poland, Spain, Switzerland, The Netherlands, United States of America, Federal Republic of Germany). The success of the workshop is due to the activeness of participants contributing to the presentations and discussions, and to the work done by referees and especially by the members of the program committee.

The editor is grateful for financial support from the Aachen University of Technology. He would like to thank the Rector, Prof. Dr. Habetha, for this valuable help. Furthermore, sponsorship by the European Association for Theoretical Computer Science (EATCS) and the Gesellschaft für Informatik (GI) is thankfully acknowledged. The friendly atmosphere of this workshop was in great part due to the attractive workshop location in Castle Rolduc. We would like to thank the personnel of this education center who looked after us so well. Last but not least the editor would like to say thanks to the members of the organization committee consisting of M. v.d. Beeck, U. Cordts, A. Fleck, M. Hirsch, Th. Janning, N. Kiesel, M. Lischewski, U. Schleef, A. Schürr, C. Weigmann, B. Westfechtel and some students of computer science at Aachen University of Technology.

Aachen, October 1989   *M. Nagl*

The following persons helped as referees to select the papers for presentation:

A. Arnold	J. G. Penaud
L. Bayer	J. Perl
R. Berghammer	G. Reich
J. Bond	W. Ries
H. Bunke	A. Scheuing
B. Courcelle	I. Schiermeyer
S. Dulucq	J. Schmid
J. Dvorak	G. Schmidt
J. Ebert	L. Schmitz
T. Glanser	H. J. Schneider
E. Gmür	J. Schreiber
D. Janssens	A. Schürr
X.-Y. Jiang	W. Thomas
P. Kropf	G. Tinhofer
M. Kunde	J. van Leeuwen
M. Ley	W. Vogler
Th. Ludwig	F. Wagner
K. Mehlhorn	B. Walter
M. Nagl	B. Westfechtel
H. Noltemeier	P. Widmayer
W. Oberschelp	

Contents

Graph Embeddings

Algorithmic Graph Theory

Graphs and Data Structures

Foreword
The Aim of the WG Workshops

The aim of this workshop series is to contribute to integration in computer science. This striving for integration is achieved by applying graph–theoretic concepts. Thereby integration appears in two ways. First, graph–theoretic concepts are applied to various fields of specialization in computer science and thus commonalities between the fields are detected. Second, the workshops aims to combine theoretical aspects with practice and applications; this is achieved either by applying theoretical concepts to practice or by taking up problems from practice and trying to solve them theoretically.

This workshop series is thus a rarity in computer science, as it is neither purely theoretical nor practical or oriented to applications. It is also a vertical cut through the different fields of computer science in which graphs and graph–theoretic concepts can be applied. On the other hand the workshop has a tradition: this volume collects the papers of its fifteenth occurrence. That alone is remarkable in a field like computer science where new topics appear and disappear quite rapidly.

Looking into this volume and its predecessors (see the list on the last page of this volume) one can see that there are many disciplines of computer science where graph–theoretic concepts can be applied and have been applied. The volumes present applications in data structures, databases, information retrieval, software engineering, design of software systems, compiler construction, construction of intelligent software tools, design of inherently parallel hardware architectures as in VLSI layout, distributed systems, process control, concurrency, communication, graphics, computational geometry, computer–aided design, computer vision, and design of algorithms and investigation of complexity, for example in operations research and optimization and in pattern recognition. In some cases applications outside computer science in such fields as chemistry, biology, geology have also been presented.

Graphs are used in these disciplines for different purposes. They are used as an appropriate level of abstract representation at which certain problems can be studied and at which certain details of representation and implementation are suppressed. Thus for a given problem a suitable graph representation has to be found. As such a problem belongs to a certain application area or to a certain class of problems, the graph is a member of a certain class of graphs.

For this class of graphs certain structural properties and/or the complexity of corresponding algorithms can be investigated. An example of a structural property of graphs and a corresponding class of graphs is given by the bounded degree graphs. An example of a particular algorithm is the Fairley algorithm. In such cases a class of graphs is given by static properties, or by algorithms with a certain complexity. We call this branch of graph–theoretic concepts *structural graph theory*.

In many application fields the graphs are not static but rather change dynamically. In these applications not only has a suitable class of graphs to be detected but also the dynamic behavior of changes has to be formally specified. Changes may belong to structural properties (for example specifying a complex insertion in a database) or they may belong to some state of execution (such as firing a transition, or reevaluating attributes of entities). In the first case graph rewriting systems (graph grammars) are a suitable formal method to describe changes, and in the second case Petri nets and attributed grammars are well known methods. Let us call this branch of graph–theoretic concepts *conceptual graph modeling*.

Solving practical problems which are regarded as problems on graphs is dramatically facilitated if suitable means of implementation are available. In our case these means should help in implementing graph classes, algorithms on graphs, changes on graphs and so on. This help may be restricted to implementing a special graph class or it may extend to implementing arbitrary classes. In this branch we find hints, mechanisms, methods, notions, tools, or basic software systems which facilitate the implementation of a general graph solution or a special solution based on graphs. Let us call this branch of graph–theoretic concepts *graph implementation mechanisms*. In this category we find graph programming languages, special data structures for graphs, special nonstandard database systems for graphs and so on.

Summing up, the importance of graphs is that we can study structural properties, special algorithms, structural or value changes, and implementation mechanisms rather separately from the problem or the application area they originate from. This level of abstraction makes it easier to get structural results independent of application or to characterize different specific occurrences of a more general problem. To give an example, phenomena like parallelism or concurrency can be adequately studied and characterized independently of applications. This formulation is abstract relative to its representation in an application area and independent of its applications in the sense that the same phenomenon is detected in different application areas. Thus the formulation helps to detect similarities of problems and solutions, or to detect uniform mechanisms which solve classes of problems rather than single problems.

The importance of graphs is growing and we have accumulated a lot of knowledge on how to handle graphs in the last fifteen years in structural graph theory, in conceptual graph modeling and in graph implementation mechanisms. On the other hand it is increasingly obvious that specialized structures like trees, even with attributes, can only serve as an adequate underlying model for limited classes of problems. The growing importance of graphs is stressed in many conferences and journals. In this sense, this workshop is quite modern despite its long tradition.

M. Nagl

DISJOINT PATHS IN THE HYPERCUBE

Seshu Madhavapeddy and I. Hal Sudborough
Computer Science Program
University of Texas at Dallas
Richardson, Texas 75083-0688

1. INTRODUCTION

In this paper we study the problem of finding disjoint paths in a hypercube. The motivation is to maximize the number of non-intersecting two way communication paths that can be kept open simultaneously between mutually disjoint pairs of nodes, irrespective of which nodes in the hypercube are chosen. That is, we consider the problem in which one is given a collection (s_1, g_1), (s_2, g_2), ..., (s_k, g_k), for some $k \geq 1$, of pairs of processors in a hypercube network and asked to determine if there are k node or edge disjoint paths connecting s_i to g_i for each i, $1 \leq i \leq k$. We thus study the possibility of routing messages in a hypercube network so that no two messages pass through the same processor or connecting wire.

A hypercube of dimension n, $Q(n)$, can be represented as an undirected graph consisting of 2^n vertices labeled from 0 to $2^n - 1$. There is an edge between two vertices i and j if and only if the binary representations of i and j differ in exactly one bit position (*Figs* (1), (2) & (3)). The hypercube $Q(n)$ has several interesting properties [SS]. It is a vertex n-connected graph and hence by Menger's theorem there exist n vertex disjoint paths between any two nodes of the hypercube [HW]. These paths can be generated quite easily in time polynomial in n or polylogarithmic in the size of the hypercube [SS]. Also $Q(n)$ can be separated into two $(n-1)$ dimensional hypercubes such that the nodes of the two are in one-to-one correspondence and the nodes thus paired up are connected by an edge.

We consider the case of finding disjoint connecting paths in a hypercube, as hypercubes are an increasingly popular architectural style for parallel interconnection networks. Often applications on such networks are hindered by the need for independent processors at some distance to communicate with each other. This communication may result in performance degradation due to too many messages being routed through the same intermediate processor. The problem of avoiding bottlenecks in communication, particularly in hypercube networks, has been studied several times before. Perhaps the most well known routing algorithm is a probabilistic one by Valiant [LGV]. However Valiant's algorithm allows collisions, meaning more than one message may be routed through an edge at the same time. This is handled by introducing queues at each node. Valiant has shown that the algorithm's performance is good in the sense that with high probability there will be few collisions. The permutation networks of Benes [VEB] can be adapted to the hypercube to achieve collision free routing of messages. Predictably this gives us an algorithm that is much slower than Valiant's. In this paper we study the problem of routing under more strict criteria. We explore the possibility of routing under the restriction that no two messages use the same (a) node or (b) edge right through the entire routing process. This would be the same as finding node/edge disjoint paths in a graph. The problem of finding node disjoint paths between pairs of nodes in a graph has long been studied. It is formally stated as:

DISJOINT CONNECTING PATHS [GJ]

Instance: Graph $G=(V,E)$, collection of disjoint node pairs (s_1,g_1), (s_2,g_2), ..., (s_k,g_k).

Question: Does G contain k mutually node disjoint paths, one connecting s_i and g_i for each i, $1 \leq i \leq k$?

This problem has been shown to be NP-complete [RMK]. However Shiloach [YS] proved that the problem can be solved in polynomial time for the case $k=2$. Watkins [MW] gave some pointers for the case $k=3$. Recently Robertson and Seymour have given a polynomial time algorithm for any fixed k. However the algorithm is impractical [RS]. In this paper we study a restricted version of the above problem, the case when graph G is a hypercube. We prove that in a hypercube of dimension n we can always find, in polynomial time, node disjoint paths between $\lfloor n/2 \rfloor$ pairs of nodes and edge disjoint paths between n pairs of nodes irrespective of which nodes of the hypercube are chosen as the source and goal nodes as long as they are all distinct.

2. NODE DISJOINT PATHS

The polynomial time algorithm to find node disjoint paths between mutually disjoint node pairs (s_1,g_1), (s_2,g_2), ...(s_k,g_k), $k \leq \lfloor n/2 \rfloor$, in $Q(n)$ is based on the fact that we can find the n node disjoint paths between any two nodes in $Q(n)$ very efficiently [SS]. First we prove the existence of these k paths. Watkins [MW] proves that the best result we can expect is $\lceil n/2 \rceil$ node disjoint paths. Hence our result is within one of optimal.

Definition: Let $n \geq 0$. A graph $G=(V,E)$ satisfies property P_n if given any $2n$ distinct vertices $s_1, s_2, ...s_n, g_1, g_2, ..., g_n \in V$, there exist n node disjoint paths $p_i[s_i, g_i]$ $(i=1,...n)$, in G such that p_i is a path from s_i to g_i.

Definition: Let $n \geq 0$. A graph $G=(V,E)$ satisfies property FP_n if given any $2n+1$ distinct vertices $s_1, s_2, ..., s_n, g_1, g_2, ..., g_n, x \in V$, there exist n node disjoint paths $p_i[s_i, g_i]$ $(i=1,2,...,n)$, in G such that none of the paths p_i $(1 \leq i \leq n)$ pass through the forbidden node x.

Lemma 2.1: $Q(2)$ satisfies P_1. $Q(3)$ satisfies P_1 and FP_1.

Proof: $Q(2)$, shown in $Fig(1)$, is a connected graph. Hence given any two vertices of $Q(2)$ as s_1, g_1, there is a path between them and it can be found in polynomial time [EWD]. $Q(3)$ is a connected graph as well. Hence by the same argument, it too satisfies P_1. $Q(3)$, shown in $Fig(2)$, is 3-connected. So, given s_1, g_1 and x, by Menger's theorem there exist three node disjoint paths from s_1 to g_1 and they can be generated efficiently [SS]. Now at most one of these paths can pass through the forbidden node x, in which case one of the other two paths is the solution.

Lemma 2.2: Given a $n \times m$ $(n \geq 2, m \geq n)$ 0/1 matrix such that all the n rows are distinct, there exists a column i $(1 \leq i \leq m)$ such that in the $n \times (m-1)$ 0/1 matrix obtained after the elimination of column i, the n rows are still distinct. Moreover, given a $n \times m$ matrix, we can find such a column in polynomial time.

Proof (by induction): Consider the case when $n=2$ as the basis. So, we have a $2 \times m$ $(m \geq 2)$ matrix where the rows are distinct. That is, they differ in at least one column. If they differ in more than one column position then we can eliminate any one of the columns and still the rows will remain distinct. If they differ in only one column then we can safely eliminate any one of the other columns. As induction hypothesis let us assume that the lemma is true for $n=k-1$. Then consider a $k \times m$ $(m \geq k)$ 0/1 matrix which has k distinct rows. In this matrix let us strike out the mth column. If the k rows are still distinct, then we have already found the column we are looking for. If they are not, then suppose p $(k/2 \leq p \leq k-1)$ distinct rows remain, the rest being duplicates. Now, by the induction hypothesis, we can find a column in the remaining $p \times (m-1)$ matrix such that the elimination of that column, say j, will still leave the p rows distinct. To this $p \times (m-2)$ matrix we now add back the $(k-p)$ duplicate rows along with column m to get a $k \times (m-1)$ 0/1 matrix in which all the rows are distinct. Thus column j is the solution. We can find this column, whose existence has been established, in time polynomial to the size of matrix by exhaustive search. That is, we single out one column at a time in the $n \times m$ matrix, throw it away and see if the n rows still remain distinct.

Lemma 2.3: If $Q(n)$ satisfies P_k $(k \leq \lfloor n/2 \rfloor)$ then for any $m > n$, $Q(m)$ satisfies P_k.

Proof: This statement simply follows from the recursive nature of the construction of a hypercube and Lemma 2.2. Given $Q(m)$ and (s_1,g_1), (s_2,g_2), ..., (s_k,g_k), form a $(2k \times m)$ 0/1 matrix with these nodes stored as rows of the matrix in the order $s_1,g_1,s_2,g_2,....,s_k,g_k$ (note that in $Q(m)$, each node is represented by a m-bit binary string). Then by Lemma 2.2, we can eliminate $(m-n)$ columns, leaving all rows in the $(2k \times n)$ matrix that remains distinct. Call the n-bit rows that remain $s'_1,g'_1,s'_2,g'_2,.......,s'_k,g'_k$. Since $Q(n)$ satisfies P_k, one can find node disjoint paths in $Q(n)$ between (s'_1,g'_1), (s'_2,g'_2), ..., (s'_k,g'_k). Then find paths (s_1,s'_1), (g_1,g'_1), (s_2,s'_2), (g_2,g'_2), ..., (s_k,s'_k), (g_k,g'_k) each in its own separate $Q(m-n)$ subcube of $Q(m)$. For example, we find the path (s_1,s_1^1) in the subcube of $Q(m)$ obtained by fixing the n-bits that are common to s_1 and s_1^1. By combining the two sets of paths we found, we get the k disjoint paths in $Q(m)$.

Lemma 2.4: $Q(n)$, the hypercube of dimension n, satisfies P_u and FP_v where $u \leq \lfloor n/2 \rfloor$ and $v \leq \lceil n/2 \rceil - 1$.

Proof (by induction): By Lemma 2.1, we know that the statement is true for $n=2,3$.

As the induction hypothesis, let us suppose it is true for $n \leq m$ where m is an even number. Then in the induction step we will have to prove that (1) $Q(m+1)$ satisfies $FP_{(m/2)}$ and (2) $Q(m+2)$ satisfies $P_{(m/2)+1}$.

Case (1): The problem $FP_{(m/2)}$ specifies $m+1$ distinct nodes $s_1, s_2, \ldots, s_{(m/2)}, g_1, g_2, \ldots, g_{(m/2)}, x$. Each of these nodes is represented by a unique binary string of length $(m+1)$ in the hypercube $Q(m+1)$. Thus, these strings can be visualized as constituting a $(m+1) \times (m+1)$ 0/1 matrix where each row of the matrix represents one of the hypercube nodes. According to Lemma 2.2, we can locate a column in this 0/1 matrix, the removal of which still leaves the $(m+1)$ rows distinct. Let this be the ith column. Now, let us split $Q(m+1)$ into two subcubes $Q_0(m)$ and $Q_1(m)$ along the ith dimension. $Q_0(m)$ consists of all the nodes in $Q(m+1)$ which have a "0" in the ith position and $Q_1(m)$ consists of all the nodes with "1" in the ith position. Consider the m nodes specified as the source and goal nodes and the nodes obtained by toggling the ith bit position of these nodes. Clearly half of these $2m$ nodes belong to $Q_0(m)$ and the other half to $Q_0(m)$. Name these $2m$ nodes as $s_1^0, s_1^1, s_2^0, s_2^1, \ldots, s_{(m/2)}^0, s_{(m/2)}^1, g_1^0, g_1^1, \ldots, g_{(m/2)}^0, g_{(m/2)}^1$ where all the nodes with superscript "0" belong to $Q_0(m)$ and superscript "1" belong to $Q_1(m)$. Also these $2m$ nodes and the node x are all distinct in $Q(m+1)$. Each s_j, $(1 \leq j \leq m/2)$ is either s_j^0 or s_j^1, the other being s_j^i, the node obtained by toggling the ith bit position in s_j. The same is true of all g_j. Under this scenario x can occur either in (a) $Q_0(m)$ or in (b) $Q_1(m)$.

(a) If x is in $Q_0(m)$ (*Fig (4)*), then find the $m/2$ disjoint paths $p_j[s_j^1, g_j^1]$ $(1 \leq j \leq m/2)$ in $Q_1(m)$. (where each s_j^1 is either s_j or s_j^i and each g_j^1 is either g_j or g_j^i). This can be done by the induction hypothesis. Then these node disjoint paths along with the edges in the ith dimension joining s_j^1 to s_j^0 $(1 \leq j \leq m/2)$ and g_j^1 to g_j^0 $(1 \leq j \leq m/2)$ give us the $m/2$ node disjoint paths

$p_j[s_j\,,\,g_j]$ $(1 \leq j \leq m/2)$. None of these paths pass through x.

(b) When x is in $Q_1(m)$ we find the $m/2$ disjoint paths $p_j[s_j^0\,,\,g_j^0]$ in $Q_0(m)$.

Case (2): The problem $P_{(m/2)+1}$ specifies $m+2$ nodes $s_1, s_2, \ldots, s_{(m/2)+1}, g_1, g_2, \ldots, g_{(m/2)+1}$. Since each of them is represented by a unique binary string of length $(m+2)$, by Lemma 2.2, we can find a position "i" $(1 \leq i \leq m+2)$, the removal of which from all the $(m+2)$ nodes will still leave them distinct from each other. Let us now divide $Q(m+2)$ into the subcubes $Q_0(m+1)$ and $Q_1(m+1)$ along the ith dimension. Then just as in case (1), we will have $2m+4$ nodes $s_1^0, s_1^1, s_2^0, s_2^1, \ldots, s_{(m/2)+1}^0, s_{(m/2)+1}^1, g_1^0, g_1^1, \ldots, g_{(m/2)+1}^0, g_{(m/2)+1}^1$. Half of them, the ones with superscript "0", belonging to $Q_0(m+1)$ and the ones with superscript "1" belonging to $Q_1(m+1)$. Then the following configurations are possible:

(a) There exists an "l" $(1 \leq l \leq (m/2)+1)$, such that $s_l^0 = s_l$ and $g_l^0 = g_l$. In this case we find the $(m/2)+1$ node disjoint paths in the following fashion. Find a path $p_l[s_l, g_l]$ in $Q_0(m+1)$ that does not pass through any of the other m marked nodes in it (note that this does not take any more time than generating the $m+1$ node disjoint paths in $Q(m+1)$ between s_l^0 and g_l^0 which are guaranteed by Menger's theorem). Then in $Q_1(m+1)$ find the $m/2$ node disjoint paths $p_j[s_j^1, g_j^1]$ $(1 \leq j \leq (m/2)+1\,,\, j \neq l)$ (by induction hypothesis). These paths along with the edges in the ith dimension joining nodes in $Q_0(m+1)$ to those in $Q_1(m+1)$ give us the $(m/2)+1$ node disjoint paths we want.

(b) All the s_l $(1 \leq l \leq (m/2)+1)$ are in $Q_0(m+1)$ and all g_l $(1 \leq l \leq (m/2)+1)$ are in $Q_1(m+1)$. Under this situation (Fig (5)) find the path $p_{(m/2)+1}[s_{(m/2)+1}\,,\, g_{(m/2)+1}^0]$ in $Q_0(m+1)$ which does not pass through any of the other m marked nodes in it. This path along with the hypercube edge in the ith dimension connecting $g_{(m/2)+1}^0$ to $g_{(m/2)+1}$ gives us the path $p_{(m/2)+1}[s_{(m/2)+1}\,,\, g_{(m/2)+1}]$. Now find $p_j[s_j^1\,,\, g_j]$ $(1 \leq j \leq m/2)$ in $Q_1(m+1)$ such that none of them pass through $g_{(m/2)+1}^1$ (possible, as proved in case (1)). These paths added with the edges in the ith dimension, (s_j, s_j^1) $(1 \leq j \leq m/2)$ give us the other $m/2$ node disjoint paths $p_j[s_j\,,\, g_j]$ $(1 \leq j \leq m/2)$.

All the possible configurations fall into one of the categories discussed above. Hence the Lemma. Based on this, we can develop a divide-and-conquer, polynomial time algorithm to find k node disjoint paths between (s_1, g_1), (s_2, g_2), ..., (s_k, g_k) in $Q(n)$ when $k \leq \lfloor n/2 \rfloor$ such that the maximum length of any of the paths is no more than $2n$.

3. EDGE DISJOINT PATHS

Node disjoint paths are also edge disjoint. So, if we are trying to find k edge disjoint paths in $Q(n)$, $k \leq \lfloor n/2 \rfloor$, we can use the same algorithm given in the previous section. However, we can guarantee the existence of up to n edge disjoint paths in $Q(n)$ $(n \geq 3)$, irrespective of which of its nodes are chosen as source and goal nodes.

Definition: Let $n \geq 0$. A graph $G = (V, E)$ satisfies property EP_n if given any $2n$ distinct vertices $s_1, s_2, ..., s_n, g_1, g_2, ..., g_n \in V$, there exist n edge disjoint paths $p_i[s_i, g_i]$ $(i = 1, .., n)$ in G, such that p_i is the path from s_i to g_i.

In this section we essentially prove that $Q(n)$ $(n \geq 3)$, satisfies EP_k $(k \leq n)$ much in the same way as in section 2 we proved $Q(n)$ satisfies P_q $(q \leq \lfloor n/2 \rfloor)$. The proof of Lemma 3.2 given below is omitted because it is similar to Lemma 2.3 and follows the same line of argument. Lemma 3.3 is proven by exhaustive search.

Lemma 3.1: Given a $2n \times m$ $(n \geq 2$, $m \geq n)$ 0/1 matrix such that all the $2n$ rows are distinct, there exists a column "i" $(1 \leq i \leq m)$ such that in the $2n \times (m-1)$ 0/1 matrix obtained after the elimination of column i, there still remain at least $(2n-2)$ distinct rows. Moreover, given a $2n \times m$ matrix, we can find such a column in polynomial time.

Proof (by contradiction): By inspection we can see that there are no 4×2 or 6×3 0/1 matrices such that all the rows in the matrix are distinct and elimination of any column yields at least 3 pairs of identical rows. So, the Lemma is true for $n = 2$, 3. However, assume that there exists a $k \times n$ matrix $(n > 3$, $k \leq 2n)$ such that elimination of any column yields at least 3 pairs of identical rows. Let n be the smallest integer for which such a matrix exists and let k be the least number of rows for that value of n. This implies that there exists a column in the matrix whose removal yields exactly 3 pairs of identical rows. Without loss of generality, let it be column "n".

Then in the $k \times n-1$ matrix obtained after eliminating column n let us suppose there are two copies of rows x, y and z (i.e; there are rows $x0$, $x1$, $y0$, $y1$, $z0$ and $z1$ in the $2n-k \times n$ matrix). After eliminating the identical rows, we will have a $k-3 \times n-1$ matrix in which all the rows are distinct. Now in this matrix there are three possibilities:

(a) Elimination of any of the $n-1$ columns yields at least 3 pairs of identical rows.

(b) There is a column "i" which would yield 3 pairs of identical rows in the original $k \times n$ matrix but now in the $k-3 \times n-1$ matrix yields only 2 pairs of identical rows. This could occur if say $y = x^i$ or any such analogous situation (x^i represents the row obtained by toggling the ith column of row x). In such a case we can ensure that column i also yields 3 identical pairs of rows by adding row z^i (unless its already present).

(c) Suppose there are two such columns, "i" and "j" such that $y = x^i$ and $z = x^j$. Then we can ensure that both columns yield 3 identical pairs by adding the row x^{ij} to the matrix (unless its already present).

So, we can definitely get a $k-2-p \times n-1$ matrix ($p = 0$ or 1) such that all the rows of the matrix are distinct and the elimination of any column yields at least 3 pairs of identical rows. But $k-2-p \leq 2(n-1)$. Hence n is not the smallest possible integer for which the Lemma is not true (contradiction).

Lemma 3.2: If $Q(n)$ satisfies EP_k $(k \leq n)$ then for any $m > n$, $Q(m)$ satisfies EP_k.

Lemma 3.3: $Q(3)$ satisfies EP_3.

Lemma 3.4: $Q(n)$ $(n \geq 3)$, the hypercube of dimension n, satisfies EP_u where $u \leq n$.

Proof (by induction): The proof of Lemma 3.4 is based on mathematical induction and depends on

Lemma 3.1 and Lemma 3.2 and is quite like the proof of Lemma 2.4. However it is complicated by the fact that upon the elimination of column "i", all the rows may not remain distinct. By Lemma 3.3, we know that the statement is true for $n = 3$. For the induction step, let us suppose it is true for any $n \leq m$. Then we prove that $Q(m+1)$ satisfies EP_{m+1}.

The problem EP_{m+1} specifies $2m+2$ distinct nodes $s_1, s_2, \ldots, s_{m+1}, g_1, g_2, \ldots, g_{m+1}$. Each of these nodes is represented by a unique binary string of length $m+1$ in the hypercube $Q(m+1)$. Thus these strings can be visualized as constituting a $(2m+2) \times (m+1)$ 0/1 matrix where each row of the matrix represents one of the hypercube nodes. According to Lemma 3.1 we can find a column, say "i", whose removal from the matrix still leaves at least $2m$ rows distinct. Now let us split $Q(m+1)$ into two subcubes $Q_0(m)$ and $Q_1(m)$ along the ith dimension. Consider the $(2m+2)$ nodes specified in EP_{m+1} and the nodes obtained by toggling the ith bit of these nodes. Half of these $2 \times (2m+2)$ nodes belong to $Q_0(m)$ and the other half to $Q_1(m)$. Let us rename these nodes as $s_1^0, s_1^1, s_2^0, s_2^1, \ldots, s_{m+1}^0, s_{m+1}^1, g_1^0, g_1^1, \ldots, g_{m+1}^0, g_{m+1}^1$ where all the nodes with superscript "0" belong to $Q_0(m)$ and those with superscript "i" belong to $Q_1(m)$. Now the following three cases could occur.

Case (1): Suppose the removal of column i leaves all the $2m+2$ rows distinct.

Then find edge disjoint paths $P_j[s_j^0, g_j^0]$ $(2 \leq j \leq m+1)$ in $Q_0(m)$, which can be done by the induction hypothesis and find a path $p_1[s_1^1, g_1^1]$ in $Q_1(m)$. These paths and the ith dimension edges which join nodes in $Q_0(m)$ and $Q_1(m)$ will together give the $m+1$ edge disjoint paths in $Q(m+1)$.

Case (2): There is one pair of identical rows after removal of column "i".

That is, suppose $s_l = g_j$, where $(1 \leq l, j \leq m+1)$. Then $s_l^0 = g_j^0$ and $s_l^1 = g_j^1$. Without loss of generality, assume $s_l = s_l^0$ and $g_j = g_j^1$. Then find edge disjoint paths $p_k[s_k^0, g_k^0]$ $(1 \leq k \leq m+1, k \neq j)$ in $Q_0(m)$ (by Induction hypothesis) and $p_j[s_j^1, g_j^1]$ in $Q_1(m)$. These paths along with the ith dimension edges give us the $m+1$ paths in $Q(m+1)$. The point is that the ith dimension edges of $Q(m+1)$ have not been used by any of the paths we found in $Q_0(m)$ and $Q_1(m)$.

Case (3): There are two pairs of identical rows after removal of column "i".

(a)   Suppose $s_l = g_j$ and $s_j = g_l$.   Then $s_l^0 = g_j^0$, $s_j^0 = g_l^0$, $s_l^1 = g_j^1$, $s_j^1 = g_l^1$.   If $s_l = s_l^0$, $g_l = g_l^0$, $s_j = s_j^1$, $g_j = g_j^1$ this simply reduces to Case (1) where we find $p_j[s_j, g_j]$ in $Q_1(m)$ and the rest of the paths in $Q_0(m)$. So, let us assume with out loss of generality that $s_l = s_l^0$, $g_l = g_l^1$, $s_j = s_j^0$, $g_j = g_j^1$ (Fig (6)).   Then   find $m-1$ edge   disjoint   paths $p_k[s_k^0, g_k^0]$ $(1 \leq k \leq m+1$, $k \neq l,j)$ in $Q_0(m)$ (by Induction hypothesis and Lemma 3.2) and two edge disjoint paths between g_j^1 and g_l^1 in $Q_1(m)$ (by a variation of Menger's Theorem). These paths along with edges in i^{th} dimension give us the $m+1$ edge disjoint paths.

(b) Suppose $s_l = g_j$, $s_j = g_k$. If $(s_j = s_j^0$, $g_j = g_j^0)$ or $(s_j = s_j^1$, $g_j = g_j^1)$. w.l.o.g assume $s_l = s_l^0$, $s_j = s_j^0$, $g_j = g_j^1$, $g_k = g_k^1$.   Then   as   in   (a),   find $m-1$ edge   disjoint   paths $p_q[s_q^0, g_q^0]$ $(1 \leq q \leq m+1$, $q \neq j,k)$ and find edge disjoint paths $(g_k^1, g_j^{1)}$ and (s_k^1, g_k^1) in $Q_1(m)$ (by a variation of Menger's theorem). Then adding the ith dimensional edges if necessary, we get $p_q[s_q, g_q]$ $(1 \leq q \leq m+1)$ in $Q(m+1)$.

(c) Suppose $s_r = g_j$, $s_k = g_l$. w.l.o.g assume $s_r = s_r^0$, $s_k = s_k^0$, $g_j = g_j^1$, $g_l = g_l^1$. Then solution is as shown in (Fig (7)).

All the possible node assignments in $Q(m+1)$ fall into one of the categories discussed above. Hence the Lemma. We can write a polynomial time recursive procedure for finding k edge disjoint paths between (s_1, g_1) , (s_2, g_2) ,......., (s_k, g_k) in $Q(n)$ $(n \geq 3$, $k \leq n)$ based on Lemma 3.4. This procedure would closely follow the proof of the Lemma and the maximum length of any path is $2n$.

4. NP-COMPLETENESS

In this section we prove that the problem of finding edge disjoint paths in a hypercube is NP-complete. Unlike the "Disjoint Connecting Paths" problem in which all the source and goal nodes are distinct, in this case each node in the hypercube can be the source or goal node for several paths.

HYPERCUBE EDGE DISJOINT PATHS (HEDP)

Instance: Hypercube $Q(n)$, collection of node pairs (s_1, g_1), (s_2, g_2), ..., (s_k, g_k). (The $2k$ nodes may not be distinct)

Question: Does $Q(n)$ contain k mutually edge disjoint paths, one connecting s_i and g_i for each i, $1 \leq i \leq k$?

We reduce the NP-complete problem 1-in-3SAT into hypercube edge disjoint paths (HEDP). 1-in-3SAT denotes the problem in which, given a predicate logic wff in conjunctive normal form such that each clause contains exactly three literals, we have to find out if the wff is satisfiable in such a manner that exactly one literal in each clause is true.

Let us suppose we have an instance of 1-in-3SAT which has n variables and m clauses. Then we construct an instance of HEDP which has a solution if and only if 1-in-3SAT has a solution. The HEDP is defined over the hypercube $Q(\lceil \log (2n+1) \rceil + \lceil \log (8m+16) \rceil)$. The number of source-goal node pairs we will define between which edge disjoint paths have to be found will be less than $(\lceil \log (2n+1) \rceil + \lceil \log (8m+16) \rceil) \times 2^{\lceil \log (2n+1) \rceil + \lceil \log (8m+16) \rceil - 1}$. This number is in fact the number of edges in the hypercube. Hence the size of the HEDP is polynomial in the size of 1-in-3SAT.

Let us consider the hypercube laid out as a grid having $2^{\lceil \log (2n+1) \rceil}$ rows and $2^{\lceil \log (8m+16) \rceil}$ columns such that each row and column is in itself a hypercube of smaller dimension ($Fig\ 3$ depicts $Q(5)$ laid out as a 4×8 grid). Now, each variable in 1-in-3SAT is assigned two consecutive rows of this grid, one representing the "true" assignment and other the "false" assignment of the variable. Each clause is represented by eight consecutive columns. We define a source-goal node pair for each variable and the path between them will have to traverse across the breadth of the grid passing through the set of columns representing each clause. Initially each "variable" path can either proceed in the direction corresponding to the "true" assignment (positive literal) of the variable or to the "false" assignment (negative literal). But once it chooses a particular path it will not be able to jump to the other path. Each path corresponding to a particular literal will be routed through the eight columns belonging to a clause as follows:

(a) If the variable is not present in the clause then it directly jumps from column 1 to column 8 (*Fig 8(a)*).

(b) If the literal is present in the clause then it is routed as shown in *Fig 8(b)*.

(c) If the opposite literal is present in the clause then it is routed as in *Fig 8(c)*.

So, essentially in each clause three paths are going to jump from column 3 to column 6 and the three paths corresponding to the literals which are opposite to the ones in the clause are going to pass through the edges joining column 4 and column 5. The rest of the paths just jump straight from column 1 to column 8. In each clause we make sure that there are only two edges joining columns 4 and 5 that are available for these "variable" paths to pass through, making it essential for at least one of the three paths corresponding to the variables in the clause to "jump" from column 3 to column 6. Which means that this variable satisfies the clause. In *Fig* (9) we can see how the clauses are built. Note that each clause "i" has source-goal node pair (s_i^0, g_i^0). This ensures that there are only two edges from column 4 to column 5 that are available for the variables. Each s_i^0 is connected by disjoint paths to the three nodes which are on the left side of the edges joining column 4 to column 5 in clause "i". Similarly g_i^0 is connected to the three right side nodes. These disjoint paths are found using the nodes in the respective columns. This can be done as long as each column has at least eight nodes, which is a fair assumption as this only means there are at least two variables in the instance of 1-in-3SAT. Similarly (s_i^1, g_i^1) and (s_i^2, g_i^2) ensure that only one "variable" path can jump from column 3 to column 6. Hence only one literal in each clause can be true.

In *Fig* (9) we show an instance of 3SAT and the corresponding HEDP problem. The source-goal node pairs (a_b, a_f), (b_b, b_f), (c_b, c_f) correspond to the variables A,B,C. The pairs (s_1^0, g_1^0), (s_1^1, g_1^1), (s_1^2, g_1^2), (s_2^0, g_2^0), (s_2^1, g_2^1), (s_2^2, g_2^2) ensure that there are only two "free" edges between column 4 and column 5 and one "free" edge between column 3 and column 6 in each of the clauses 1 and 2. The hypercube edges not shown in the picture are rendered unavailable for any of the paths mentioned above by defining the two nodes connected by each of these edges as a source-goal node pair. We can very easily verify that this method of construction gives us an instance of HEDP which has a solution if and only if the corresponding instance of 1-in-3SAT has a solution. Moreover the transformation takes polynomial time. Hence HEDP is NP-complete.

5. CONCLUSION

This paper represents preliminary work done in this area. Watkins [MW] proves that a graph can possibly satisfy P_n only if it is vertex $(2n-1)$ connected. Hence it would be interesting to see if we can improve upon Lemma 2.4 and prove that $Q(2n-1)$ satisfies P_n (In fact we have recently proved this). It is also our conjecture that the general problem of finding k node disjoint paths in $Q(n)$, $(k \leq 2^{n-1})$ is also NP-complete. Further work could be done in improving the results obtained regarding edge disjoint paths as well. There is scope for improving the algorithms to include situations where the existence of the paths is not guaranteed. We can also work at reducing the bound on the length of the paths, so they would be a function of the Hamming distance between the source and goal nodes rather than the size of the hypercube.

6. REFERENCES

[LGV] L.G. Valiant, "A Scheme for Fast Parallel Communication", SIAM J. Comput., May 1982, pp 350-361.

[VEB] V.E. Benes, "Mathematical Theory of Connecting Networks and Telephone Traffic", Academic Press, New York 1965.

[GJ]   Michael R. Garey & David S. Johnson, "Computers and Intractability", W.H. Freeman and Company.

[RMK] R.M. Karp, "On the Computational Complexity of Combinatorial Problems", Networks, 5: pp 45-68.

[YS]   Y. Shiloach, "The Two Paths Problem is Polynomial", Technical Report, Stanford University, STAN-CS-78-654.

[MW] Mark E. Watkins, "On the Existence of Certain Disjoint Arcs in Graphs", Duke Math. Journal, 1968.

[HW] H. Whitney, "Congruent Graphs and the Connectivity of Graphs", Amer. J. Math. 54 (1932), 150-168.

[EWD] E.W. Dijkstra, "A note on two Problems in connection with Graphs", Numerische Mathematik, 1 (1959) pp 269-271.

[SS] Youcef Saad and Martin H. Schultz, "Topological Properties of Hypercubes", Research report, YALE/DCS/RR-389.

[RS] N. Robertson and P. Seymour, "Graph Minors XIII. The Disjoint Paths Problem", 1986 Manuscript.

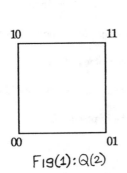

Fig(1): Q(2)

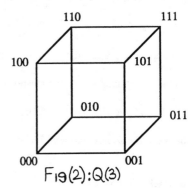

Fig(2): Q(3)

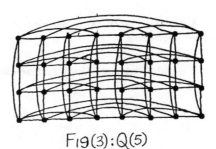

Fig(3): Q(5)

17

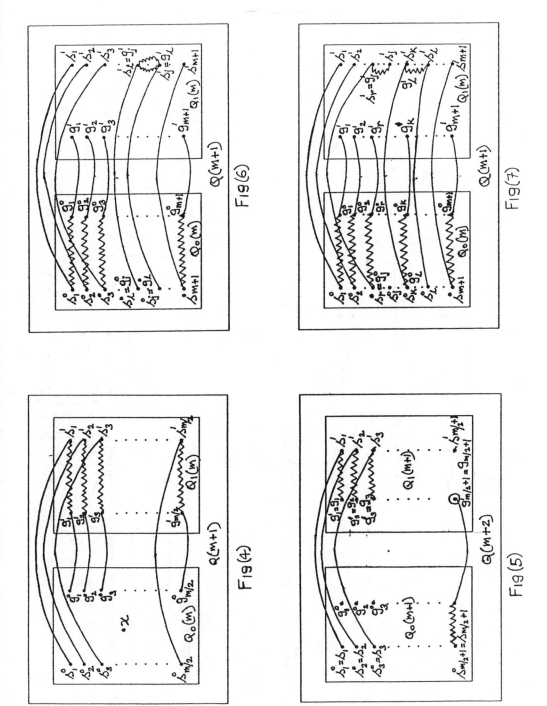

Fig(6)

Fig(7)

Fig(4)

Fig(5)

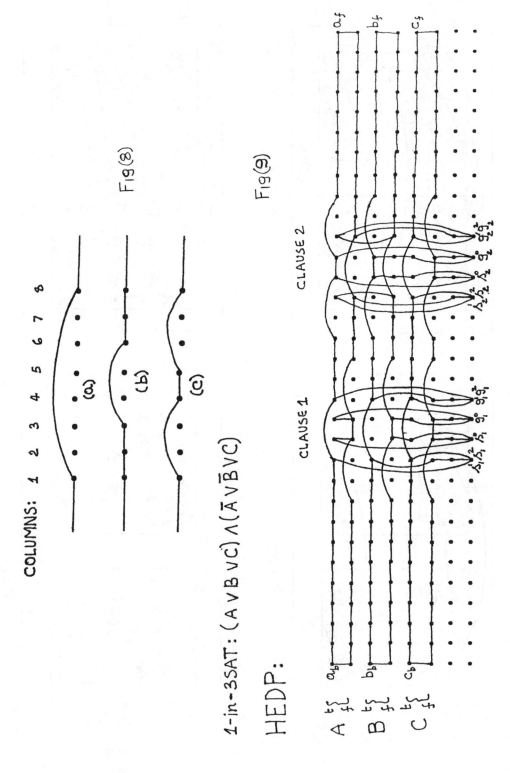

COLUMNS: 1 2 3 4 5 6 7 8

(a)

(b)

(c)

Fig(8)

1-in-3SAT: $(A \vee B \vee \bar{C}) \wedge (\bar{A} \vee \bar{B} \vee C)$

HEDP:

Fig(9)

Time Bounds for Broadcasting
in Bounded Degree Graphs*

Renato M. Capocelli
Dipartimento di Matematica G. Castelnuovo
Università di Roma "La Sapienza", 00185 Roma, Italy

Luisa Gargano and Ugo Vaccaro
Dipartimento di Informatica ed Applicazioni
Università di Salerno, 84081 Baronissi (SA), Italy

Abstract

Broadcasting is the process of transmitting a message from a node to all other nodes in a network. The problem of constructing sparse graphs in which the broadcast could be accomplished in minimum time has been investigated in several papers. However, the proposed graphs may have nodes of high degree that could be unacceptable from a network designer's point of view. In this paper we consider the problem of broadcasting on bounded degree graphs. We provide a lower bound on the time necessary to accomplish the broadcast when the maximum node degree is fixed. Moreover, we propose a method to construct bounded degree graphs in which the time to complete the broadcast from any node is of the same order as the lower bound. We also obtain the only known families of graphs having exactly minimum broadcast time.

1. Introduction

Broadcast refers to the process of information dissemination from a node (*originator*) to all the other nodes in the network. This concept finds wide application in the control of distributed systems. For instance, in computer networks there are many tasks, such as scheduling, updating distributed data bases, etc. [14], where a processor is required to send a message to all the other processors. In general, the broadcast is executed by a series of message transmissions (*calls*) through the communication lines of the network. The goal is to complete the process as quickly as possible.

As usual, the network is modeled as a graph $G = (V, E)$, where the set of nodes V represents the processors of the network and the set of edges E corresponds to the communication

---

* This work was supported in part by the Italian Ministry of Education, Project: Algoritmi e Sistemi di Calcolo

lines between pairs of processors. Nodes connected by a line are called *neighbors*. In this paper we assume that communication is synchronous and that each call requires one unit of time, that is, one unit of time is long enough for the longest packet that could be generated to be transmitted by the slowest processor. Moreover, we assume that a node may call at most one other node at time, and may call only neighbors. Many authors have considered this model for local broadcast, see for example [4-8],[12-13]. For a complete list of references on the argument the reader is referred to the recent survey by Hedetniemi *et al.* [9].

Two important measures of the goodness of a network are the time needed to perform the broadcast and the number of communication lines. It is easy to see that the minimum number of time units required to broadcast a message on a network of n nodes (*broadcast time of the network*) is $\lceil \log n \rceil$ † since the number of informed nodes can at most double during each time unit. A *minimal broadcast network* on n nodes is a network in which any node can broadcast in $\lceil \log n \rceil$ time units [5]. Let $B(n)$ be the minimum number of edges in any minimal broadcast network with n nodes. No general method is known to construct minimal broadcast networks with $B(n)$ edges, unless n is a power of 2 [7]. Values of $B(n)$ for few values of n are known [7,12, 15] and the problem of constructing minimal broadcast networks with $B(n)$ edges, for arbitrary n, is thought to be NP-complete [7]. Notice that already the problem of recognizing whether a given network is a minimal broadcast network is NP-complete [7]. Since it difficult to find minimal broadcast networks with minimum number of edges, much work has been done in order to obtain minimal broadcast networks with a relatively small number of edges. The problem was first considered by Farley [5] who gave algorithms for constructing minimal broadcast networks with n nodes and approximatively $(n/2)\log n$ edges. Chau and Liestman [4] and Gargano and Vaccaro [8] gave algorithms which improve Farley's results. Recently, Peleg [13] showed that $B(n) \in \Theta(L(n-1)n)$, where $L(n)$ is the number of leading ones in the binary representation of n. Unfortunately, Peleg's construction can produce graphs with vertices of degree $O(n)$, whereas the algorithms of Farley, Chau and Liestman and Gargano and Vaccaro produces graphs with nodes of degree $O(\log n)$.

From a practical point of view, it would be interesting to have graphs of maximum node degree independent from n and in which the broadcast could be still accomplished "quickly". Liestman and Peters were the first to study the problem of broadcasting on networks with fixed

---

† Unless otherwise specified, all logarithms in this paper are in base 2.

maximum degree. In particular, they studied the broadcast time for networks of maximum node degree $d = 3$ and $d = 4$ [10]. Broadcast in De Bruijn graphs has been recently considered in [2]. These results have been improved and extended by Bermond *et al.* [1]. More precisely, they gave a lower bound on the time necessary to accomplish the broadcast in a network of fixed node degree d and proposed a method to construct graphs with broadcasting time of same order as the lower bound. Unfortunately, their graphs can be constructed only when the number of nodes n is of the form $n = pq$, where p is the size of a De Bruijn graph and q is an integer depending on d and p. Moreover the method proposed in [1] requires, for every suitable n, the search of the appropriate p and q and of an appropriate graph of size q. Therefore, the problem of finding graphs of bounded degree having broadcasting time close to the optimal is still open for general n and d.

In this paper we afford two problems: lower bounding the broadcast time in graphs with fixed maximum degree and constructing graphs with broadcast time of the same order as our derived lower bound.

The paper is organized as follows. In Section 2 we study the rate of growth of the number of informed nodes in a network of bounded degree. We give un upper bound on the maximum number of informed nodes in the networks t time units after the beginning of the broadcast. This will allow us to obtain a lower bound on the time necessary to perform the broadcast in the network. We point out that our lower bound is better than the one given by Bermond *et al.* [1] in that they use an evaluation of the rate of growth of the number of informed nodes less precise than ours.

In Section 3 we introduce a new family of graphs that can be constructed for any number of nodes n and any value of d, the maximum node degree. These graphs have broadcast time $(c(d) + 1/\log(d - 3)) \log n$. Note that the lower bound is of the form $c(d) \log n + O(1)$, for a constant $c(d) > 1$ depending on d.

In Section 4 we consider the broadcast time in graphs obtained as products of rings. Even though they do not have the best possible asymptotic time, they result interesting for various reasons. Indeed, their regular structure allows a very simple broadcast protocol. Moreover, from some of these graphs we can obtain the only known families of graphs with fixed node degree having optimum broadcast time.

2. A lower bound on the broadcast time

In this section we establish a lower bound on the time necessary to complete the broadcast in a graph having nodes of degree not greater than d. The result is obtained by finding an upper bound on the maximum number of nodes, n_t^d, that can be informed in t time units in a graph with n nodes and maximum degree d.

Since each informed node can call at most one neighbor at each time unit, it follows that $n_t^{(d)} \leq 2^t$ for each t. Moreover, one can make the reasonable assumption that any informed node is not idle unless all its neighbors have been called. Therefore, after time unit $t + d - 1$, all nodes that were called at time t have informed all their neighbors and must be idle. It then follows that $n_{t+d}^{(d)} \leq n_{t+d-1}^{(d)} + (n_{t+d-1}^{(d)} - n_t^{(d)}) = 2n_{t+d-1}^{(d)} - n_t^{(d)}$. From this inequality one gets that an upper bound on $n_{t+d}^{(d)}$ is the solution of the recurrence relation

$$b_t^{(d)} = 2^t, \text{ for } t = 0, \ldots, d$$

$$b_{t+d}^{(d)} = 2b_{t+d-1}^{(d)} - b_t^{(d)} \text{ for each } t \geq 1. \tag{1}$$

Therefore, remembering that the maximum number of nodes that can be informed within t time units satisfies $n_t^{(d)} \leq b_t^{(d)}$ one gets that the number $t(n,d)$ of time units necessary to broadcast on n nodes satisfies

$$t(n,d) \geq \min\{t | b_t^{(d)} \geq n\}. \tag{2}$$

Bermond et al. [1], derived an estimate of the solution of (1) and showed that for each constant $c > \log e$, and for all sufficiently large n and d it results

$$\left(1 + \frac{\log e}{2^d}\right) \log n \leq \min\{t | b_t^{(d)} \geq n\} \leq \left(1 + \frac{c}{2^d} \log n\right). \tag{3}$$

The main result of this section is the following theorem.

Theorem 1. *For each d and t, the maximum number of nodes that can be informed in t time units in a graph of maximum degree d satisfies*

$$n_t^{(d)} \leq b_t^{(d)} = \left\lfloor \frac{2(\phi_0^{(d-1)})^{t-2}}{d\phi_0^{(d-1)} - 2(d-1)} - \frac{2}{d-2} + 0.5 \right\rfloor \tag{4}$$

where $\phi_0^{(d-1)}$ is the unique positive real root of the polynomial $x^{d-1} - x^{d-2} - \ldots - x - 1$.

Equation (4) allow us to improve the lower bound (3). Indeed, from (4) and (2), we get

$$t(n,d) \geq \left\lceil \frac{\log((d\phi_0^{(d-1)} - 2(d-1))(2(d-2)n - d + 6)) - 2 - \log(d-2)}{\log \phi_0^{(d-1)}} \right\rceil + 2 \tag{5}$$

The rest of this section is devoted to the proof of Theorem 1. To this aim we need some preliminary results. Let $k = d - 1$, the recurrence relation (1) is equivalent to

$$b_t^{(k+1)} = 2^t, \text{ for } t = 0, \ldots, k+1$$

$$b_t^{(k+1)} = b_{t-1}^{(k+1)} + \ldots + b_{t-k}^{(k+1)} + 2 \text{ for each } t > k+1. \tag{6}$$

As it is well known, a solution of (6) can be obtained by subtracting the constant $2/(k-1)$ from the solution of the corresponding homogeneous relation $y_t = y_{t-1} + \ldots + y_{t-k}$, with $y_t = b_t^{(d)} + 2/(k-1)$, $i = 0, \ldots, k+1$. Since we have two extra initial conditions we also shift the sequence by 2. Then we need to solve the recurrence relation

$$x_t^{(k+1)} = 4(2^t) + 2/(k-1), \text{ for } t = 0, \ldots, k-1$$

$$x_t^{(k+1)} = x_{t-1}^{(k+1)} + \ldots + x_{t-k}^{(k+1)} \text{ for } t \geq k. \tag{7}$$

where

$$x_t^{(k+1)} = b_{t+2}^{(k+1)} + 2/(k-1). \tag{8}$$

As it is well known [11], the solution of (7) has the form

$$x_t^{(k+1)} = \sum_{i=0}^{k-1} \alpha_i^{(k)} (\phi_i^{(k)})^t,$$

where the $\phi_i^{(k)}$'s are the roots of the polynomial $x^k - x^{k-1} - \ldots - x - 1$ and the $\alpha_i^{(k)}$'s are determined by the initial conditions. Where it is not confusing, we write α_i and ϕ_i for $\alpha_i^{(k)}$ and $\phi_i^{(k)}$, respectively.

Recurrence relation (7) is, apart from the initial conditions, the same that arises when considering generalized Fibonacci numbers [11]. We can then partially apply the results presented in [3], where the authors study generalized Fibonacci numbers. In the following we report without proof those results shown in [3] which do not depend upon the initial condition, i.e., from the values of the α_i's.

Fact 1. [3] For each k, the polynomial $x^k - x^{k-1} - \cdots - x - 1$ has only one positive root $\phi_0 > 1$ satisfying $2 - 2/2^k \leq \phi_0 \leq 2 - 1/2^k$. Any other root satisfies $|\phi_i| < 1$. □

Fact 2. [3] For each $i = 0, \ldots, k-1$

$$\alpha_i = \frac{(\mathbf{R_i} \circ (x_{k-1}^{(k+1)}, \ldots, x_0^{(k+1)}))(\phi_i - 1)}{\phi_i^{k-1}((k+1)\phi_i - 2k)} \tag{9}$$

where $\mathbf{R_i} = (1, \phi_i - 1, \phi_i^2 - \phi_i - 1, \ldots, \phi_i^{k-1} - \phi_i^{k-2} - \ldots - \phi_i - 1)$, and $\circ$ is the vector product. □

Fact 3. For each $i = 0, \ldots, k-1$

$$\alpha_i = \frac{2\phi_i^2}{(k+1)\phi_i + 2k} \tag{10}$$

Proof. Noticing that each ϕ_i satisfies the relation

$$\phi_i^k = \phi_i^{k-1} + \ldots \phi_i + 1 = (\phi_i^k - 1)/(\phi_i - 1) \tag{11}$$

one has

$$
\begin{aligned}
(\mathbf{R_i} \circ (x_{k-1}^{(k+1)}, \ldots x_0^{(k+1)})) &= \sum_{j=0}^{k-1} \phi_i^j \left(4 - \frac{2(k-j-2)}{k-1} \right) = \sum_{j=0}^{k-1} \phi_i^j \left(\frac{2(k+j)}{k-1} \right) \\
&= \frac{2k\phi_i^k}{k-1} + \left(\frac{2}{(k-1)} \right) \left(\frac{\phi_i}{(\phi_i-1)^2} \right) ((k-1)\phi_i^k - k\phi_i^{k-1} + 1) \\
&= \frac{2\phi_i}{(k-1)(\phi_i-1)^2} (k\phi_i^{k-1}(\phi_i-1)^2 + (k-1)\phi_i^k - k\phi_i^{k-1} + 1) \\
&= \frac{2\phi_i}{(k-1)(\phi_i-1)^2} (k\phi_i^{k+1} - (k+1)\phi_i^k + 1)
\end{aligned}
$$

From which, noticing that (11) is equivalent to $\phi_i^k = 1/(2 - \phi_i)$, one has

$$(\mathbf{R_i} \circ (x_{k-1}^{(k+1)}, \ldots x_0^{(k+1)})) = \frac{2\phi_i(k\phi_i - (k+1) + 2 - \phi_i)}{(k-1)(\phi_i - 1)^2(2 - \phi_i)} = \frac{2\phi_i}{(\phi_i - 1)(2 - \phi_i)}$$

Then from (9), one gets that (10) holds. □

In order to get the desired result, we need to show that $x_i^{(k+1)}$ differs from $\alpha_0 \phi_0^t$ for less than $1/2$. Define the deviations, d_t, as the differences

$$d_t = x_i^{(k+1)} - \alpha_0 \phi_0^t. \tag{12}$$

Fact 4. [3] For the recurrence relation (7), the d_t's satisfy the following properties

1) at most $k-1$ deviations have the same sign;

2) the deviations satisfy the recurrence relation $d_{t+1} = 2d_t - d_{t-k}$;

3) if $|d_t| \geq 1/2$, then $|d_{t-j}| > 1/2$ for some j such that $k \geq j \geq 2$. □

Fact 5. For each $t \geq 1$, $|d_t| < 1/2$.

Proof. From Fact 4 it follows that in order to prove that each deviation has absolute value less then $1/2$, it is sufficient to find k consecutive deviations each having absolute value less then $1/2$. Going backwards one has that $x_{-1} + x_0 + \ldots x_{k-2} = x_{k-1}$ and then $x_{-1} = 2 + 2/(k-1)$. We show now that

$$1/2 > d_{-1} > d_0 > \ldots > d_{k-2} > -1/2. \tag{13}$$

In order to prove that $1/2 > d_{-1} = 2 + 2/(k-1) - \alpha_0\phi_0^{-1}$, consider the function $f(x) = 2x/((k+1)x - 2k)$. The function f is decreasing in $(2k/(k+1), 2)$. One has then $f(\phi_0) > f(2) = 2$, which implies $1/2 > d_{-1}$ for each $k > 4$. For smaller values of k direct computation shows that this holds, too.

We show now that $d_i < d_{i-1}$, for $i = 0, \ldots, k-2$. From (10) and (12), this is equivalent to show that

$$\frac{2\phi_0^{i+1}(\phi_0 - 1)}{(k+1)\phi_0 - 2k} > 2^{i+1}. \tag{14}$$

Let the function $g_i(x)$ defined as $g_i(x) = 2x^{i+1}(x-1)/((k+1)x - 2k)$. Inequality (14) is equivalent to $g_i(\phi_0) > 2^{i+1}$. Since $g_i(x)$ is decreasing in the interval $(2k/(k+1), 2]$, one has that $g_i(\phi_0) > g_i(2) = 2^{i+1}$ and (14) holds. In order to prove (13), it remains to show that $d_{k-2} > -1/2$. From (10) and (12), this is equivalent to show that

$$\frac{2\phi_0^k}{(k+1)\phi_0 - 2k} < 2^k + \frac{k+3}{2(k-1)} = A_k. \tag{15}$$

Let $h(x) = A_k/x^k((k+1)x - 2k) - 2$. Inequality (15) is equivalent to $h(\phi_0) > 0$.

The function h is increasing and convex $\cap$ in the interval $[2k/(k+1), 2]$. One has then

$$h\left(\frac{2k+1}{k+1}\right) = h\left(\frac{1}{2}\frac{2k+1}{k+1} + \frac{1}{2}2\right) \geq \frac{1}{2}\left(f\left(\frac{2k+1}{k+1}\right) + f(2)\right) = \frac{1}{2}\frac{A_k - 2^k}{2^{k-1}} > 0$$

Since, from Fact 1, $\phi_0 > (2k+1)/(k+1)$ for each $k > 2$ and h is increasing, one has $h(\phi_0) > 0$ and (15) is proved. For $k = 2$, direct computation shows that (15) holds. Then $|d_i| < 1/2$, $i = -1, \ldots, k-2$, and Fact 5 is proved. $\quad\square$

From the above result, one has that $x_t^{(k+1)}$ differs from $\alpha_0\phi_0^t$ for less than $1/2$. From equality (8), it follows that $b_t^{(k+1)}$ differs from $\alpha_0\phi_0^{t-2} - 2/(k-1)$ for less than $1/2$. Using equality (10) Theorem 1 is proved.

3. An upper bound on the broadcast time

In this section we shall describe an algorithm for constructing graphs with n nodes of degree $\leq d+1$ and broadcast time not greater than $t(n,d) + \lceil \log_{d-2} n \rceil + 2$. These graphs can be constructed for any value of n and d.

The basis of our construction is a tree in which the root can broadcast in exactly $t(n,d)$ time units. Such a tree, indicated by $T_{n,d}$, is constructed as follows. Label the nodes with the integers from 1 to n. The root is node 1. As first step, connect node 2 to the root. In general, at step $t \geq 2$, let $b_t^{(d)}$ defined as in (1). If $n > b_{t-1}^{(d)}$ then insert the edges

$$(b_t^{(d)} - i, b_{t-1}^{(d)} - i) \quad \text{for } i = \begin{cases} 0, \ldots, b_t^{(d)} - b_{t-1}^{(d)} - 1 & \text{if } n \geq b_t^{(d)} \\ b_t^{(d)} - n, \ldots, b_t^{(d)} - b_{t-1}^{(d)} - 1 & \text{otherwise} \end{cases}$$

The resulting tree for $d = 3$ and $n = 20$ is represented in Fig.1.

Fact 6. Each node in $T_{n,d}$ has degree not greater than d

Proof. Consider a node i, with $b_{r-1}^{(d)} < i \leq b_r^{(d)}$ for some integer r. Node i receives its first connection at step $t = r$. For $t \geq r + d$, nodes $b_{t-1}^{(d)} + 1, b_{t-1}^{(d)} + 2, \ldots$ are connected to nodes $b_{t-d}^{(d)} + 1, b_{t-d}^{(d)} + 2, \ldots$, respectively. But $t - d \geq r$ and therefore $b_{t-d}^{(d)} + 1 \geq b_r^{(d)} + 1 > i$. Then node i can receive a new connection only for $t = r, \ldots, r + d - 1$, and therefore has degree not greater than d. □

Consider the following broadcast scheme from the root of $T_{n,d}$ to all nodes in the network: each informed node sends the message to the root of its subtree of maximum size; at the successive time unit, among the remaining sons, calls the root of the subtree of maximum size, and so on until all sons have been informed. In this way at time unit t there are exactly $b_t^{(d)}$ informed nodes (n at the last time unit) and the broadcast takes time $t(n,d)$. We want now to add edges to this tree in such a way that the degree of each node is not greater than $d+1$ and each node can broadcast in $t(n,d) + \lceil \log_{d-2} n \rceil + O(1)$ time units. To this aim we need the following result.

Fact 7. For each $d \geq 3$ and n, the number of leaves in $T_{n,d}$ is not less than $n/3$

Proof. Consider first the case $n = b_t^{(d)}$ for some t. One has $b_t^{(d)} = 2b_{t-1}^{(d)} - b_{t-d}^{(d)}$ and the number of leaves in $T_{b_t^{(d)},d}$ is $b_{t-1}^{(d)} - b_{t-d}^{(d)}$. From (1) one has $b_{t-1}^{(d)} \geq 2b_{t-d}^{(d)}$. It follows that $3(b_{t-1}^{(d)} - b_{t-d}^{(d)}) \geq 2b_{t-1}^{(d)} - b_{t-d}^{(d)} = n$, and the proof is complete in this case. Suppose now that $b_t^{(d)} + m = n \leq b_{t+1}^{(d)}$, for some t and $m > 0$. If the remaining $n - b_t^{(d)}$ nodes are connected only to internal nodes of $T_{b_t^{(d)},d}$, then the total number of leaves in $T_{n,d}$ is equal to the number of leaves in $T_{b_t^{(d)},d}$ plus m, that is, the number of leaves is $b_{t-1}^{(d)} + b_{t-d}^{(d)} + m$. One has

$$b_{t-1}^{(d)} + b_{t-d}^{(d)} + m \geq b_t^{(d)}/3 + m \geq (b_t^{(d)} + m)/3 = n/3.$$

Consider now the case that nodes are connected also to leaves of $T_{b_t^{(d)},d}$ (i.e., to nodes numbered from $b_{t-1}^{(d)} + 1$ to $b_t^{(d)}$) Since connecting a new node to a leaf does not increase the number of leaves, one gets that the number of leaves in $T_{n,d}$ is equal to the number of leaves in $T_{b_{t+1}^{(d)},d}$. Being $n < b_{t+1}^{(d)}$, this fact is proved. □

From the above fact, it follows that it is possible to connect each internal node to a leaf in such a way that each leaf has no more than 2 incident edges: the one that belongs to the tree plus an eventual connection to another internal node which is not connected to any leaf by edges in $T_{n,d}$. Let v be the leaf which is connected to the root. Add edges to construct a $(d-2)$–ary tree T rooted in v and whose nodes are the leaves of $T_{n,d}$. Call $G_{n,d+1}$ the resulting graph. Notice that each leaf has no more than $(d-1)+2 = d+1$ edges. The height of this tree is the minimum h such that the number of its nodes $((d-2)^{h+1} - 1)/(d-3)$ is not less than the number, f, of leaves in $T_{n,d}$; then $h = \lceil \log_{d-2} f(d-3) + 1 \rceil - 1 \leq \lceil \log_{d-2} n \rceil$. The broadcast process in $G_{n,d+1}$ proceeds as follows. The node that has to perform the broadcast sends the message to the leaf to which it is connected. The leaf sends the message toward the root v of the $(d-2)$–ary tree T constructed on the leaves of $T_{n,d}$, (at most $h \leq \log_{d-2} n$ time units are needed); v sends the message to the root of $T_{n,d}$ which broadcasts on $T_{n,d}$. Then the graph $G_{n,d+1}$ has maximum degree $d+1 \geq 5$ and the maximum time to complete the broadcast from any node is $1 + \lceil \log_{d-2} n \rceil + 1 + t(n,d)$. By using (5), we get then the following result.

Theorem 2. *For each $d > 4$ and n, the graph $G_{n,d}$ with n nodes of maximum degree d has broadcast time*

$$\lceil \log_{d-3} n \rceil + t(n,d-1) + 2 = \left(\frac{1}{\log \phi_0^{(d-2)}} + \frac{1}{\log d - 3} \right) \log n + O(d)$$

□

The graph $G_{26,5}$ constructed according to the method outlined above is shown in Fig.2.

If one allows that leaves, once they have called the node in the path toward v (the root of the tree constructed on the leaves of $T_{n,d}$), continue calling their uninformed neighbors, some time units can be saved to complete the broadcast. In particular additional edges can be introduced between nodes whose total degree is less than $d+1$ until each node has exactly degree $d + 1$. In this way, a node x informed at time t can inform in the following d time units d new nodes, unless a collision happens. Here collision means that at a certain time $t + d' \leq t + d$ all the neighbors of x have be previously informed by other nodes.

4. Broadcasting in product of rings

In this section we consider a class of graphs derived from product of rings that can be constructed for any value n of the number of nodes and for any even value d of the maximum degree allowed for a node, and on which broadcasting can be accomplished by a very simple protocol. Moreover, even though the obtained broadcasting time is not asymptotically the best known, for many values of n and d these graphs allow to obtain the best known broadcasting time. In particular some of these graphs have broadcast time equal or very near to the lower bound. In section 4.2, we show that from some of these graphs it is possible to obtain the only families of graphs having provably optimal broadcast time, i.e. equal to the lower bound.

4.1 Product of rings

The description of the graphs and of the correspondent broadcast schemes is divided into two parts: We first consider the simpler case in which the number n of nodes is $n = r^k$, $k = d/2$, for some integer r and, subsequently, the general case is considered.

Graphs with $n = r^k$ nodes. The graphs considered in this section are obtained as a product of rings. We recall that the product of two graphs [16] $G_1 = (V_1, E_1)$ and $G_2 = (V_2, E_2)$ is defined as $G_1 \times G_2 = (V, E)$, with $V = \{(v_1 v_2) | v_1 \in V_1, \ v_2 \in V_2\}$ and $E = \{((v_1 v_2), (w_1 w_2)) | (v_1, w_1) \in E_1 \text{ and } v_2 = w_2 \text{ or } (v_2, w_2) \in E_2 \text{ and } v_1 = w_1\}$

A ring with r nodes is the graph with node set $V = \{0, \ldots, r-1\}$ and edge set $E = \{(i,j) | j = i + 1 \bmod r\}$. The graph $C_{r^k, r}$ with r^k nodes and degree $d = 2k$ is defined as

$$C_{r^k, r} = \underbrace{R_r \times R_r \times \cdots \times R_r}_{k \text{ times}}$$

Notice that for $r = 2$ the resulting graph is an hypercube. For $d = 4$ we get the torus network defined in [17].

It is easy to see that the broadcast in the ring R_r requires time $\lceil r/2 \rceil$. In $C_{r^k, r}$ each node belongs to k rings. We can individuate k directions for the rings (e.g., in $C_{r^2, r}$ we can individuate 2 directions: horizontal and vertical), let them be $d_1, \ldots, d_k$. Consider a node $w = w_1 \ldots w_k$. Let $R^1, \ldots, R^k$ be the rings containing w, nodes in R^i have the form $w_1 \ldots w_{i-1} a w_{i+1} \ldots w_k$, $0 \le a \le r-1$, and R^i has direction d_i, for $i = 1, \ldots k$. Let $a_1, \ldots, a_k$ be the originator of the broadcast. If w receives the message for the first time along an edge of R^i then for $t = i, \ldots, k$ the node w calls the nodes $w_1 \ldots w_{t-1} b w_{t+1} \ldots w_k$ of R^t, where the b's are the nodes called (in the same order) by w_t during the broadcast from a_t in the ring R^t.

It is easy to verify that after $k\lceil r/2 \rceil$ time units, each node is informed. Actually, one could prove the following more general result.

Fact 8. If the graph G is the product of two graphs G_1 and G_2, then the broadcast time of G is equal to the summ of the broadcast times of G_1 and G_2. □

Graphs with $n \ne r^k$ nodes. Suppose now that $n = hr^k + m < r^{k+1}$, with $m < r^k$. If $m = 0$ we can consider the graph $C_{r^k, r} \times R_h$. Suppose $m > 0$. Proceed by induction on k. If $k = 0$ then $n < r$ and we can take $C_{n,r} = R_n$.

Suppose one can construct $C_{n',r}$ for each $n' \le r^k$. Indicate by G_0 the graph having nodes $a0$ such that a is a node of $C_{m,r}$ and there is an edge between nodes $a0$ and $b0$ if and only if there is an edge between a and b in $C_{m,r}$. Let G_i, $i = 1, \ldots, h$, be the graph having ai such that a is a node of $C_{r^k,r}$ and there is an edge between nodes ai and bi if and only if there is an edge between a and b in $C_{r^k,r}$. Connect the G_i's as follows. For each a such that $a0$ is in G_0 construct a ring on $a0, \ldots, ah$. If $a0$ is not in G_0, G_0 construct a ring on $a1, \ldots, ah$. Each node in $C_{n,r}$ has degree at most $2(k+1)$. Following the lines of the broadcast scheme in case n is a power of r, it is easy to verify that each node in $C_{n,r}$ can perform the broadcast in $k\lceil r/2 \rceil + \lceil (h+1)/2 \rceil$.

One has then the following theorem.

Theorem 3. For each $k \ge 1$ and n the graph $C_{n,r}$, where r is an integer such that $r^k < n \le r^{k+1}$, has maximum degree $d = 2(k+1)$ and each node can perform the broadcast on $C_{n,r}$ in $\lfloor \log_r n \rfloor \lceil r/2 \rceil + \lceil n/(2r^{\lfloor \log_r n \rfloor}) \rceil$ time units. □

4.2 Optimal graphs

When d is fixed, the value obtained in Theorem 3 grows faster than the lower bound and than the best known upper bounds. An interesting point is that for many values of d and n the broadcast time of Theorem 3 is the best known. In particular the graph $C_{36,6}$ with $n = 36$ and

$d = 4$ has optimal broadcast time 6. From this graph we get an infinite family of graphs having minimum broadcast time. Consider the graphs $\mathcal{G}_i = C_{36,6} \times C_{2^i,2}$, for $i > 1$, with $n = 2^i \cdot 36$ and $d = 4 + i$. We recall that $C_{2^i,2}$ is the hypercube with 2^i nodes. Since broadcasting in $C_{2^i,2}$ requires i time units [8], and broadcasting in $C_{36,6}$ requires 6 time units, from Fact 8 one can see that the broadcast in $\mathcal{G}_i$ can be completed in $i + 6$ time units. From (1), one gets that the obtained time is optimal. We then obtain that $t(36 \cdot 2^i, 4 + i) = i + 6$, for each i. We observe now that $n = 36$ and $d = 4$ are not the only values for which the graphs $C_{n,r}$ and their product with hypercubes have optimal broadcast time. There are many other values for which it is possible to prove the optimality and then to derive the exact expression of $t(n2^i, d)$.

Fact 9. For all integers d, n, and Δ, with $2^{d+\Delta} \leq n \leq b_{d+\Delta+1}^{(d)}$ one has $t(n2^i, d + i) \geq d + \Delta + i + 1$.

Proof. Immediate from (1). □

Let n satisfy the hypothesis of the above fact. If a graph $C_{n,r}$ has broadcast time equal to the lower bound, then each member of the family of graphs $C_{n,r} \times C_{2^i,2}$ has optimal broadcast time. We can therefore establish that $t(2^i n, d + i) = d + \Delta + i + 1$. Some values of n and d for which this holds are given in Table 1(a). Finally we notice that an analogous reasoning can be done, for n as in Fact 9, also in case the broadcast time of $C_{n,r}$ is not provably optimal, i.e., equal to the lower bound plus some t. In this case, using $C_{n,r} \times C_{2^i,2}$, we can establish that $t(2^i n, d + i) \in [d + \Delta + i + 1, d + \Delta + i + 1 + t]$ for each i. Consider for example $n = 64$ and $d = 4$. One has $t(2^i 64, 4 + i) \geq 7 + i$. For each i the graph $C_{64,2} \times C_{2^i,2}$ has broadcast time $8 + i$, thus giving $7 + i \leq t(2^i 64, 2) \leq 8 + i$. A few other values of n and d for which the value of $t(2^i n, d + i)$ is contained in a range of size not greater than 2 are given in Table 1b.

References

[1] Bermond, J.C., Hell, P., Liestman, A.L., and Peters, J.G., "Broadcasting in Bounded Degree Graphs", *Technical Report 88-5*, Simon Fraser University (1988).

[2] Bermond, J.C., and Peyrat, C., "Broadcasting in De Bruijn Networks", *Technical Report 88-4*, Simon Fraser University (1988).

[3] Capocelli, R.M. and Cull, P., "Generalized Fibonacci Numbers Are Rounded Powers", to appear in *Proceedings of Third Int. Conf. on Fibonacci Numbers and Their Applications*, Pisa, Italy, (1988)

[4] Chau, S. and Liestman, A. L., "Constructing Minimal Broadcast Networks", *J. Combin., Inform. & System Sci.* **10** (1985), 110–122.

[5] Farley, A. M., "Minimal Broadcast Networks", *NETWORKS* **9** (1979), 313–332.

[6] Farley, A.M., "Broadcast Time in Communication Networks", *SIAM J. Appl. Math.* **39** (1980), 385–390.

[7] Farley, A. M., Hedetniemi, S.T., Proskurowski, A. and Mitchell, S., "Minimum Broadcast Graphs", *Discrete Math.* **25** (1979), 189–193.

[8] Gargano, L., and Vaccaro, U.,"On the Construction of Minimal Broadcast Networks", *NETWORKS*, to appear.

[9] Hedetniemi, S.T., Hedetniemi, S.M., and Liestman, A.L., "A Survey of Broadcasting and Gossiping in Communication Networks", *NETWORKS*, **18**, (1988), 319– 351.

[10] Liestman, A.L., and Peters, J.G.,"Broadcast Networks of Bounded Degree", *SIAM J. Disc. Math.*,**1**, (1988), 531–540.

[11] Miles, E.P., Jr., "Generalized Fibonacci Numbers and Matrices", *Amer. Math. Monthly.* **67**, (1960), 745–757.

[12] Mitchell, S. and Hedetniemi, S. T., "A Census of Minimum Broadcast Graphs", *J. Combin., Inform. & Systems Sci.* **9** (1980), 119–129.

[13] Peleg, D., "Tight Bounds on Minimum Broadcast Graphs", *SIAM J. Disc. Math.*, to appear.

[14] Tanebaum, A. S., "*Computer Networks*", Prentice-Hall, Englewood Cliffs, N.J., (1981).

[15] Bermond, J.C., Hell, P., Liestman, A.L., and Peters, J.G., "New Minimum Broadcast Graphs and Sparse Broadcast Graphs", *Technical Report 88-4*, Simon Fraser University (1988).

[16] G. Sabidussi, "Graph Multiplication", *Math. Zeitschr.*, **72**, (1960), 446–457.

[17] C. von Conta, "Torus and Other Networks as Communication Networks with up to some Hundred Points", *IEEE Trans. Comp.*, **C-32**, (1983), 657–666.

d	n	$t(n2^i, d+i)$
4	$5 \cdot 6 + 1 \quad \div \quad 6 \cdot 6$	$6+i$
5	$6^2 + 28 \quad \div \quad 6^2 + 36$	$7+i$
6	$3 \cdot 6^2 + 20 \quad \div \quad 3 \cdot 6^2 + 36$	$8+i$

(a)

d	n	$t(n2^i, d+i)$	
		lower bound	upper bound
4	$4 \cdot 8 + 5 \quad \div \quad 6 \cdot 8$	$6+i$	$7+i$
4	$6 \cdot 10 + 4 \quad \div \quad 8 \cdot 10$	$7+i$	$9+i$
5	$8 \cdot 8 + 1 \quad \div \quad 8 \cdot 8 + 56$	$7+i$	$9+i$
6	$4 \cdot 6^2 + 1 \quad \div \quad 6^3$	$8+i$	$9+i$
7	$6^3 + 1 \quad \div \quad 6^3 + 38$	$8+i$	$10+i$
8	$2 \cdot 6^3 + 70 \quad \div \quad 4 \cdot 6^3$	$10+i$	$11+i$
8	$4 \cdot 6^3 + 1 \quad \div \quad b_{10}^{(8)} (= 4 \cdot 6^3 + 152)$	$10+i$	$12+i$
9	$6^4 + 752 \quad \div \quad 2 \cdot 6^4$	$12+i$	$13+i$
10	$3 \cdot 5^4 + 173 \quad \div \quad b_{12}^{(10)} (= 3 \cdot 6^4 + 200)$	$12+i$	$14+i$

(b)

TABLE 1

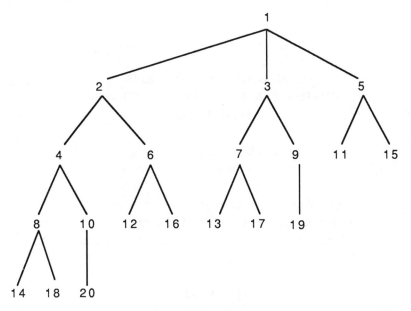

FIGURE 1

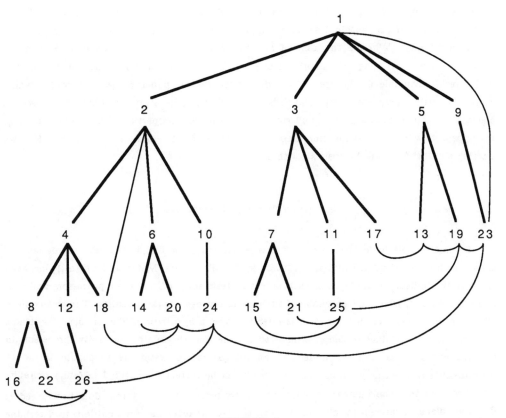

FIGURE 2

t/s-DIAGNOSABLE SYSTEMS: A CHARACTERIZATION AND DIAGNOSIS ALGORITHM

A. Das (1), K. Thulasiraman (1),

V.K. Agarwal (11), K.B. Lakshmanan (111)

(1) E.E. Dept., Concordia University, Montreal

(11) E.E. Dept., McGill University, Montreal

(111) Dept. of Math. and Comp. Sc., SUNY, Brockport, N.Y.

ABSTRACT

A classical PMC multiprocessor system [1] composed of n units is said to be t/s-**diagnosable** [6] if, given a syndrome (complete collection of test results), the set of faulty units can be isolated to within a set of at most s units, assuming that at most t units in the system are faulty. Whereas the issue of t/s-diagnosability has been resolved by Sullivan in [11], the efficient t/t-diagnosis algorithm due to Yang, Masson and Leonetti [9], is the only major result available in the area of t/s-diagnosis. In this paper we present results which represent further progress in the diagnosis of t/s-diagnosable systems. First, we present a characterization of such a system and establish several new properties of allowable fault sets with respect to a given syndrome. Based on these properties and the characterization, we develop an $O(n^{3.5})$ $t/t+1$-diagnosis algorithm.

I. INTRODUCTION

Several models have been proposed in the literature for diagnosable system design. Of these, the now well-known PMC model introduced by Preparata, Metze and Chien [1] has been extensively studied. In this model, each processor tests some of the other processors and produces test results, which are unreliable if the testing processor is itself faulty. The collection of all test results over the entire system is referred to as a **syndrome**. The classical constraint used in the study of diagnosable systems is to assume that the number of faulty processors in the entire system is upper bounded by an integer t. A system is then said to be t-**diagnosable** if given a syndrome, all processors can be correctly identified as faulty or fault-free, provided that the number of faulty processors present in the system does not exceed t. Three problems of interest in this context are : the t-**characterization problem** to determine the necessary and sufficient conditions for the system test assignment to be t-diagnosable, the *t*-**diagnosability problem** to determine the largest value of *t* for which a given system is *t*-diagnosable, and finally the *t*-**diagnosis problem** to locate the faulty units present in a *t*-diagnosable system, using a given syndrome.

Hakimi and Amin [2] gave a solution to the *t*-characterization problem. An $O(|E|n^{3/2})$ algorithm for the *t*-diagnosability problem was presented by Sullivan [3]. Dahbura and Masson [4] published an $O(n^{2.5})$ algorithm for the *t*-diagnosis problem; a *t*-diagnosis algorithm with complexity $O(|E|+t^3)$ was presented by Sullivan [5].

The requirement that all the faulty processors in a multiprocessor system be identified exactly is rather restrictive. Friedman [6] introduced the concept of *t/s*-**diagnosability** which allowed the possible replacement of fault-free processors, whereas in *t*-diagnosability only the replacement of faulty processors is considered. A multiprocessor system S is said to be *t/s*-**diagnosable**, if given a syndrome, the set of faulty processors can be isolated to within a set of at most *s* processors provided that the number of faulty processors does not exceed *t*. Allowing some fault-free processors to be possibly identified as faulty permits the system to have far fewer tests. It has been shown that *t/t*-diagnosable systems with $n*\lceil(t+1)/2\rceil$ tests can be constructed [7]. *t/t*-diagnosable systems have been studied extensively in the literature. Chwa and Hakimi [8] gave a characterization of *t/t*-diagnosable systems, Sullivan [3] presented a polynomial time algorithm for the *t/t*-diagnosability problem, and Yang et al. [9] presented an $O(n^{2.5})$ algorithm for the *t/t*-diagnosis problem. Sullivan also presented in [10] a *t/t+k*-diagnosability algorithm which runs in polynomial time for each fixed integer *k*. This diagnosability algorithm is based on a characterization of *t/t+k*-diagnosable systems also developed in [10].

The objective of this work is to develop diagnosis algorithms for *t/s*-diagnosable systems. The paper is organised as follows. First in section II, we present certain basic definitions, notations and results. In section III, we present characterizations of *t/s*-diagnosable systems which generalize those given earlier in [8,9]. With the objective of determining an effecient test for a vertex *v* to be in an allowable fault set of size at most *t*, we establish in section IV several properties of allowable fault sets. Using these properties and the characterizations given in section III, we develop in section V an $O(n^{3.5})$ algorithm for diagnosis of a *t/t+1*-diagnosable system.

II. PRELIMINARIES

A multiprocessor system S consists of *n* units or processors, denoted by the set $U = \{u_1, u_2, ..., u_n\}$. Each unit is assigned a subset of other units for testing. Thus the test interconnection can be modeled as a directed graph $G = (U, E)$. The **test outcome** a_{ij}, which results when unit u_i tests unit u_j, has value 1 (0) if u_i evaluates unit u_j to be faulty (fault-free). Since all faults considered are permanent, the test outcome a_{ij} is reliable if and only unit u_i is fault-free. The collection of all test results over the entire system is referred to as a **syndrome**.

If $a_{ij} = 0$ (1) then u_i is said to have a 0-link (1-link) to u_j and u_j is said to have a 0-link (1-link) from u_i.

Given a syndrome, **the disagreement set** $\Delta_1(u_i)$ of $u_i \in U$ is defined as

$$\Delta_1(u_i) = \{u_j \mid a_{ij} = 1 \text{ or } a_{ji} = 1\}.$$

For a subset $W \subseteq U$,

$$\Delta_1(W) = \bigcup_{u_j \in W} \Delta_1(u_j).$$

Given a syndrome, the set of **0-descendents** of u_i is represented by the set

$$D_0(u_i) = \{u_j : \text{there is a directed path of 0-links from } u_i \text{ to } u_j\}$$

and for a set $W \subseteq U$, the **0-ancestors** of W denotes the set

$$A_0(W) = \{u_i : u_j \in D_0(u_i) \text{ and } u_j \in W\}.$$

For $u_i \in U$, $H_0(u_i)$ corresponds to the set $A_0(u_i) \cup \{u_i\}$.

Definition 1 [4]: Given a system S and a syndrome, a subset $F \subseteq U$ is an **allowable fault set** (AFS) if and only if

c1: $u_i \in (U-F)$ and $a_{ij} = 0$ imply $u_j \in (U-F)$, and

c2: $u_i \in (U-F)$ and $a_{ij} = 1$ imply $u_j \in F$.

In other words, F is an AFS for a given syndrome if and only if the assumption that the units in F are faulty and the units in $U-F$ are fault-free is consistent with the given syndrome. A **minimum allowable fault set** (MAFS) is an allowable fault set of minimum cardinality.

Definition 2 [4]: Given a system S and a syndrome, the **implied faulty set** $L(u_i)$ of $u_i \in U$ is the set of all units of S that may be deduced to be faulty under the assumption that u_i is fault-free.

It follows that

$$L(u_i) = \Delta_1(D_0(u_i)) \cup A_0(\Delta_1(D_0(u_i))).$$

Note that if $u_j \in L(u_i)$ then there exists a 1-link (u_k, u_l) or (u_l, u_k) such that there is a directed path of 0-links from u_i to u_k and a directed path of 0-links from u_j to u_l. Such a path will be refered to as an **implied-fault path** between u_i and u_j.

If $u_i \in L(u_i)$ then clearly the unit u_i is faulty. Such a unit will be in every AFS. Without loss of generality, we assume in this paper that $u_i \notin L(u_i)$ for any $u_i \in U$.

The following lemmas determine a few properties of AFS's and implied faulty sets.

Lemma 1 [4]: Given a system S and a syndrome, each of the following statements holds:

(1) for $u_i, u_j \in U$, $u_i \in L(u_j)$ if and only if $u_j \in L(u_i)$.

(2) for $u_i, u_j \in U$, if $a_{ij} = 0$ then $L(u_j) \subseteq L(u_i)$.

(3) If $F \subseteq U$ is an AFS, then $\bigcup_{u_i \in U-F} L(u_i) \subseteq F$. //

Lemma 2 [10]: Given a system S and a syndrome, if F_1 and F_2 are AFS's then so is $(F_1 \cup F_2)$. //

Lemma 3 : Given a system S and a syndrome, let $F \subseteq U$ be an AFS containing $u_i \in U$. Then $H_0(u_i) \subseteq F$.

Proof : Suppose that $u_j \in H_0(u_i)$ is not a member of F. Since $u_j \in H_0(u_i)$ there exists a directed path of 0-links from u_j to u_i. Since $u_j \in U-F$ and $u_i \in F$, there exists a 0-link from

$U-F$ to F on this path, contradicting the assumption that F is an AFS. //

For what follows, let $G'=(U',E')$ denote a general, undirected graph.

Definition 3 : A subset $K \subseteq U'$ is called a **vertex cover set** (VCS) [11] of G' if every edge in G' is incident to at least one vertex in K. A **minimum vertex cover set** (MVCS) is a VCS of minimum cardinality in G'.

Definition 4 : A subset $M \subseteq E'$ is called a **matching** [11] if no vertex in U' is incident to more than one edge in M. A **maximum matching** is a matching of maximum cardinality in G'.

A **bipartite graph** , with bipartition (X,Y), is one whose vertex set can be partitioned into two subsets X and Y such that every edge is incident to a vertex in X and a vertex in Y. Finally, for $u_i \in U'$, $N(u_i)$ denotes the set of all vertices which are adjacent to u_i.

III. A CHARACTERIZATION OF t/s-DIAGNOSABLE SYSTEMS

As stated earlier in Section I, Friedman [6] introduced the concept of t/s-diagnosability which allowed the possible replacement of fault-free units in system repair. In this section, we present a characterization of a t/s-diagnosable system. This characterization generalizes the ones given in [8] and [9] for t/t-diagnosable systems.

Definition 5 [6]: A system S is t/s-**diagnosable** if and only if, given a syndrome, all faulty units can be isolated to within a set of at most s units, provided that the number of faulty units in the system does not exceed t.

From the above definition, $t \leq s$ for a system to be t/s-diagnosable. It should be observed that a system is trivially t/s-diagnosable if $n=s$. Thus, in this paper, it is required that $0<t\leq s<n$. It should also be noted that under these conditions $n \geq 2t+1$ for t/s-diagnosable systems.

Definition 6 : Given a system S and a subset $X \subseteq U$, a set $A = \{X_1,...,X_r\}$ is said to be a t-**decomposition** of X if and only if $\bigcup_{1\leq i \leq r} X_i = X$ and $0< |X_i| \leq t$ for $1\leq i \leq r$. The set P_X is the collection of all t-decompositions of X.

Theorem 1 : A system S with test interconnection graph $G=(U,E)$ is t/s-diagnosable if and only if for all $X \subseteq U$, $|X| >s$, and for all t-decompositions $A \in P_X$, there exist subsets $X_i,X_j \in A$ such that there is a test from $U-X_i-X_j$ to $X_i \oplus X_j$.

Proof : (Necessity) Assume that the system S is t/s-diagnosable and that the condition of the theorem does not hold. Then there exists $X \subseteq U$ with $|X| >s$ and $A \in P_X$ such that for all $X_i,X_j \in A$ there is no test from $U-X_i-X_j$ to $X_i \oplus X_j$. Consider now the syndrome where for each edge $(u_k,u_l) \in E$ the outcome is defined as follows :

Case 1. $u_k \in U-X$ or $u_l \in U-X$

    1.1. $u_k, u_l \in U-X$; then set $a_{kl} = 0$

    1.2. $u_k \in U-X$ and for all $X_i \in A$, $u_l \in X_i$; then set $a_{kl} = 1$

    1.3. $u_k \in X$ and $u_l \in U-X$; then set $a_{kl} = 0$

Case 2. $u_k, u_l \in X$

    2.1. $u_k, u_l \notin X_i$ for some $X_i \in A$; then set $a_{kl} = 0$

It is observed that in this case, there is no subset $X_j \in A$ such that $u_l \in X_j$ and $u_k \notin X_j$. For otherwise, there would be a test from $U-X_i-X_j$ to $X_i \oplus X_j$.

    2.2. For all $X_i \in A$ either $u_k \in X_i$ or $u_l \in X_i$; then set $a_{kl} = 1$

It can be seen that for the above syndrome each $X_i \in A$ is an allowable fault set of size less than or equal to t. Since the union of all these allowable fault sets is X and $|X| > s$, no subset of units of U of size at most s can isolate the faulty units for the above syndrome. Hence S is not t/s-diagnosable, a contradiction.

(Sufficiency). Proof is given by using a contrapositive argument. Assume S is not t/s-diagnosable. Then there exists a syndrome, say θ, and subsets $X_1, X_2, ..., X_m$ of size at most t such that these subsets are allowable fault sets with respect to θ and that $|X| > s$, where X is the union of $X_1, X_2, ..., X_m$.

Suppose that for some pair X_i, X_j, $1 \le j < k \le m$ there is a test from $U-X_i-X_j$ to $X_i \oplus X_j$. Let (u_k, u_l) be such a test edge. Without loss of generality, let $u_l \in X_i$. If X_i is the fault set then the test outcome $a_{kl} = 1$; if X_j is the fault set then $a_{kl} = 0$. This contradicts the assumption that both X_i and X_j are allowable fault sets for θ. This shows that the condition of the theorem is not satisfied. //

Combining the above Theorem with Lemmas 2 and 3 of [9], we obtain the following equivalent characterizations of t/t-diagnosable systems.

Theorem 2 :

1. S is t/t-diagnosable.

2. For all $X \subseteq U$ with $|X| > t$ and for all t-decompositions $A \in P_X$, there exist subsets $X_i, X_j \in A$ such that there is a test from $U-X_i-X_j$ to $X_i \oplus X_j$.

3. For any two sets of units $X_1, X_2 \subseteq U$ where $|X_1| = |X_2| = t$ and $X_1 \ne X_2$, there is a test from $U-X_1-X_2$ to $X_1 \oplus X_2$.

4. For any two sets of units $X_1, X_2 \subseteq U$ where $|X_1|, |X_2| \le t$ and $|X_1| \nsubseteq |X_2|$, there is

a test from $U-X_1-X_2$ to $X_1 \oplus X_2$.

5. For each integer p with $0 \leq p \leq t$ and each $X' \subseteq U$ with $|X'|=2(t-p)$, $|\Gamma^{-1}(X')| > p$, where $\Gamma^{-1}(X')$ denotes the set containing every unit u such that $u \notin X'$ and u tests a unit in X'.

IV. BASIC PROPERTIES OF ALLOWABLE FAULT SETS

In this section we we establish certain properties of allowable fault sets with respect to a given syndrome. Our study is directed towards investigating conditions for a vertex v to be in an allowable fault set of size at most t. For this purpose we use the notion of implied-fault set and the implied-fault graph used by Dahbura and Masson [4] in their study.

Given a syndrome for a system S, define the **implied-fault graph** $G'=(U',E')$ to be an undirected graph such that $U'=U$ and $E'= \{(u_i,u_j) : u_i \in L(u_j)\}$. For $u \in U$, let G'_u denote the subgraph of G' obtained after all units in $H_0(u)$ and all edges incident on these units have been removed from G'. Let K_u represent a MVCS of G'_u and let $G-H_0(u)$ denote the subgraph of G where all vertices in $H_0(u)$ along with all edges incident on these vertices have been removed from G. Finally, we define $G'(F)$ to be the subgraph of G' such that all edges which connect vertices entirely inside F have been deleted.

Recall that we have assumed that $u_i \notin L(u_i)$ for any $u_i \in U$. This means that G' has no self-loops.

The results of the following lemma can be found in [4]. We present this lemma for the sake of completeness.

Lemma 4 : Given a syndrome for a system S, we have the following:

(i)   Every AFS of G is a VCS of G'.

(ii) If $F \subseteq U$ is a minimal VCS of $G*$, then F is an AFS of G.

(iii) $F \subseteq U$ is a MAFS of G if and only if F is a MVCS of G'.

Proof:

(i)   Let F be an AFS of G for the given syndrome. Assume F is not a VCS of G'. Then there exist $u_i,u_j \in U-F$ such that (u_i,u_j) is an edge in G'. Since all edges from $U-F$ into F in G are 1-links (F is an AFS of G) and an implied-fault path between u_i and u_j can contain only one 1-link, all vertices which lie on an implied-fault path between u_i and u_j must belong to $U-F$. But this implies that there is a 1-link between two vertices in $U-F$, contradicting the assumption that F is an AFS of G. This shows that (i) holds.

(ii) Let F be a minimal VCS of G'. Assume (ii) does not hold. Then at least one of the following conditions is satisfied.

(a) There exist $u_i, u_j \in U-F$ with $a_{ij}=1$

(b) There exist $u_j \in F$, $u_i \in U-F$ with $a_{ij}=0$

Assume (a) holds. Then the edge (u_i, u_j) is in G'. But this contradicts the fact that F is a VCS of G' since neither u_i nor u_j is a member of F.

Now assume (b) holds and (a) does not hold. Since F is a minimal VCS of G' there exists a unit u_k in $U-F$ such that (u_j, u_k) is an edge in G'; for otherwise $F-\{u_j\}$ will be a VCS contradicting the minimality of F. Hence $u_j \in L(u_k)$. Since $a_{ij}=0$, it follows that $u_i \in L(u_k)$ and so (u_i, u_k) is an edge in G'. Since neither u_i nor u_k is a member of F, this contradicts the fact that F is a VCS of G'.

(iii) Statement (iii) follows from (i) and (ii). //

Lemma 5 : F is an AFS in G of minimum size containing unit v if and only if $H=F-H_0(v)$ is a MAFS of $G-H_0(v)$.

Proof : (Necessity) We first show that H is an AFS of $G-H_0(v)$. Since $U-F = (U-H_0(v))-H$ and F is an AFS of G all edges within $(U-H_0(v))-H$ are 0-links and all edges from $(U-H_0(v))-H$ into H are 1-links. Hence H is an AFS of $G-H_0(v)$. To show that H is a MAFS of $G-H_0(v)$, assume H_1 is a AFS of $G-H_0(v)$. Clearly all edges with both vertices incident on vertices in $(U-H_0(v))-H_1$ are 0-links and all edges from $U-H_0(v)-H_1$ into H_1 are 1-links. Now consider edges from $(U-H_0(v))-H_1$ into $H_0(v)$. These edges must all be 1-links, otherwise the vertices incident on these edges would all belong to $H_0(v)$. This shows that the set $H_1 \bigcup H_0(v)$ is an AFS of G. Hence if $H_1 < H$ then $H_1 \bigcup H_0(v)$ is an AFS of smaller size than F, contradicting the fact that F is an AFS of minimum size containing v. Hence $H_1 \leq H$ and H is an MAFS of $G-H_0(v)$.

(Sufficiency) If $F-H_0(v)$ is an MAFS of $G-H_0(v)$ then as we have shown in the proof of necessity, F is an AFS of G. If F is not an AFS in G of minimum size containing v, then let F_1 be an AFS of G containing v with $|F_1| < |F|$. But then $F_1-H_0(v)$, from the necessity part, would be an AFS of $G-H_0(v)$ of smaller size than $F-H_0(v)$, a contradiction.

Lemma 6 : For $v \in U$, $(G-H_0(v))'=G_v'$.

Proof : Since the vertex sets of both graphs are the same, we need only show that the edge sets are identical. Clearly every edge in $(G-H_0(v))'$ is in G_v'. Now assume that there is an edge (u_i, u_j) in G_v' which is not in $(G-H_0(v))'$. Then every implied-fault path in G between u_i and u_j must contain at least one vertex from $H_0(v)$. But this implies that either u_i or u_j is a member of $H_0(v)$, contradicting the assumption that both vertices are members of $G-H_0(v)$. Hence the two edge sets are also identical. //

Lemma 7 : Given a syndrome for a system S, let $F \subseteq U$ be an AFS containing $v \in U$. Then $F-H_0(v)$ is a VCS of G_v'.

Proof : Let $F_1 = F - H_0(v)$. From Lemma 3 and the proof of Lemma 5, it follows that F_1 is an AFS of $G - H_0(v)$. Then from Lemma 4, F_1 is a VCS of $(G - H_0(v))'$. Thus, by Lemma 6, F_1 is a VCS of G_v'. //

Theorem 3 : Given a syndrome for a system S, F is an AFS of minimum size among all allowable fault sets that contain unit $u \in U$ if and only if $F - H_0(v)$ is a MVCS of G_v'.

Proof : Proof follows from Lemmas 4, 5 and 6. //

The condition in the above theorem can be used to test if a unit belongs to an AFS of size at most t for a given syndrome. However this condition requires determining a MVCS for a general undirected graph, a problem which is known to be NP-Complete. So, we would like to develop a test which requires determining a MVCS of a bipartite graph. With this objective in mind, we now define a bipartite graph for each vertex v. This bipartite graph is derived from G_v'. We then relate an MVCS of this graph to an AFS containing vertex v and establish certain properties of this AFS which will be used in the following sections to develop appropriate diagnosis algorithms.

Given a system S and a syndrome, define $B = (U_B, E_B)$ to be the undirected bipartite graph with bipartition (X, Y) where

$$X = \{x_1, ..., x_n\}, \ Y = \{y_1, ..., y_n\}$$

and

$$E_B = \{(x_i, y_j) : u_i \in L(u_j) \text{ in } S\}$$

For $v \in U$, define the undirected bipartite graph $B_v = (U_v, E_v)$ with bipartition (X_v, Y_v) to be the vertex induced subgraph of B such that

$$X_v = \{x_i : u_i \in U - H_0(v)\}, \ Y_v = \{y_i : u_i \in U - H_{0(v)}\}.$$

Let t_v correspond to the value $t_v = t - |H_0(v)|$.

Theorem 4 : Given a syndrome for a system S, a unit $v \in U$ does not belong to any AFS of size at most t if B_v has a MVCS of size greater than $2t_v$.

Proof : Let the size of a MVCS of B_v be greater than $2t_v$. Assume $v \in U$ belongs to an AFS F such that $|F| \leq t$. Let $H = F - H_0(v)$. Clearly $|H| \leq t_v$. Define $B_X(H) = (U_X, E_X)$ to be the vertex induced subgraph of B_v, where

$$U_X = \{x_i : u_i \in H\} \cup \{y_i : u_i \in U - H\}.$$

$B_Y(H) = (U_Y, E_Y)$ is defined to be the vertex induced subgraph of B_v, where

$$U_Y = \{x_i : u_i \in U - H\} \cup \{y_i : u_i \in H\}.$$

Clearly $F_X = \{x_i : u_i \in H\}$ and $F_Y = \{y_i : u_i \in H\}$ are VCS's of $B_X(H)$ and $B_Y(H)$ respectively. It follows that $F_B = F_X \cup F_Y$ is a VCS of $B_X(H) \cup B_Y(H)$. Since F is an AFS, in G' there are no edges connecting vertices of $U - F$. From this it follows that every edge in $B_v - B_X(H) - B_Y(H)$ connects vertices in F_B. Therefore F_B is a VCS of B_v, contradicting our assumption that the size of a MVCS of B_v is greater than t_v. Hence v is not contained in any AFS of size at most t. //

In the following we use $F_B(v)$ to denote a MVCS of B_v. For a given $F_B(v)$ let

$$F_I = \{u_i \mid x_i \in F_B(v) \text{ and } y_i \in F_B(v)\}$$

and

$$F_v = \{u_i \mid x_i \in F_B(v) \text{ or } y_i \in F_B(v)\}.$$

We now proceed to establish certain properties of F_v.

Lemma 8 : $F_v \bigcup H_0(v)$ is an AFS of G.

Proof: Assume the contrary. Then at least one of the following conditions is satisfied.

(a) There exist $u_i, u_j \in U - F_v - H_0(v)$ with $a_{ij} = 1$

(b) There exist $u_j \in F_v \bigcup H_0(v)$ and $u_i \in U - (F_v \bigcup H_0(v))$ with $a_{ij} = 0$

Assume (a) holds. Then the edge (u_i, u_j) is in G_v'. Hence (x_i, y_j) is an edge in B_v. But this contradicts the fact that $F_B(v)$ is a VCS of B_v since neither x_i nor y_j is a member of $F_B(v)$.

Now assume (b) holds and (a) does not hold. Clearly $u_j \notin H_0(v)$; for otherwise u_i would also belong to $H_0(v)$. Thus $u_j \in F_v$. Hence either x_j or y_j is a member of $F_B(v)$. Without loss of generality let $x_j \in F_B(v)$. Since $F_B(v)$ is a MVCS of B_v, there exists y_k in B_v with $y_k \notin F_B(v)$ such that (x_j, y_k) is an edge in B_v. Hence $u_j \in L(u_k)$. Since $a_{ij} = 0$, $u_i \in L(u_k)$. Hence (x_i, y_k) is an edge in B_v. Since neither x_i nor y_k is a member of $F_B(v)$, this contradicts the fact that $F_B(v)$ is a VCS of B_v. //

Lemma 9 : Given a syndrome for a system S and a unit $v \in U$, we have the following:

(1) In G', there is no edge (u_i, u_j) with $u_i \in U - (F_v \bigcup H_0(v))$ and $u_j \in F_v - F_I$.

(ii) In G, there is no edge (u_i, u_j) with $u_i \in U - (F_v \bigcup H_0(v))$ and $u_j \in F_v - F_I$.

Proof:

(1) Assume the contrary. Let (u_i, u_j) be an edge from $U - (F_v \bigcup H_0(v))$ into $F_v - F_I$ in G'. Then either x_j or y_j is not a member of $F_B(v)$. Thus in B_v either the edge (x_i, y_j) or the edge (x_j, y_i) is not incident on any vertex in $F_B(v)$, contradicting the fact that $F_b(v)$ is a VCS of B_v.

(ii) By Lemma 8, the set $F_v \bigcup H_0(v)$ is an AFS of G. Thus every edge from $U - (F_v \bigcup H_0(v))$ into $F_v - F_I$ in G must be a 1-link. So if such an edge (u_i, u_j) exists in G, then (u_i, u_j) is an edge in G'. Thus from (1) it follows that there is no edge (u_i, u_j) in G with $u_i \in U - (F_v \bigcup H_0(v))$ and $u_j \in F_v - F_I$. //

Proof: To show that every AFS of G contained in $F_v \bigcup H_0(v)$ contains the subset F_I it suffices to show that every VCS of G' contained in $F_v \bigcup H_0(v)$ contains F_I. The above assertion holds if every vertex in F_I is incident on some vertex of $U - (F_v \bigcup H_0(v))$ in G'. Assume the contrary. Let u_k be a vertex in F_I which is not incident on any vertex of the set $U - (F_v \bigcup H_0(v))$. Then let $W_B(v) = \{x_i \mid u_i \in F_v\} \bigcup \{y_i \mid u_i \in F_I - \{u_k\}\}$. From Lemma 9(1) and the construction of B_v it follows that in B_v, there is no edge (x_i, y_j) with $u_i \in U - (F_v \bigcup H_0(v))$ and $u_j \in F_v - F_I$. So $W_B(v)$ is a VCS of B_v. But $|W_B(v)| = |F_B(v)| - 1$. This contradicts the assumption that $F_B(v)$ is a MVCS of B_v. Thus every vertex in F_I is incident on some vertex of $U - (F_v \bigcup H_0(v))$ in G'. This implies that every VCS of G' contained in $F_v \bigcup H_0(v)$ contains the subset F_I. By Lemma 4, it follows that every AFS of G contained in $F_\bullet \bigcup H_0(v)$ contains the subset F_l. //

V. $O(n^{3.5})$ ALGORITHM FOR DIAGNOSIS OF
A $t/t+1$-DIAGNOSABLE SYSTEM

In this section we establish that in the case of a $t/t+1$ diagnosable system the condition of Theorem 4 is both necessary and sufficient for a vertex v to be in an AFS of size at most t. This will lead to an $O(n^{3.5})$ diagnosis algorithm to isolate all faulty units to within at most $t+1$ units in a $t/t+1$-diagnosable system. First we derive a necessary condition for a system to be $t/t+k$-diagnosable.

Lemma 11 :If S, a multiprocessor system with test interconnection $G=(U,E)$, is $t/t+1$-diagnosable then for all $X_i, X_j \subseteq U$ with $|X_i| > t$, $X_j \nsubseteq X_i$, and $|X_i| + |X_j| \leq 2t$, there exists a test from $U-X_i-X_j$ into $X_i \oplus X_j$.

Proof: Assume S is $t/t+1$-diagnosable but the condition does not hold. Then there exist $X_i, X_j \subseteq U$ with $|X_i| > t$, $X_j \nsubseteq X_i$, $|X_i| + |X_j| \leq 2t$ such that there is no test from $U-X_i-X_j$ into $X_i \oplus X_j$.

Since $|X_i| > t$ and $X_j \nsubseteq X_i$, $|X_i \bigcup X_j| > t+1$, we construct two sets W_i and W_j from X_i and X_j by moving elements from X_i-X_j into X_j-X_i until W_i and W_j have size at most t. Thus we obtain two sets W_i and W_j with $|W_i| \leq t$, $|W_j| \leq t$ such that there is no test from $U-W_i-W_j$ into $W_i \oplus W_j$. By Theorem 1, this contradicts the assumption that S is $t/t+1$-diagnosable. //

Recall from the previous section that $F_B(v)$ is a MVCS of $B_\bullet$ and $F_\bullet$ and F_l are sets derived from $F_B(v)$.

Theorem 5 : Given a syndrome for a $t/t+1$-diagnosable system S, a unit $u \in U$ belongs to an AFS of size at most t if and only if $|F_B(v)| \leq 2t_\bullet$. //

Proof : If $|F_B(v)| > 2t_\bullet$ then, by Theorem 4, G does not contain an AFS of size at most t containing the unit v.

Now assume $|F_B(v)| \leq 2t_\bullet$. If $F_\bullet \bigcup H_0(v)$ contains an AFS of G of size at most t containing the unit v then we are through. So assume $|F_B(v)| \leq 2t_\bullet$ and $F_\bullet \bigcup H_0(v)$ does not contain any AFS of G of size at most t containing the unit v. From Lemma 8, $F_\bullet \bigcup H_0(v)$ is an AFS of G containing the unit v. If $F_\bullet = F_l$ then $|F_\bullet \bigcup H_0(v)| \leq t$ since $|F_B(v)| \leq 2t_\bullet$. So we further assume that $F_\bullet \neq F_l$. Since G' does not contain any units with self-loops and $F_\bullet \neq F_l$, the subset $|F_\bullet - F_l| \geq 2$. Let F_α be an AFS of smallest size containing unit v such that $F_\alpha \subseteq F_\bullet \bigcup H_0(v)$. Clearly $|F_\alpha| > t$.

By Lemma 10 every AFS of G contained in $F_\bullet \bigcup H_0(v)$ contains the subset F_l, and since $v \in F_\alpha$, it follows that $F_l \bigcup H_0(v) \subseteq F_\alpha$.

We next show that $F_\alpha \neq F_\bullet \bigcup H_0(v)$. Let $u_i \in F_\bullet - F_l$. Then $W = F_\bullet - \{u_i\}$ is a VCS of G'.

because, by Lemma 9(1), in G_v', there is no edge (u_i, u_j) with $u_i \in U - (F_v \bigcup H_0(v))$ and $u_j \in F_v - F_l$. Hence by Lemma 6, W is a VCS of $(G - H_0(v))'$. This means, by Lemma 4(11), W contains an AFS of $G - H_0(v)$. Thus $W \bigcup H_0(v)$ has an AFS of G containing unit v and of size less than that of $F_v \bigcup H_0(v)$. Since F_α is an AFS of smallest size containing v such that $F_\alpha \subseteq F_v \bigcup H_0(v)$, it follows that $F_b = (F_v \bigcup H_0(v)) - F_\alpha$ is non-empty and the set $F_\beta = F_b \bigcup F_l \bigcup H_0(v)$ is not contained in F_α.

Now $|F_\alpha| + |F_\beta| = |F_B(v)| + 2|H_0(v)|$
$$\leq 2t_v + 2|H_0(v)| \leq 2t$$

Thus we have $|F_\alpha| > t$, $F_\beta \not\subseteq F_\alpha$, $|F_\alpha| + |F_\beta| \leq 2t$, and there is no test from $U - F_\alpha - F_\beta$ into $F_\alpha \oplus F_\beta$. This, by Lemma 11, contradicts our assumption that the system S is $t/t+1$-diagnosable. //

Given a valid syndrome for a $t/t+1$-diagnosable system S and a unit v in S, we have shown that the bipartite graph B_v has a MVCS of size at most $2t_v$ if and only if G has an AFS of size at most t containing the unit v. Thus we have the following algorithm to isolate all faulty units in a $t/t+1$-diagnosable system.

Algorithm I: Diagnosis of a $t/t+1$-Diagnosable System

Step 1. Given a $t/t+1$-diagnosable system S and a syndrome, construct the bipartite graph $B = (U_B, E_B)$ with bipartition (X, Y).

Step 2. Set $F = \phi$; for all $v \in U$, label v unmarked.

Step 3. **While** there exists an unmarked $v \in U$
  begin
    3.1. Label v marked.
    3.2. Set $t_v = t - |H_0(v)|$.
    3.3 Construct B_v from B.
    3.4. Compute a maximum matching K_v of B_v
        using the Hopcroft/Karp algorithm [12].
    3.5. **If** $|K_v| \leq 2t_v$ then add v to F.
  end
Step 4. **If** $|F| \leq t+1$ **then** F is the required set
  else the given syndrome is not valid.

The bipartite graph in Step 1 can be constructed in $O(n^{2.5})$ operations [4]. Step 2 requires $O(n)$ operations. The computation within Step 3 is dominated by the computation of a maximum matching which requires $O(n^{2.5})$ operations. Since Step 3 is performed for each unit in U, the complexity of the entire algorithm is $O(n^{3.5})$.

VI. CONCLUSIONS

In this paper we have studied the problem of diagnosing t/s-diagnosable systems. First we presented a characterization of these systems which generalize those given in [8,9] for t/t-diagnosable systems. We then presented a diagnosis algorithm for $t/t+1$-diagnosable systems which runs in $O(n^{3.5})$ time. This algorithm is based on certain properties derived from the characterization of t/s-diagnosable systems (Section III) and the structure of allowable fault sets (Section IV).

REFERENCES

[1] Preparata, F.P., Metze, G., Chien. R.T., "On the Connection Assignment Problem of Diagnosable Systems," *IEEE Trans. Electr. Compt.* , vol EC-16, pp. 848-854, Dec 1967.

[2] Hakimi, S.L., Amin, A., "Characterization of the Connection Assignment of Diagnosable Systems," *IEEE Trans. Comp.* , vol C-23, pp 86-88, Jan 1974.

[3] Sullivan, G.F., "A Polynomial Time Algorithm for Fault Diagnosability," in *Proc. 25th Annu. Symp. Foundations Comp. Sc.* , pp. 148-156, Oct 1984.

[4] Dahbura, A.T., Masson, G.M., "A Practical Variation of the $O(n^{2.5})$ Fault Diagnosis Algorithm," in *14th Int. Symp. Fault-Tolerant Comput.* , 1984.

[5] Sullivan, G.F., "An $O(t^3+|E|)$ Fault Identification Algorithm for Diagnosable Systems," *IEEE Trans. Comp.* , vol C-37, pp. 388-397, April 1988.

[6] Friedman, A.D., "A New Measure of Digital System Diagnosis," in *Dig. 1975 Int. Symp. Fault-Tolerant Comput.* , pp. 167-170, June 1975.

[7] Kavianpour, A., Friedman, A.D., "Efficient Design of Easily Diagnosable Systems," in *Proc. 3rd USA-Japan Comput. Conf., pp. 251-257, 1978.*

[8] Chwa, K.Y., Hakimi, S.L., "On Fault Identification in Diagnosable Systems," *IEEE Trans. Comp.* , vol. C-30, pp. 414-422, June 1981.

[9] Yang, C.L., Masson, G.M., Leonetti, R.A., "On Fault Identification and Isolation in t_1/t_1-Diagnosable Systems," *IEEE Trans. Comp.* , vol C-35, pp. 639-643, July 1986.

[10] Sullivan, G., "The Complexity of System-Level Fault Diagnosis and Diagnosability," *Ph.D. Dissertation* , Yale University, 1986.

[11] Bondy, J.A., Murty, U.S.R., *Graph Theory with Applications* , New York: Elsevier North-Holland 1976.

[12] Hopcroft, J.E., Karp, R.M., "A $n^{2.5}$ algorithm for maximum matching in bipartite graphs," *SIAM J. Comput.* , vol. 2, Dec 1973.

Toward a complete representation of graphoids in graphs
- Abridged Version

Robert Y. Geva & Azaria Paz†
Technion IIT - Haifa Israel

Abstract: We consider 3 place relations $I(X,Z,Y)$ where X, Y and Z are sets of propositional variables and $I(X,Z,Y)$ stands for the statement "Knowing Z renders X independent of Y". These relations are called *graphoids*. The theory of graphoids uncovers the axiomatic basis of informational dependencies and ties it to vertex separation in graphs. In this paper we advance towards a characterization of graphoids by families of graphs. Given two graphs R and B, let M be the set of all independencies which are implied by R and B under closure by the 5 graphoid axioms (defined in the text). We show the following results:

1   An algorithm which generates a family of graphs that represent M.

2   The number of graphs needed to represent M might be exponential.

3   A polynomial time algorithm for the following problem: given an independency t is $t \in M$?

4   We define annotated graphs and show an efficient representation of M by this model.

1. Introduction and motivation

Any system that reasons about knowledge and beliefs must make use of information about dependencies and relevancies. If we have acquired a body of knowledge Z and wish to assess the truth value of a proposition X, it is important to know whether it would be worthwhile to consult another proposition Y, which is not in Z. In other words, before we examine Y, we need to know if it's value can potentially generate new information relative to X, information not available from Z.

Many *AI* systems approach this problem in ad - hoc ways. These systems, though computationally convenient, are semantically sloppy. They often yield surprising and counterintuitive conclusions ([Pearl 88]).

The other approach to the problem of dealing with irrelevance, as with any other notion involving uncertainty, is to handle it within probability theory, which is the appropriate mathematical framework. The problem with this approach is that it cures the problem of lack of semantics, but introduces computational inefficiency.

The goal of the theory of *graphoids* is to make probabilistic systems operational by making relevance relationships explicit.

The representation has to be made in a way which will make it easy to identify the facts which are irrelevant and therefore can be neglected, or, even better, make it easy to identify the relevant facts, which must be considered. This is done via representation in graphs: belief networks encode relevancies as neighbouring vertices in a graph, thus ensuring that by consulting the neighbourhood one gains a license to act: what you don't see locally doesn't matter. This graph representation will be explained in the sequel.

† The results shown in this report are part of the MSc Thesis of the first author done under the supervision of the second author - to be presented to Graduate School of the Technion, Israel Institute of Technology. Contribution of the second author supported by the fundation for the promotion of research at the Technion.

The relationship of conditional irrelevance was given mathematical definition in probability theory by Lauritzen [Lauritzen 82]. One can interpret conditional irrelevance as conditional independence. Given a joint probability distribution $P()$, the random variables X and Y are irrelevant when Z is known if

$$P(X,Y\,|Z)=P(X\,|Z){\cdot}P(Y\,|Z).$$

It has been shown [Pearl & Paz 85] that any relation I induced by a probability distribution is closed under the following 4 independent axioms: (the relation symbol I is omitted for clearity, and the notation YW stands for the union $Y \cup W$).

(1)	$(X,Z,Y) \rightarrow (Y,Z,X)$	Symmetry
(2)	$(X,Z,YW) \rightarrow (X,Z,Y) \wedge (X,Z,W)$	Decomposition
(3)	$(X,Z,YW) \rightarrow (X,ZY,W)$	Weak Union
(4)	$(X,Z,Y) \wedge (X,ZY,W) \rightarrow (X,Z,YW)$	Contraction

In addition if for every X, $P(X){>}0$ then the relation I is closed under a 5'th axiom:

(5)	$(X,ZY,W) \wedge (X,ZW,Y) \rightarrow (X,Z,YW)$	Intersection

This set of 5 axioms is termed *graphoid axioms*, and a set of triplets closed under these axioms is a *graphoid*. For example consider axiom (2). The left hand side is equivalent (under the given interpretation) to

$$P(X\ Y\ W\ |\ Z)=P(X\,|Z){\cdot}P(Y\ W\,|Z)$$

Summing the above equation over all possibles values w of W we get

$$\sum_{w}[P(X\ Y\ W\,|Z)=P(X\,|Z){\cdot}P(Y\ W\,|Z)] \equiv$$

$$P(X\ Y\,|Z)=P(X\,|Z){\cdot}P(Y\,|Z) \equiv$$

$$I(X,Z,Y).$$

Similarly summing over Y we get $I(X,Z,W)$ obtaining the right hand side of axiom (2).

Consider an undirected graph as a model of representation for an independency relation I. The vertices of the graph represent the variables in U. A triplet (X,Z,Y) (X,Y and Z are subsets of of the vertex set of the graph) is *represented* by the graph if the set of vertices Z is a cutset between the sets of vertices X and Y.

Pearl and Paz have managed to characterize the properties of cutset - separation in graphs by very similar axioms. A set of triplets (a relation) can be represented by an undirected graph if and only if it is closed under the following independent axioms:

(1)	$(X,Z,Y) \rightarrow (Y,Z,X)$	Symmetry
(2)	$(X,Z,YW) \rightarrow (X,Z,Y) \wedge (X,Z,W)$	Decomposition
(5)	$(X,ZY,W) \wedge (X,ZW,Y) \rightarrow (X,Z,YW)$	Intersection
(6)	$(X,Z,Y) \rightarrow (X,ZW,Y),\ W \subset U$	Strong Union
(7)	$(X,Z,Y) \rightarrow (X,Z,\gamma) \vee (\gamma,Z,Y),\ \gamma \in U$	Transitivity

Remarks:

1. γ is a singleton element of U and all three arguments of $I()$ must represent disjoint subsets.

2. The axioms are clearly satisfied for vertex separation in graphs. Eq. (7) is the contrapositive form of connectedness transitivity, stating that if X is connected to some vertex γ and γ is connected to Y then X is connected to Y. Eq. (6) states that if Z is a vertex cutset separating X from Y then adding more vertices W from

the graph to Z leaves X and Y still separated. Eq. (5) states that if X is separated from Y with W removed and X is separated from W with Y removed then X must be separated from both Y and W.

3.   Eqs. (5) and (6) imply the contraction axiom (no (4)), and also the converse of Eq. (2) which is:

(8) $$(X, Z, Y) \wedge (X, Z, W) \rightarrow (X, Z, YW) \qquad\qquad\qquad \text{composition}$$

meaning that I is completely defined by the set of triplets (a, Z, b) in which a and b are individual elements of U. Note that the union axiom, Eq. (6) is unconditional and therefore strictly stronger then Eq. (3) which is required for probabilistic dependencies.

While axioms 1,2,5,6 and 7 imply the set of axioms 1,2,3,4, and 5 as is easy to see, the two sets of axioms are logically inequivalent. Therefore, not every graphoid can be represented by an undirected graph.

One way out is to settle for approximation. Pearl and Paz defined an approximation in terms of I- mapness: a graph G is an I- map of a graphoid I if there is a 1 - 1 correspondence between the universe of the random variables of the graphoid and the vertices of the graph, and if a triplet is represented by the graph then it is in the graphoid (but not necessarily the other way around).
Note that the complete graph is always a trivial I- map as no triplet is represented by it. Indeed, it has been shown that there are graphoids for which the only I- map is the complete graph.

Another possibility is to extend the representation of relations as defined above for *sets of graphs*. This approach is followed in this paper.

The main advantages of graph representation are that the size of the graph is polynomial in the size of the set of random variables and it can represent exponentially many triplets.

An illustration

Consider the relation $I = \{(1,2,3),(3,2,1),(1,4,3,),(3,4,1)\}$. The relation is a graphoid, as is easy to verify. (It is closed under the five graphoid axioms). The relation is represented by the two graphs shown below.

Figure 1.1

Thus (1,2,3) and (3,2,1) are represented by the first graph since the vertex 2 is a cutset between the vertices 3 and 1. However, no single graph can represent this relation. To see this, assume that there exists a graph that represents this relation. Then it's vertex set must contain all four vertices, to represent the elements 1,2,3 and 4. If the graph is complete then no triplet is represented by it (no cutset exists), therefore it can not represent the given graphoid. If some edge e.g. (1, 3) is missing, then by definition the triplet $(1,\{2,4\},3)$ is represented by the graph. But no triplet in the graphoid has two elements in it's middle entry, and therefore the graph can not represent the graphoid. Thus no single graph can represent this graphoid, and moreover, the only I-map for this graphoid is the trivial complete graph.

2. Definitions and previous results

Any three - place relation $I(\ )$ that satisfies Eqs. 1,2,3 and 4 is called a *semi - graphoid*. If it satisfies Eqs. 1,2,3 and 5 it is called a *pseudo - graphoid* and a relation that satisfies all 5 Eqs. is a *graphoid*.

A triplet $t=(X,Z,Y)$ is *represented* by a set of graphs $\{G_1,\ldots,G_n\}$ if it is represented by at least one of the graphs. The triplet is represented by one of the graphs, say $G_i=(V_i\ E_i)$ if $X \cup Y \cup Z \subset V_i$ and the vertex set Z is a cutset in G_i between the vertex sets X and Y.

It has been shown in [Pearl & Paz 85] that not every graphoid can be represented by an undirected graph. On the other hand it has been shown [Geiger 87] that a three place relation is closed under the axioms symmetry, decomposition and weak union if and only if it can be represented by a set of graphs.

Definition: Let $G_i(V_i)$ denote a graph over V_i. If α, β, are elements of V_i then "(α, β) is a *nonedge* of G_i" means that (α, β) is not in the edge set of G_i.

Definition: For $V_j \subsetneq V_i$ define (α, β) as a *nonedge* of G_i mod V_j if α is not connected to β in $G_i(V_i/V_j)$ i.e., removing the vertices V_j and the incident edges from $G_i(V_i)$ will render α and β disconnected.

Definition: A graph $G_1=(V_1,E_1)$ *implies* another graph $G_2=(V_2,E_2)$ if $V_2 \subset V_1$ and every nonedge (a,b) of G_2 is a nonedge of G_1 mod $V_2-\{a,b\}$.

Remark: A graph G_1 implies another graph G_2 if and only if every triplet represented by the graph G_2 is represented also in the graph G_1.

Definition (of the $\otimes$ operation on graphs): Given two graphs $G_i(V_i)$ and $G_j(V_j)$ let $V_k = V_i \cap V_j$ and assume $V_k \neq \varnothing$. Define the graph $G_k(V_k)= G_i \otimes G_j$ as follows:

1   Every pair (α, β) over V_k which is a nonedge of G_i mod $(V_k - \alpha - \beta)$ *or* is a nonedge of G_j mod $(V_k - \alpha - \beta)$ is a nonedge of G_k

2   Every pair (α, β) to which 1 above does not apply is an edge of G_k

    Example:

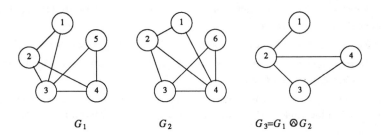

$$G_1 \qquad\qquad G_2 \qquad\qquad G_3{=}G_1 \otimes G_2$$

Figure 2.1

G_3 is not implied by either G_1 nor G_2 e.g. $(1,2,\{3,4\})$, $(1,2,3)$ and $(1,2,4)$ represented by G_3 are not represented by G_1 nor in G_2.

Notice that $(1, 3)$ is a nonedge in the graph G_2 and that $(1, 4)$ is a nonedge in G_1 mod $\{2,3\}$. ∎

It also has been shown [Paz 87] that a three place relation is closed under the axioms symmetry, decomposition, weak union and intersection if and only if it can be represented by a set of graphs closed under the $\otimes$ (intersection) operation.

The if part is easy. The approach taken in [Paz 87] to prove the only if case is the following: assume a basis $B=\{t_1,\ldots,t_n\}$ is given, consisting of triplets $t_i=(X_i,Z_i,Y_i)$. The goal is to develop an algorithm for representing the *pseudo - graphoid* closure by a set of graphs such that a triplet $t=(X,Z,Y)$ is in the closure if and only if Z is a cutset between X and Y in some graph in the set of graphs.

It is shown first that the set of graphs $\{G_i, G_j, G_i \otimes G_j\}$ is the pseudo graphoid closure of the graphs G_i and G_j. The algorithm represents the triplets in the basis B by an initial set of graphs. Then, based on the above mentioned property of the operation $\otimes$, it generates new graphs via the $\otimes$ operation applied on previously generated graphs. When the algorithm halts, the set of graphs is closed under the $\otimes$ operation, and the set of triplets represented by the set of graphs is the pseudo - graphoid closure. After each graph generation, a better approximation (in terms of I- mapness) is achieved.

This approach expresses the scenario in which an expert gives intuitively some irrelevancy statements (triplets) and the knowledge engineer has to find their graphoid closure.

A computational problem involved in this approach is that the number of graphs generated may be exponential in the size of the set of random variables.

A solution to this problem is the following: do not generate the graphs. Construct only the representation of the basis. In order to test membership of a given triplet to the closure execute the following steps:

1. Let $G_1, G_2, \ldots, G_l$ be the graphs that contain $V_t = X \cup Y \cup Z$ in their vertex set.
2. Construct $G_t = G_1 \otimes G_2, \ldots, \otimes G_l$.
3. The triplet t is in the closure if and only if it is represented in the graph G_t.

3. New results and paper organization

In this paper we advance towards extending the results from pseudographoids to graphoids. We assume an initial set of triplets, which we represent in an initial set of graphs. Then we develop an algorithm which, given two graphs, produces a set of graphs which represents the graphoid closure of the triplets represented by the two given graphs. Unlike the pseudographoid closure algorithm which creates a simple new graph in order to achieve the closure of the two given graphs, the graphoid closure operation may require the creation of exponentially many graphs (in the number of vertices) in order to achieve the closure of the two given graphs (as will be shown in the sequel).

On the other hand we will show polynomial algorithms for the following two problems:

1.   The membership problem: given a triplet t and two graph, decide whether t is in the graphoid closure of the triplets in the two given graphs.

2.   Represent the closure of two given graphs in two "annotated" graphs (the notion of an annotated graph will be defined in the sequel). This result is extended to a more general case involving a "chain" of graphs.

The rest of the paper is organized as follows:

In the next section we present some definitions and technical lemmas.

In section 5 we prove that the graphoid closure of two graphs may require exponentially many graphs for it's representation.

In section 6 we investigate the emptiness problem which is the following: are there triplets in the closure of two given graphs R and B which are not represented by any one of the three graphs R, B and $R \otimes B$? We show that this problem is decidable in polynomial time.

In section 7 we present the contraction algorithm: given two graphs, the contraction algorithm generates a set of graphs that represent the graphoid closure of the triplets in the given graphs.

In section 8 we define the membership problem and prove that the problem is decidable in polynomial time.

In section 9 we define annotated graphs, and show how to represent the closure of the triplets represented by any two given graphs in two annotated graphs.

In section 10 we generalize the result of section 9 from the case of two graphs to the case of a "chain" of graphs (defined in the text).

In the last section we draw conclusions and state some open problems.

4. Technical lemmas

(The proofs are omitted here but will be included in the full paper).

When talking about pairs of graphs we will usually call them R and B - "red" and "blue", for convenience. The vertices common to both graphs will be called V_{BR}. In the left hand side of the contraction axiom there are two triplets: one of the form (X,YZ,W) and the other (X,Z,Y). The first one will sometimes be referred to as the 'big' one, and the second one as the 'small' one.

Let S be a set of graphs. We define

1. $[S]$ is the set of triplets represented by the graphs in the set S.

2. $CL_G(S)$ is the set of all triplets in the closure under the 5 graphoid axioms of the triplets represented by the graphs in S.

3. $cont(R,B)$ is defined as $cont(R,B)=CL_G(\{R,B,R \otimes B\})-[\{R,B,R \otimes B\}]$.

Remark: We will say that t is *in* S when the triplet t is represented by one of the graphs in the set of graphs S.

Let $G=(V,E)$ be a graph. Let U be a subset of the vertex set V. Let (a,b) be two vertices in U. Then $V_G^U(a,b)$ is the set of all vertices in $V-U$ such that discarding $U-a$ from the graph will render them connected to a *and* discarding $U-b$ from the graph G will render them connected to b.

These are the vertices on the paths between the two vertices a and b, outside the vertex set U.

Example: in the graph shown in figure 4.1 below, $V_G^z(a,b)=\{u,v\}$.

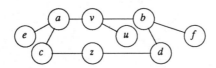

Figure 4.1

4.1 An equivalent axiom

Define the symmetric contraction (notation : SC) axiom as below:

SC: $(Q,XZ,YW) \wedge (QX,ZY,W) \wedge (X,Z,Y) \rightarrow (QX,Z,YW)$.

Lemma 4.1 The contraction axiom and SC are equivalent modulo the decomposition axiom.

Note: In order to get the right hand side of the SC axiom one has to apply twice the contraction axiom. This observation is valuable since in the development of a contaction algorithm (see section 7 in the sequel) the use of the SC axiom instead of the contraction axiom may save computation time.

4.2 The restriction lemma

Let $G = (V, E)$ be a graph, and V' a subset of V. Let $C(V')$ be the complete graph over the vertices V'. $\otimes G(V')$ is defined to be the graph resulting from G and $C(V')$ by the $\otimes$ operation. In the two graphs R and B shown below, $B = \otimes R(\{x, y, z, w\})$.

Figure 4.2

Lemma 4.2: (The restriction lemma): Let $G = (V, E)$ be any graph, V' is a subset of V. Then the triplet (X, Z, Y) is represented by $\otimes G(V')$ if and only if (X, Z, Y) is represented by G and $X \cup Y \cup Z \subset V'$.

5. The complexity of $cont(R, B)$

As opposed to the pseudographoid case, in which one additional graph is enough in order to represent the closure of a set of triplets given by two graphs, we show here that the graphoid closure is a much more complicated problem.

Theorem 5.1: For every integer $n \geq 2$ there are two graphs R_n and B_n over $3n + 1$ vertices such that $cont(R, B)$ requires at least $2^n - 2$ additional graphs for its graphical representation.

The theorem and it's proof are illustrated here for the case $n=2$. The general case is similar. The two graphs R_2 and B_2 are shown in figure 5.1 below.

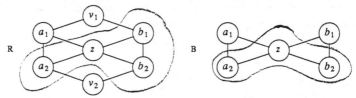

Figure 5.1

The basic idea is the following: $2^n - 2$ triplets are shown to exist in $CL(R, B)$, such that if any two of them are represented by the same graph, then some triplet can be found, which is represented by the same graph, but is not in $CL(R, B)$. Therefore all the $2^n - 2$ triplets must be represented by $2^n - 2$ different graphs.

In the case where $n=2$ there are two pairs of triplets implying two additional triplets:

1. $(a_1, a_2 z b_2 b_1, v_2)$ in R and $(a_1, a_2 z b_2, b_1)$ in B imply $(a_1, a_2 z b_2, b_1 v_2)$.

2. $(a_2, a_1 z b_1 b_2, v_1)$ in R and $(a_2, a_1 z b_1, b_2)$ in B imply $(a_2, a_1 z b_1, b_2 v_1)$.

The representation of the first pair of triplets is illustrated in the above figure. The vertices in the cutsets are encircled.

These two implied triplets must be represented by two different additional graphs. If the two of them are represented by the same graph, then by transitivity and decomposition, one of the two triplets (a_1, a_2zb_2, v_1) or (v_1, a_2zb_2, b_1) is represented by that same graph. One can prove that neither one of these two triplets is in $CL(R,B)$ by showing that the contraction algorithm, described in section 7 below, does not generate their representation.

6. The Emptiness Problem

Since the general problem of representing the graphoid closure is hard, it is natural to try to single out easy cases. The most general easy case is the case where the size of the set of graphs needed to represent the closure is polynomial. In this section we investigate such a case, the case where the graphoid closure is equal to the pseudographoid closure. We present an algorithm which decides this problem in polynomial time.

Formally, the emptiness problem for two given graphs R and B is to decide whether $cont\ (R,B) \neq \emptyset$.

We need first a few lemmas. If $cont\ (R,B) \neq \emptyset$ then there exist 3 triplets t_1, t_2 and t_3 such that t_1 and t_2 are in two different graphs of the set $S = \{R, B, R \otimes B\}$ and t_3, which is implied by contraction from t_1 and t_2, is not in S.

Lemma 6.1: If $cont\ (R,B) \neq \emptyset$ then we may assume the following configuration w.l.o.g.: there are triplets t_1, t_2 and t_3 such that

$t_1 = (X,YZ,W), t_1 \in R, t_1 \notin B, t_1 \notin R \otimes B$
$t_2 = (X,Z,Y), t_2 \in B, t_2 \notin R$
$t_3 = (X,Z,YW), t_3 \notin S, S = \{R, B, R \otimes B\}$
$X, Y, Z \subset V_{BR}, W \not\subset V_B, W \neq \emptyset$.

The proof of this lemma is easy and left to the reader.

Lemma 6.2: Given that triplets t_1, t_2 and t_3 as in lemma 6.1 exist, another set of triplets t_1', t_2', t_3' can be found such that $t'_1 = (a, V_{BR} - a, W')$ $t'_2 = (a, V_{BR} - a - b, b)$ $t'_3 = (a, V_{BR} - a, bW')$ a, b are vertices; $W' = W - V_{BR} \neq \emptyset$
$t'_1 \in R, t'_1 \notin B, t'_1 \notin R \otimes B; t'_2 \in B, t'_2 \notin R; t'_3 \notin S$.

The proof of the lemma is based on the following argument:

Given that $t_2 = (X,Z,Y) \notin R$ there must be two vertices $a \in X$ and $b \in Y$ which are connected in R (outside Z) according to one of the following 4 alternatives

1   (a,b) is an edge in R

2   a is connected to b in R via a path inside $V_{BR} - Z$

3   a is connected to b in R via a path outside V_{BR}

4   a is connected to b in R via an alternating path, inside $V_{BR} - Z$ and outside V_{BR}

For any of the above alternatives, triplets t'_1, t'_2, t'_3 as required can be found - based on the given triplets t_1, t_2, t_3. Notice that since $R \otimes B \in S, t'_3 \notin S$ implies that t'_3 is not implied by the intersection of any two triplets in S.

This configuration is illustrated in figure 6.1 below.

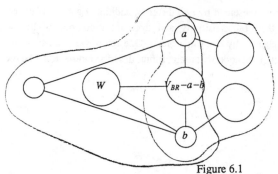

Figure 6.1

The left hand side circle is the graph R, the right hand side circle is the graph B and in the intersection are the vertices V_{BR}, common to both graphs.

We are now ready to present the algorithm.

Emptiness Algorithm, Empty (B,R).

Procedure check (B,R)

1   Construct the two graphs: $\otimes B\ (V_{BR})$ and $\otimes R\ (V_{BR})$.

2   Check if $\otimes R\,(V_{BR})$ implies $\otimes B\,(V_{BR})$ (i.e., check whether all the independencies represented by $\otimes B\,(V_{BR})$ are represented by $\otimes R\,(V_{BR})$, due to graph properties it is enough to check the independencies represented by nonedges in $\otimes B(V_{BR})$, (a,V_B-a-b,b)). If $\otimes R\,(V_{BR})$ implies $\otimes B\,(V_{BR})$ return 'false'

3   Else -- collect all witnesses to that fact, namely all edges (a,b) in $\otimes R\,(V_{BR})$ which are nonedges in $\otimes B\,(V_{BR})$.

4   For each witness (a,b) check if there exists a nonempty set $U \subset V_R - V_{BR}$ such that either $(a,V_{BR}-a,U)$ or $(b,V_{BR}-b,U)$ holds in R. If this test succeeds for at least one of the witnesses -- return 'true', otherwise return 'false'.

End of procedure.

Set Empty $(B,R)=no$ iff Check (B,R) or Check (R,B).

Example: Consider the two graphs R and B shown below.

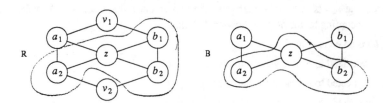

Figure 6.2

B is a subgraph of R and therefore when witnesses are collected from R, there will be no sets U found in the graph B in step 4 of the procedure. The witnesses in B are the two pairs of vertices (a_1,b_1) and (a_2,b_2). In step 4, the vertex v_2 which is in V_R-V_B will be found disconnected from a_1, (also from b_1) and the vertex v_1 will be found disconnected from a_2 (and also from b_2). Therefore the algorithm's answer is yes, and indeed $(a_1,za_2,b_1b_2v_2) \in cont\,(R,B)$. The representation of the pair of triplets implying this triplet is illustrated in figure 6.2, as cutsets are encircled.

Lemma 6.3: The emptiness algorithm is polynomial (in the maximal number of vertices in R or B).

Proof: The complexity of the algorithm equals the complexity of the procedure involved in it. As to the procedure:

1   Step 1 is polynomial by definition of the $\oslash$ operation.

2   Step 2 is polynomial (implication is determined by edges comparison, since both graphs have the same set of vertices).

3   Step 3 is carried out as a part of step 2. The distinction between them is only for clearity.

4   Step 4 can be carried out in the following way: for every witness (a, b) remove the set $V_{BR}-a$ from the graph and check whether there is some vertex set disconnected from the vertex a, then remove the set $V_{BR}-b$ from the graph and check whether there is some vertex set disconnected from the vertex b. (This check can be done using the BFS algorithm [Even 79]).

Lemma 6.4: The emptiness algorithm is correct.

Proof: If the algorithm outputs "yes" then it means that it has found a triplet in $cont(R, B)$ and therefore $cont(R, B) \neq \varnothing$.
On the other hand, if the algorithm answer is "no", then if it was obtained in step 2 then every small triplet in B is also in R. If it was obtained in step 4, then due to lemma 6.2 only triplets of the form $a, V_{BR}-a-b, b)$ have to be tested, it means that no small triplet has a corresponding big triplet and therefore $cont(R, B) = \varnothing$.

7. A Contraction Algorithm

In this section we present an algorithm that takes as input two graphs R and B, and generates graphs to represent $CL(R, B)$. It follows from theorem 5.1 that this algorithm may generate exponentially many graphs, and hence it may run for an exponential time. However it is polynomial in the length of the output. Therefore if it is run on an input which belongs to an easy case, such that a polynomial set of graphs is output, then the algorithm will be polynomial for such a case.

The algorithm generates graphs to represent $CL(R, B)$. Atfer each generation of a graph, a better appoximation (in terms of I-mappness) is achieved. Therefore, in the general case, the algorithm may be run with a time bound to generate an I-map as good as one can afford.

Another benefit of this algorithm is that given R and B, it provides a characterization of the set of graphs that represents $CL(R, B)$. As was mentioned in section 5, we had to show that certain triplets were not in $CL(R, B)$. This was done by showing that the contraction algorithm does not generate their representation.

Unlike the the problem of the pseudographoid closure, solved by the $\oslash$ operation, the contraction axiom involves some intrinsic difficulties which must be taken into account.
Specifically consider the following example: Let R and B be the graphs shown in figure 7.1 below:

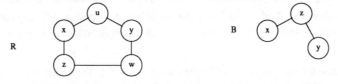

Figure 7.1

Then (x,yz,w) is represented by R and (x,z,y) is represented by B. By contraction we get (x,z,yw) which is not represented by R nor in B.

If we disconnect u from y or u from x in R (which are the only possibilities for representing the new triplet in R) we will get a new graph, say the one in figure 7.2.a, replacing R, in which (x,z,yw) is represented. But also the triplet (ux,z,yw) is represented, and this triplet is not implied (by the graphoid axioms) from the triplets represented by R and B, as one can check explicitly.

Therefore the only way out is to represent the new triplet (x,z,yw) in a new graph, the one in figure 7.2.b, and add this new graph to the set of graphs B and R.

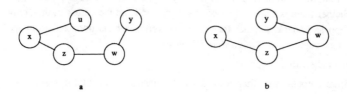

Figure 7.2

To overcome the above mentioned difficulties we will use the following construction: We will find all pairs of vertices (a,b) which are nonedges in B modulo V_{BR} and not nonedges in R modulo V_{BR}. These are witnesses in terms of the emptiness algorithm. Each witness induces a partition on the set of vertices $V_R - V_B$ such that one subset of $V_R - V_B$ is the set of vertices which are directly connected (outside the set V_{BR}) to both vertices a and b. These vertices should not appear in the graph resulting from R by removing the edge (a,b). Therefore they will be discarded (vertex u in the example). All other vertices are either not connected to a, they will play the role of W is the SC axiom, or not connected to b - they will play the role of Q in the axiom. These vertices should stay in the graph.

Every witness may induce a different partition, and therefore it will follow from the construction that the algorithm is exponential in it's input. On the other hand it is easy to see that the algorithm is polynomial in its' output.

A Contraction Algorithm

1. Construct $P = R \otimes B$.

2. For each graph H in $\{R, B\}$ do:

   2.1. Insert to a set S_H all pairs $(a,b) \in V_P \times V_P$, $a \neq b$, of vertices such that: (a,b) is a nonedge of P and (a,b) is not a nonedge of H modulo $V_P - a - b$.

   2.2. For each $(a,b) \in S_H$ find the set $V_H^P(a,b)$.

   2.3. For each $(a,b) \in S_H$ do:

      2.3.1 If $V_H^P(a,b) \cup V_P = V_H$ then set $S_H = S_H - \{(a,b)\}$.

      2.3.2 If $V_H^P(a,b) = \varnothing$ then remove the edge (a,b) from H and set $S_H = S_H - \{(a,b)\}$.

   2.4. For each subset S' of S_H do: Let U be the union of all the sets $V_H^P(a,b)$ such that $(a,b) \in S'$. Construct the graph $\otimes H(V - U)$ and remove all the edges $(a,b) \in S'$ from the graph.

End.

Consider again the two graphs R_2 and B_2 shown in figure 5.1. The contraction algorithm, applied on this two graphs, generates the two additional graphs shown in figure 7.3 below.

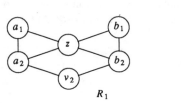 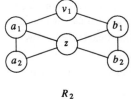

$$R_1 \qquad\qquad R_2$$

Figure 7.3

Theorem 7.1: Let D be the set of graphs output by the contraction algorithm. Then $[D]=CL(R,B)$.

proof: We prove equality by double inclusion.

Lemma 7.1: $[D] \subset CL(R,B)$

Idea of proof: The Idea is the following: we show that whenever an edge (a,b) is removed from a graph $G=(V,E)$, which is the only operation in the algorithm that generates the representation of new triplets, the only new triplet of the form $(u,V-u-v,v)$ (u and v are single vertices) is the triplet $(a,V-a-b,b)$. We show that this triplet is in the closure by deriving it via the axioms. Then we apply a proof in [Pearl & Paz 85] to show that all the triplets in the graph are in the closure.

To prove the other direction, which is $CL(R,B) \subset [D]$ we show that $[D]$ includes all the triplets in $[R]$ and $[B]$, and is closed under the five graphoid axioms.

Lemma 7.2 $\{[R] \cup [B]\} \subset [D]$.

Proof: The two graphs R and B are in D. Edges may only be deleted from them, not added to them.

Lemma 7.3: $[D]$ is closed under symmetry, decomposition and weak union.

Proof: Trivial.

Lemma 7.4: $[D]$ is closed under intersection.

Idea of proof: The idea is to show that the set of graphs D is closed under the $\otimes$ operation. The lemma then follows from Paz's theorem, mentioned in section 2 above.

Lemma 7.5: $[D]$ is closed under contraction.

Idea of proof: The proof is by way of contradiction. The set $[D]$ is not closed under contraction if and only if there are two triplets, t_1 and t_2, represented by two different graphs in the set D, implying a third triplet by contraction, and the third triplet is not in $[D]$. As in the proof of the emptiness algorithm we may assume that certain elements in the triplets are singletons. Then we show that the small triplet is always in the graph P. Then we show explicitly how the algorithm takes care of this situation, and derive a contradiction to the assumption that the set $[D]$ is not closed under contraction.

8. The membership problem

A possible way out of the need to generate exponentially many graphs is not to generate an explicit representation of the closure while still having a polynomial membership testing algorithm. Such a solution is provided in this section.

Definition: The membership problem for two graphs R and B and a triplet t is to determine whether $t \in CL_G(R,B)$.

In this section we show that the membership problem is decidable in polynomial time. This result is somewhat surprising, given the fact that the graphical representation of $CL_G(R,B)$ may require exponential space. Consider the algorithm defined below.

Membership Algorithm:

Let V_t be the set of nodes appearing in the triplet t.

1. If $V_t \not\subset V_R$ and $V_t \not\subset V_B$ - return no.

2. If $V_t \subset V_R \cap V_B$ then construct $P = R \otimes B$. $t \in CL_G(R,B)$ if and only if t is in P.

3. Else - assume $V_t \subset V_R$ (otherwise exchange the names of the graphs).
    Let $V_{tP} = V_t \cup V_P$.

3.1. Construct $P = R \otimes B$ and $R'_t = \otimes R(V_{tP})$.

3.2. Collect all pairs (a,b) of vertices in P which are nonedges in P and edges in R'_t.

3.3. For each pair (a,b) collected in step 3.2, find the set of vertices $V^P_{R't}(a,b)$. If $V(a,b) = \emptyset$ then remove the edge (a,b) from R'_t.

4. Let R_t be the graph obtained from R'_t in step 3.3. $t \in CL_G(R,B)$ if and only if t is in R_t.

Theorem 8.1 The membership algorithm is correct.

The idea of the proof is to show that correctness is implied by the contraction algorithm. The given triplet t is in the graph R_t if and only if it is in some graph in the set of graphs D, generated by the contraction algorithm.

Remark: The algorithm is obviously polynomial, since there are no iterations in it, and every step is polynomial.

9. An efficient representation of $CL(R,B)$

As we have seen in section 5, the representation of $CL(R,B)$ may require exponential space. The question is whether this limitation is inevitable. Obviously, as long as we represent triplets in graphs, using the cutset separation semantics, the answer is yes.

However, a closer look at the contraction algorithm shows that the information, according to which the graphs are constructed, can be stored in polynomial space and generated in polynomial time. This information is the set of pairs of vertices S, and for each element (a,b) in S the set of vertices $V^P_R(a,b)$ (or $V^P_B(a,b)$).

If R and B are two graphs over a set of n vertices then the number of pairs of vertices is S is of order n^2 and for each pair of vertices, the size of the vertex set associated to it is of order n and therefore the size of the necessary information is of order n^3.

We will show next that this reduced representation is also efficient in the sense that a polynomial membership algorithm can be based on it.

We now give the formal definitions.

Definition: An *annotated graph* is a triple $\tilde{G} = (V, E, K)$ where V is a set of vertices, E is a set of edges and K is a function $K: V \times V \to P(V)$.

Definition: The *underlying graph* of an annotated graph $\tilde{G} = (V, E, K)$ is $G = (V, E)$.

The representation of the graphoid closure of the triplets in two given graphs R and B can be set in two annotated graphs constructed by the following algorithm:

1. Construct $P = R \otimes B$.

2. For each graph H in $\{R, B\}$ do:

   2.1 Insert to a set S_H all pairs of vertices $(a, b) \subset V_P \times V_P$, $a \neq b$, such that (a, b) is a nonedge of P and (a, b) is not a nonedge of H modulo $V_P - a - b$.

   2.2 For each $(a, b) \in S$ find the set $V_H^P(a, b)$.

   2.3 For each $(a, b) \in S$ do:

   2.3.1 If $V_H^P(a, b) \cup V_P = V_H$ then set $S = S - \{(a, b)\}$.

   2.3.2 If $V_H^P(a, b) = \emptyset$ then set $E_H = E_H - \{(a, b)\}$; set $S_H = S_H - \{(a, b)\}$.

   2.4 Define the function K_H as follows: for each $(a, b) \in S_H$ set $K_H(a, b) = V_H^P(a, b)$.

3. Output: $\bar{R} = (V_R, E_R, K_R)$ and $\bar{B} = (V_B, E_B, K_B)$.

End.

Example: Consider again the two graphs R_2 and B_2 defined in section 5. Since the graph B_2 is a subgraph of R_2, their graphoid closure is represented by one annotated graph $\bar{R}$ whose underlying graph is the original graph R_2 and the function K is as follows:

$$K(a_1, b_1) = \{v_1\} \; ; \; K(a_2, b_2) = \{v_2\}$$

We now show a membership testing algorithm for the two resulting annotated graphs and a triplet t.

Let V_t be the set of elements in the triplet t.

1. If $V_t \not\subset V_R$ and $V_t \not\subset V_B$ then return no.

2. If $V_t \subset V_{BR}$ then construct $P = R \otimes B$. Check if t is in P and answer accordingly.

3. Otherwise, V_t is a subset of one of the vertex sets and not of the other. Say $V_t \subset V_R$. Set $U = \{\cup K_R(a, b)\}_{K_R(a, b) \cap V_t = \emptyset}$ and set $S = \{(a, b): K_R(a, b) \subset U\}$.

4. Construct the graph $R'_t = \otimes R (V_R - U)$ and remove all the edges (a, b) in S from the graph.

5. Let R_t be the resulting graph. $t \in CL(R, B)$ if and only if t is in R_t.

End.

Theorem 9.1 A triplet t is in the graph R_t constructed by the membership testing algorithm if and only if it is in $CL_G(R, B)$.

   Idea of proof: The idea of the proof is to show that the triplet t is in the graph R_t if and only if in in is in the set of graphs D generated by the contraction algorithm applied on the two graphs R and B.

10. Representation of the Graphoid Closure of a Chain of Graphs

We say that $G_1 = (V_1, E_1), \ldots, G_n = (V_n, E_n)$ is a chain of graphs if every graph is a subgraph of the next graph i.e. $V_i \subset V_{i+1}$ and $E_i \subset E_{i+1}$ for all $1 \leq i \leq n-1$.

   In this section we generalize the results shown in the previous section and show how to represent in one annotated graph, the graphoid closure of triplets given in a chain of graphs.

Example: a chain of five graphs in shown in figure 10.1 below.

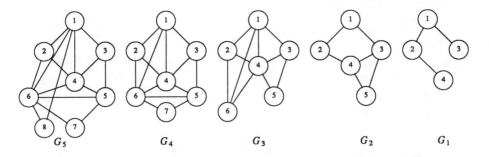

Figure 10.1

A generalization of the annotation shown in the previous section is to annotate the second graph according to the first graph, then to annotate the third graph according to the second graph *and it's annotation,* and so on.

Applying this procedure on the chain of graphs in our example would yield the following annotation functions:

$$K_2(3, 4)=\{5\}$$

$$K_3(3, 4)=\{5\} \; ; \; K_3(1, 4)=\{6\}$$

$$K_4(3, 4)=\{5, 7\} \; ; \; K_4(1, 4)=\{6, 7\} \; ; \; K_4(5, 6)=\{7\}$$

$$K_5(3, 4)=\{5, 7\} \; ; \; K_5(1, 4)=\{6, 7, 8\} \; ; \; K_5(5, 6)=\{7\}$$

The next step, as in the previous section, is to define when a given triplet is represented by a given annotated graph. This has to be done carefully. As opposed to the previous case, the two sets of vertices, the one in the domain of the function K and the one in the range of the function, are not necessarily disjoint. In order to test the representation of a triplet by an annotated graph we will discard sets of vertices which are in the range of the function K, and then remove the appropriate edge in the domain of the function. However, we have to discard these sets of vertices in an order which will make sure that we do not discard a set of vertices which includes an endpoint of an edge on which the function K is defined. The problem is that the removal of an edge may be necessary in order to discover the representation of a triplet, even if the endpoints do not appear in the triplet. This requirement induces a partial order on the sets of vertices in the range of the function. We will discard the sets of vertices and remove the corresponding edges according to this partial order.

We develop a membership testing algorithm which is polynomial and which serves as the definition of the representation of a triplet by an annotated graph (the membership testing algorithm described in the previous section is a special case). Under this definition, we can show that all the triplets represented by the annotated graph $\tilde{G}_i$ are also represented by the annotated graph $\tilde{G}_{i+1}$. Therefore all the triplets represented by the annotated chain are represented by the last annotated graph $\tilde{G}_n$. Then we can show that the set of triplets represented by the annotated graph $\tilde{G}_n$ is the graphoid closure of the set of triplets represented by the initial chain of graphs.

To conclude the example, let us assume now that we have to test whether the triplet $t=(3\,8, 1\,6, 2\,4)$ is represented by $\tilde{G}_5$. The membership testing algorithm will generate first the graph shown in figure 10.2.a and then the graph in figure 10.2.b.

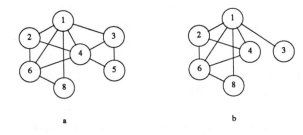

Figure 10.2

The triplet t is represented by the second graph, therefore the algorithm halts and outputs "yes".

Directed acyclic graphs (DAGs) have also been studied in the literature [Pearl &Verma 87} as a model for representation of graphoids. It has been shown that not every graphoid can be represented by a DAG. It has also been shown that there are graphoids that can be represented be an undirected graph but can not be represented by a DAG, and conversely there are graphoids that can be represented by a DAG and can not be represented by an undirected graph. Thus no one of these two models of representation is more expressive then the other.

Verma [Verma 86] has shown that any DAG can be uniquely defined by a structure which can be represented by a chain of graphs. Therefore it follows from our result that every graphoid which can be represented by a DAG can be represented by an annotated graph.

11. Conclusions

We have shown that the graph representation of $CL(R, B)$ when R and B are two given graphs, can be constructed, but the construction is not efficient since the representation may require exponentially many graphs.

We have shown a polynomial time algorithm for the membership problem.

We have defined annotated graphs as triples of vertices, edges and a function that maps pairs of vertices to sets of vertices. We have shown how to represent the graphoid closure of two given graphs in two annotated graphs. This result is generalized to chains of graphs.

The main goal remains open: find an efficient representation of any graphoid in graphs. Based on our results, we can suggest two ways: the first one is to represent a subset of the graphoid in graphs, and construct efficient membership testing algorithms. The second one is to construct a representation by an annotated graph.

Bibliography

[Even 79]

Even, S., 1979. Graph algorithms. Potomac, md.: Computer science Press.

[Geiger 87]

Geiger, D., 1987. "Towards the Formalization of Informational Dependencies," UCLA Cognitive Systems Laboratory, *Technical Report 880053 (R-102)*, (Based on the author's MS thesis).

[Lauritzen 82]

Lauritzen, S.L., 1982 Lectures on contingency tables. 2nd ed. Aalborg Denmark: University of Aalborg press.

[Paz 87]

Paz, A., 1987 "A Full Characterization of Pseudographoids in Terms of Families of Undirected Graphs," UCLA Cognitive Systems Laboratory, *Technical Report 870055 (R-95)*.

[Pearl & Paz 85]

Pearl, J., and Paz, A., 1985. "GRAPHOIDS: A Graph-Based Logic for Reasoning about Relevance Relations," UCLA Computer Science Department, *Technical Report 850038 (R-53); In Advances in Artificial* Intelligence-II, Edited by B. Du Boulay et al. Amsterdam: North-Holland

[Pearl 88]

Pearl, J., 1988. *Probabilistic Reasoning in Intelligent Systems: Networks of Plausible* Inference. Morgan-Kaufmann, San Mateo, CA.

[Pearl & Verma 87]

Pearl, J. and Verma, T.S., 1987. The logic of representing dependencies by directed graphs. Proc. 6th Natl. Conf. on *AI* (AAAI - 87), Seattle, 374 - 79.

[Verma 86]

Verma, T.S., 1986. "Causal networks: Semantics and expressiveness. Technical report R-65, Cognitive Systems Laboratory, University of California, Los Angeles.

CADULA - A Graph-Based Model for Monitoring CAD-Processes

Detlev Ruland

Informatik I, Universität Würzburg, D-8700 Würzburg, W.-Germany

Abstract

CADULA is a graph-based model for describing and monitoring CAD-processes. Today, the design of product consists of several subtasks, which are realized by functions using different and loosely coupled CAD-systems. The functions are not independent, since they commonly design the same product. The dependencies are primarily given by the information flow between the functions, i.e. the data flow between the CAD-systems realizing these functions. In CADULA, a CAD-process is described by a CAD-process scheme and a CAD-process realization. A process scheme describes the functions by specifying their preconditions, actions, and postconditions. Furthermore, the dependencies between the functions are described by a network. A process realization describes the development of a design process for a specific product. Since design processes are evolutionary processes, for each design function several versions and alternatives are developped. A process realization is based on a process scheme, i.e. a process realization must satisfy the structural properties given by the process scheme. Furthermore, CADULA supports hierarchical refinements and decompositions of functions into a number of related smaller functions. The decomposition methodology is widely accepted as the most powerful methodology for breaking down the design complexity. CADULA is currently validated by industrial applications (software engineering and electronic design).

1 Introduction

In today's marketplace the design and manufacturing of a product from the first idea to its series production consists of several stages. Each stage is responsible for a certain specified subtask (subfunction) of the overall design task (design function). Designing products in a single stage is unrealistic, because of the complexity of the products as well as the used design tools and methods, respectively. In many companies these subtasks are often performed independently using different and loosely coupled CAD-systems. This is a direct consequence of the **taylorism** often applied in company organization. But, there are strong dependencies and relationships between the various functions, because they commonly realize the design and development of the same product. The dependencies between the functions are mainly given by the necessary information exchange between the functions, i.e. by the exchange of results. Two functions F_i and F_j depend on each other, if the (suceeding) function F_j needs the results of the (preceding) function F_i. Thus, the dependencies between the functions primarily describe the **information flow** of the CAD/CAM-process.

It is widely accepted, that the fundamental goal of each **CIM-strategy** must be the integration of all design and manufacturing functions by integrating the **information flow** among these functions.

The **decomposition** of a function into a set of subfunctions is a very powerful tool for breaking down the complexity of a design task (cf. e.g. [KaAnCh86], [KeBe87], [Neu83], [NeuHo82], [TotenH87] etc.). The refining set of subfunctions must be in some sense *equivalent* to the decomposed function, i.e. the refining subfunctions must perform the same task as the decomposed function. The decomposition can be iteratively applied such that the final functions are small enough and can easily be performed. Iterative decompositions yield **decomposition hierarchies**.

Summarizing these requirements, a universal and powerful tool for describing and controlling CAD-processes is necessary. CADULA is introduced in this paper. Since CADULA is graph-based, it is universal and covers a wide range of applications. Furthermore, it is simple and clear. CADULA supports the widely used and accepted methods for desribing dependencies between functions as well as for decomposing functions into subfunctions.

CADULA supports two kinds of dependencies between design functions. First, **horizontal dependencies** describe the dependencies between functions in a set of functions yielding a **network of functions**. Second, **vertical dependencies** describe the dependencies between functions resulting from the **decomposition** of functions.

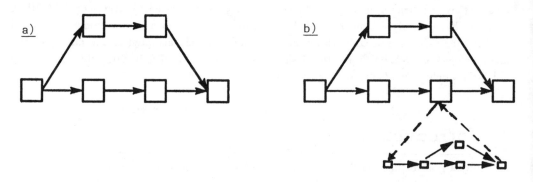

Figure 1: a) Network of Functions, b) Decompositions of Functions

CADULA describes CAD-processes on two different **abstraction levels**. First, **CAD-process schemes** describe the structural properties of CAD-processes. Especially, dependencies and decompositions of functions are described. Second, **CAD-process realizations** describe design processes of certain products by **realization graphs**. Process realizations are based on a process scheme. Since design processes are **evolutionary processes**, for each function of the process scheme several **versions** are developped. The process realization describes the dependencies of the versions, such that these dependencies are *compatible* with the higher level dependencies specified by the process scheme. Thus, each process realization must be an instance of a process scheme. A process scheme describes a set of process realizations.

2 CADULA

2.1 CAD-Environments

A CAD-environment describes the overall scope and frame of a design task (CAD-task).
A CAD-environment consists of a set of **design subtasks**, a **specification space**, a
description space, and a **CAD-function**.
A **CAD-function** is a mapping from a **specification space** into a **description space**.
The **specification space** consists of the **specification objects** of products to be designed. Well-known and widely used methods for describing specification objects are
calculi of the first (or higher) order predicate logic, temporal, and modal logic, respectively. The **description space** (or implementation space) contains the descriptions of
the final design results of products to be designed, i.e. **design objects**. A CAD-function
maps specification objects on design objects, such that the design objects satisfy the
requirements described by the specification models.

Definition:
A **CAD-environment** is a 4-tuple $CAD\text{-}env = (\mathcal{T}, \mathcal{S}, \mathcal{D}, \mathcal{C})$, where:

- $\mathcal{T} = \{T_1, \ldots, T_n\}$ is a set of **design subtasks**,

- $\mathcal{S} = S_1 \times \ldots \times S_n$ is a **specification space**,

- $\mathcal{D} = D_1 \times \ldots \times D_n$ is a **description space**, and

- $\mathcal{C} : \mathcal{S} \to \mathcal{D}$ is called **CAD-function**.

Examples:
The subtasks of the **electronic product design process** are functional capture, logical design (schematic capture), layout (placement and routing), manufacturing planning
(NC-programming) etc. The subtasks of the **software engineering process** are requirement analysis, preliminary design, detailed design, code and debug, test, operation,
and maintenance.

Up to now, no dependencies between design subtasks are considered. There is a direct
dependency between design subtask T_i and T_j, if the results of design subtask T_i are
necessary for design subtask T_j, i.e. the results of design subtask T_i are necessary for
performing design subtask T_j. Hence, T_i must be performed before T_j can be performed.
The dependencies of the design subtasks are expressed by a **dependency relation**,
which is a irreflexive and non-transitive relation on the index set of all subtasks.

Definition:
An irreflexive and non-transitive relation

$$\Theta \subset \{1, \ldots, n\} \times \{1, \ldots, n\}$$

is called a **dependency relation** of CAD-env, iff the transitive closure Θ^* defines an irreflexive ordering on $\{1, \ldots, n\}$, and it exists exactly one minimal and maximal element $min, max \in \{1, \ldots, n\}$ with respect to Θ^*.

Notation:
It is assumed, that the common $<$-ordering is always a Θ-compatible ordering, i.e. $(i, j) \in \Theta^* \Rightarrow i < j$. Hence, $(1, 2, \ldots, n-1, n)$ is always a Θ-compatible sequence. Especially, the minimal (maximal) element in the index set with respect to the $<$-ordering is also the minimal (maximal) element with respect to Θ^*.
The design subtask indexed by the minimal / maximal element (with respect to Θ^*) is called **starting / goal** design subtask.

The dependency relation determines for each design subtask T_i its relevant factors D_j of the description space $\mathcal{D}$. A factor D_j is relevant for T_i, if T_j is a direct predecessor of T_i, i.e. $(j, i) \in \Theta$. D^i consists of all factors of $\mathcal{D}$, which are relevant for T_i, $1 \le i \le n$:

$$D^i := \times_{(j,i) \in \Theta} D_j, \quad \pi_i : \mathcal{D} \to D^i \text{ is the related } \textbf{projection function.}$$

For each design subtask $T_i \in \mathcal{T}$ a **design function**

$$C_i : S_i \times D^i \to D_i$$

is defined. $S_i \times D^i$ is called **requirement space** of the design function C_i, $1 \le i \le n$. An element $(s_i, d^i) \in S_i \times D^i$ is called **requirement object** of the design function C_i.

The design functions $C_1, \ldots, C_n$ must be in a certain sense a horizontal decomposition of a CAD-function C. This is expressed by the following

Horizontal Decomposition Property:
For all $s = (s_1, \ldots, s_n) \in \mathcal{S}$ and all $d = (d_1, \ldots, d_n) \in \mathcal{D}$:
$$C(s_1, \ldots, s_n) = (d_1, \ldots, d_n) \Rightarrow (\forall i \in \{1, \ldots, n\}) : (d_i = C_i(s_i, \pi_i(d))).$$

A CAD-network describes the **horizontal dependencies** between the design functions C_i, which are given by the dependency relation Θ.

Definition:
The **CAD-network** CAD-$net(CAD$-$env, \Theta)$ of a CAD-environment CAD-env and a dependency relation Θ is defined as follows:

$$
\begin{aligned}
CAD\text{-}net(CAD\text{-}env, \Theta) &:= (V, E): \\
V &:= \{C_1, \ldots, C_n\} \\
E &:= \{(C_i, C_j) \mid (i, j) \in \Theta\}.
\end{aligned}
$$

2.2 Refinements and Hierarchical Decompositions

The **refinement** of design function in a CAD-network by a CAD-network is called **hierarchical decomposition**. The refining CAD-network must be in a certain sense compatible with the refined design function. Therefore, the refined function is called **interface function** and the refining CAD-network is called **implementation network** (of the interface function). The network refinement mechanism of CADULA is defined by the **network refinement operator**.

The refinement of a CAD-network induces a refinement of the considered CAD-environment and dependency relation, respectively. The parent CAD-environment and parent dependency relation are refined in such a way, that the refined CAD-network is a CAD-network of the refined CAD-environment and refined dependency relation.

Notation:

$CAD\text{-}net(CAD\text{-}env, \Theta) = (V, E)$ is a CAD-network of the CAD-environment $CAD\text{-}env = (\mathcal{T}, \mathcal{S}, \mathcal{D}, \mathcal{C})$ and dependency relation Θ, where $\mathcal{T} = \{T_1, \ldots, T_n\}$. $T_i \in \mathcal{T}$ is the design subtask to be refined. Hence, the design function C_i is the interface function.

First, a CAD-environment $CAD\text{-}env_i$ is defined, such that the CAD-networks of this CAD-environment are possible implementation networks for the interface function C_i. A CAD-environment $CAD\text{-}env_i := (\mathcal{T}_i, \mathcal{S}_i, \mathcal{D}_i, \mathcal{C}_i)$, where $\mathcal{T}_i := \{T_{i,1}, \ldots, T_{i,n_i}\}$, $\mathcal{S}_i := S_{i,1} \times \ldots \times S_{i,n_i}$, $\mathcal{D}_i := D_{i,1} \times \ldots \times D_{i,n_i}$, and $\mathcal{C}_i : \mathcal{S}_i \to \mathcal{D}_i$ is an **implementation CAD-environment** for the interface function C_i in the parent CAD-environment $CAD\text{-}env$, if the following **vertical decomposition property** holds.

Vertical Decomposition Property:

- $\mathcal{T}_i \cap \mathcal{T} = \{T_{i,n_i}\} = \{T_i\}$

- $\mathcal{S}_i = S_i \times D^i$

- $D_{i,n_i} = D_i$

- $\mathcal{C}_i : \mathcal{S}_i \to \mathcal{D}_i \quad : \quad proj_i \circ \mathcal{C}_i = C_i$, where
  $proj_i : \mathcal{D}_i \to D_i \quad : \quad proj_i(d_{i,1}, \ldots, d_{i,n_i}) := d_{i,n_i}.$

Thus, the CAD-function $\mathcal{C}_i$ of the implementation CAD-environment $CAD\text{-}env_i$ represents in a certain sense the interface function C_i. Each CAD-network $CAD\text{-}net_i$ of the implementation CAD-environment $CAD\text{-}env_i$ is an **implementation CAD-network** for the interface function C_i.

The CAD-network refinement is expressed by the **CAD-network refinement operator** yielding a **refined CAD-network**. The refined network is the suitable union of the parent and implementation CAD-network.

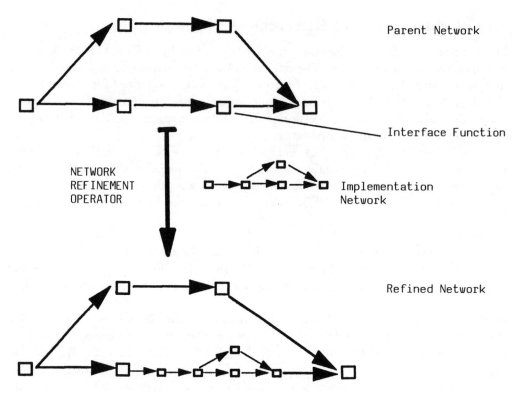

Parent Network

Interface Function

NETWORK
REFINEMENT
OPERATOR

Implementation
Network

Refined Network

Figure 2: Network Refinement Operator

Definition:
Given a CAD-network $CAD\text{-}net_i(CAD\text{-}env_i, \Theta_i) = (V_i, E_i)$ of $CAD\text{-}env_i$
The **network refinement operator**

$$RN(CAD\text{-}net(CAD\text{-}env, \Theta), C_i, CAD\text{-}net_i(CAD\text{-}env_i, \Theta_i))$$
$$:= CAD\text{-}net' = (V', E')$$

is defined as follows:

$$V' := V \cup V_i$$
$$E' := E \cup E_i - \{(C_j, C_i) \in E\} \cup \{(C_j, C_{i,1}) \mid (C_j, C_i) \in E_i\}.$$

The refinement of the CAD-network $CAD\text{-}net$ induces a refinement of the related parent CAD-environment $CAD\text{-}env$ and of the parent dependency relation Θ, respectively.

Definition:
The **environment refinement operator**

$$RE(CAD\text{-}env, T_i, CAD\text{-}env_i) := CAD\text{-}env' = (\mathcal{T}', \mathcal{S}', \mathcal{D}', \mathcal{C}')$$

is defined as follows:

$T' := T \cup T_i,$

$S' := S_1 \times \ldots \times S_{i-1} \times S_{i,1} \times \ldots S_{i,n_i-1} \times S_i \times S_{i+1} \times \ldots S_n,$

$D' := D_1 \times \ldots \times D_{i-1} \times D_{i,1} \times \ldots \times D_{i,n_i-1} \times D_i \times D_{i+1} \times \ldots \times D_n,$

$C' : S' \to D' \quad : \quad proj_D \circ C' := C \circ proj_S,$ where

$proj_D : D_i \to D \quad :$

$proj_D(d_1, \ldots, d_{i-1}, d_{i,1}, \ldots, d_{i,n_i-1}, d_i, d_{i+1}, \ldots, d_n) := (d_1, \ldots, d_{i-1}, d_i, d_{i+1}, \ldots, d_n)$

$proj_S : S_i \to S \quad :$

$proj_S(s_1, \ldots, s_{i-1}, s_{i,1}, \ldots, s_{i,n_i-1}, s_i, s_{i+1}, \ldots, s_n) := (s_1, \ldots, s_{i-1}, s_i, s_{i+1}, \ldots, s_n).$

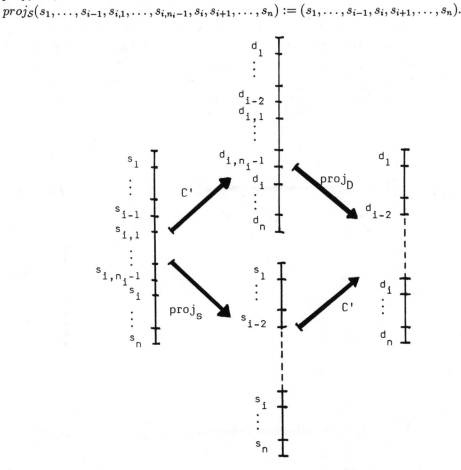

Figure 3: Refined CAD-Function

Definition:

The **dependency relation refinement operator**

$$RD(\Theta, i, \Theta_i) := \Theta'$$

is defined as follows:

$$\Theta' \subset (\{1, \ldots, n\} \cup \{(i, 1), \ldots, (i, n_i - 1)\})^2$$

$$\Theta' := \Theta \cup \Theta_i - \{(j,i) \in \Theta\} \cup \{(j,(i,1)) \mid (j,i) \in \Theta\}.$$

Now, a major result is, that the refined CAD-network CAD-net' is a CAD-network of the refined CAD-environment CAD-env' and the refined dependency relation Θ'.

Lemma:
$$RN(CAD\text{-}net, C_i, CAD\text{-}net_i) = CAD\text{-}net(RE(CAD\text{-}env, T_i, CAD\text{-}env_i), RD(\Theta, i, \Theta_i)).$$

Now, each function $C_j \neq C_i$ of the refined CAD-network CAD-net' can be decomposed again. The interface function C_j can be either a function of the original parent CAD-network CAD-net or of the implementation CAD-network CAD-net_i. Thus, a sequence of network refinement operator applications corresponds to a decomposition hierarchy. Consider the following sequence $\mathcal{REF}$ of n network refinement operator applications:
$$\mathcal{REF} := RN(\ldots RN(Net_0, C_{0,i_0}, Net_{0,i_0})\ldots, C_{n-1,i_{n-1}}, Net_{n-1,i_{n-1}}).$$

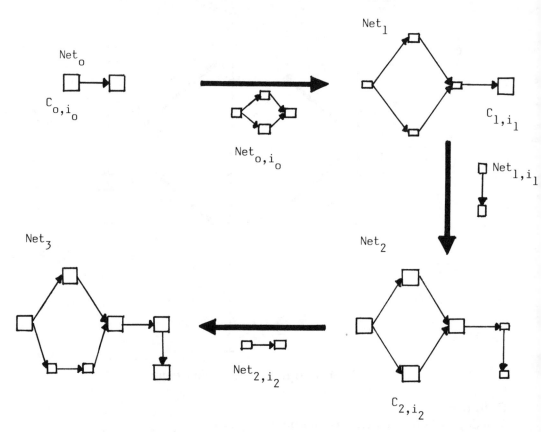

Figure 4: Sequence of Refinement Operator Applications

The interface function C_{0,i_0} in Net_0 is refined by the implementation network Net_{0,i_0}. In the resulting refined network the interface function C_{1,i_1} is refined by Net_{1,i_1}. The interface function C_{1,i_1} is either in Net_0 or in Net_{0,i_0}. The further refinement operators are applied in the same way. Net_j is the network resulting from the j-th refinement operator application, $j = 1,\ldots,n$. Furthermore, $Net_0 = (V_0, E_0)$ and $Net_{j,i_j} = (V_{j,i_j}, E_{j,i_j}), j = 0,\ldots, n-1$. The relationships between refined and refining CAD-networks is described by the **refinement graph**. The edges are the CAD-networks. There is an edge between two CAD-networks, if the second network refines a function of the first network. The edges are marked by the interface functions. Hence, the refinement graph describes in an abstract way the vertical dependencies.

Definition:
The **refinement graph of** $\mathcal{REF}$ is an edge-marked graph $RG_{\mathcal{REF}} := (V, E, m)$:
$V := \{Net_0\} \cup \{Net_{0,i_0}, \ldots, Net_{n-1,i_{n-1}}\}$,
$E := \{(Net, Net') \mid Net'$ is implementation network of an interface function in $Net\}$
$m : E \rightarrow V_0 \cup \bigcup_{j=1}^{n-1} V_{j,i_j} : m(Net, Net') = C :\Leftrightarrow$
(Net' is implementation network of interface function C in Net).

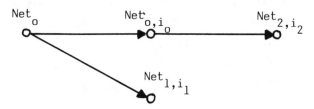

Figure 5: Network Refinement Graph

Suitbale network refinement sequences correspond to decomposition hierarchies.

Lemma:
If $V_0 \cap V_{j,i_j} = \emptyset, 0 \le j \le n-1$ and $V_{j,i_j} \cap V_{k,i_k} = \emptyset, 0 \le j < k \le n-1$, then $RG_{\mathcal{R}}$ is a tree with root Net_0.

Definition:
Each CAD-network is a **CAD-process scheme**.
A refined CAD-network, resulting from a sequence of network refinement operator applications is a **hierarchical CAD-process scheme**.

3   CAD-Process Realizations

A CAD-process scheme describes a set of CAD-process realizations. A CAD-process realization for a CAD-process scheme describes the development of a design object, i.e. a product. Since design processes are **evolutionary processes**, several **versions** of design objects are developped. Consider a certain design subtask T_i and a design subobject $d_i = C_i(s_i, d^i)$. The realization of the design subtask does not produce directly the considered design subobject d_i. Moreover, several versions are sucessively developped. These versions satisfy the same requirements, the desired design object d_i satisfies, i.e. all versions fullfill the requirement object (s_i, d^i). This is expressed by **CAD-relations** and **design relations**, respectively. A **CAD-relation** is a relation among the specification space and the description space, such that the related CAD-function is preserved.

Definition:
A **CAD-relation** of $CAD\text{-}env$ is a relation

$$\mathcal{R} \subset \mathcal{S} \times \mathcal{D} \ : \ (\forall\, s \in \mathcal{S}) : ((s, \mathcal{C}(s)) \in \mathcal{R}).$$

Similary, a **design relation** R_i is defined for each design subtask $T_i \in \mathcal{T}$.

Definition:
A **design relation** of the design subtask T_i is a relation

$$R_i \subset (S_i \times D^i) \times D_i \ : \ (\forall\, (s_i, d^i) \in S_i \times D^i) : (((s_i, d^i), C_i(s_i, d^i)) \in R_i).$$

All design objects, which are related by the CAD-relation to the same specification object / requirement object are called **versions**, i.e.

$$Ver(s) := \{d \in \mathcal{D} \mid (s, d) \in \mathcal{R}\},$$
$$Ver_i(s_i, d^i) := \{d_i \in D_i \mid (s_i, d_i) \in R_i\}, \ 1 \le i \le n.$$

Furthermore, design relations must satisfy the following

Version Property:
For all $((s_1, \dots, s_n), (d_1, \dots, d_n)) \in \mathcal{R}$ and all $i \in \{1, \dots, n\}$:

$$Ver_i(s_i, \pi_i(d)) = Ver_i(s_i, \bar{d}^i), \text{ if}$$
$$\pi_i(d) = (d^i_{i_1}, \dots, d^i_{i_{m_i}}), \ \bar{d}^i = (\bar{d}^i_{i_1}, \dots, \bar{d}^i_{i_{m_i}}) \text{ and}$$
$$\bar{d}^i_{i_j} \in Ver_{i_j}(s_{i_j}, \pi_{i_j}(d)), \ 1 \le j \le m_i.$$

A CAD-process realization for a specification object describes the dependencies of the successively developped versions. Two versions depend on each other, if the second one is derived from the first one.
A CAD-process realization is a **network realization graph**, which consist of exactly one **function realization graph** for each design function in the considered CAD-network. A **function realization graph** for design function C_i describes the dependencies between versions. A **network realization graph** additionally describes the **submission** of design objects between the function realizations graphs. The submission dependencies in the network realization graph must be in a certain sense compatible with the horizontal dependencies in the CAD-network.

Notations:
$CAD\text{-}env = (\mathcal{T}, \mathcal{S}, \mathcal{D}, \mathcal{R})$ is a CAD-environment, where $\mathcal{T} = \{T_1, \dots, T_n\}$. $CAD\text{-}net(CAD\text{-}env, \Theta) = (N, R)$ is a related CAD-network.

Definition:
Given a requirement object $(s_i, d^i) \in S_i \times D^i$.
$C_i\text{-}RG(s_i, d^i) = (V_i, E_i)$ is called C_i-**function realization graph** for (s_i, d^i), iff
$V_i \subset Ver_i(s_i, d^i)$ and (V_i, E_i) is a forest.

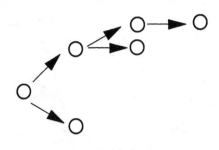

Figure 6: Function Realization Graph

A root design subobject in a C_i-function realization graph is called C_i-**starting subob-ject**. A leaf design subobject is called C_i-**ending subobject**. A path from a C_i-starting subobject to a C_i-ending subobject is called C_i-**development path.**

In a **network realization graph** the submission of design objects between the function realization graphs is described by **submission edges**.

Definition:
Given a design object $d = C(s) = (d_1, \ldots, d_n) \in \mathcal{D}$.
Given a function realization graph $C_i\text{-}RG(s_i, \pi_i(d)) = (V_i, E_i)$ for each design function $C_i, 1 \leq i \leq n$.
$Net\text{-}RG = (V, E)$ is called **network realization graph** of $CAD\text{-}net(CAD\text{-}env, \Theta)$, iff:

- $V := V_1 \cup \ldots \cup V_n,$

- $E := E_1 \cup \ldots \cup E_n \ \cup \ \bigcup_{1 \leq i < j \leq n} E_{i,j},$ such that the following property holds:

$$E_{i,j} := \begin{cases} \emptyset & \text{if } (i,j) \notin \Theta \\ \Delta_{i,j} & \text{if } (i,j) \in \Theta \end{cases}$$

where:

$$\Delta_{i,j} \subset \{(e, f) \mid e \text{ is } C_i\text{-ending subobject} \wedge \\ f \text{ is } C_j\text{-starting subobject} \wedge (C_i, C_j) \in R\}$$

- $(\forall\,(C_i, C_j) \in R):$
  $((\forall\, f \in V_j\ ,\ f$ is C_i-starting subobject$):$
  $((\exists!\ e \in V_i):(e$ is C_j-ending subobject $\wedge\ ((e, f) \in \Delta_{i,j}))).$

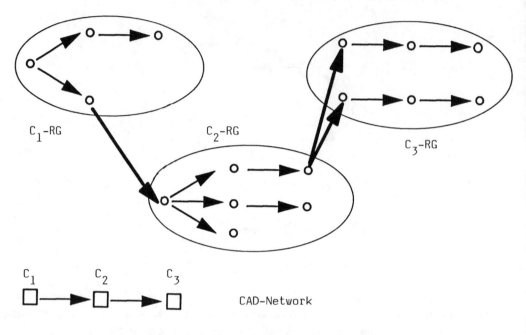

Figure 7: Network Realization Graph

A design subobject $d_1 \in D_1$ in *Net-RG* with $fan\text{-}in(d_1) = 0$ is called **network starting object**. A design subobject $d_n \in D_n$ with $fan\text{-}out(d_n) = 0$ is called **network ending object**. A path from a starting object to an ending object is called **network development path**.

The submission of design subobjects among the function realization graphs must be compatible with the dependencies of the design functions described by the CAD-network. Therefore, a **submission graph** is derived from the network realization graph. Using the submission graph, the correct submission of design objects between function realization graphs can be decided. The submission graph is defined such that the submission of design objects is correct, if the CAD-network can be in some way identified in the submission graph. The nodes in the submission graph represent development paths of the function realization graphs (**dp-nodes**), which are labeled by the starting design object and the goal design object, as well as by the index of the design function. The edges are derived from the submission edges in the function realization graph. If there is a C_i-development path with starting subobject e and ending subobject f, then edges in the submission graph are defined from any dp-node labeled by the ending subobject *e* to any dp-node labeled by the starting subobject *f*.

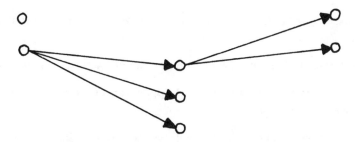

Figure 8: Submission Graph

Definition
An directed triple edge-labeled graph

$$SG(Net\text{-}RG) \;=\; (W, F, m_1, m_2, m_3)$$
$$m_1, m_2 : W \to \mathcal{D}, \quad m_3 : W \to \{1, \ldots, n\}$$

is called **submission graph** of the network realization graph $Net\text{-}RG$, iff the following properties hold:

- For each C_i-development path graph in $Net\text{-}RG$ with C_i-starting subobject e and C_i-ending subobject f, it exists a dp-node $w \in W$ with $m_1(w) := e$, $m_2(w) := f$, and $m_3(w) := i$.

- For each submission edge (e, f) in $Net\text{-}RG$:
  $\{(w_1, w_2) \mid m_2(w_1) = e \text{ and } m_1(w_2) = f\} \subset F$.

Given a set $DP = \{dp_1, \ldots, dp_n\} \subset W$ of n dp-nodes, such that there exists exactly one dp-node for each design subtask, i.e. $m_3(dp_i) = i, 1 \leq i \leq n$. Consider the subgraph $SG(Net\text{-}RG)_{DP}$ of $SG(Net\text{-}RG)$ induced by DP. The ending network object $m_2(dp_n)$ is correctly developped (i.e. the submission of design objects in the network realization graph is correct), if for each edge (C_i, C_j) in the CAD-network there exists exactly one edge in $SG(Net\text{-}RG)_{DP}$ from a node with a m_3-label i to a node with a m_3-label j.

4   Conclusions

CADULA is a powerful model for representing, managing and monitoring design automation processes, i.e. CAD-processes. CADULA is graph-based and consists of two submodels: (1) the CAD-process scheme model for representing the structure of a design process, and (2) the CAD-process realization model for representing the development of versions of design objetcs following a CAD-process scheme.

Furthermore, CADULA can be used as basis of a transaction mechanism in CAD-databases. The benefits of CADULA can be briefly summarized as follows:

- CADULA supports a dynamic refinement or decomposition mechanism for design task or functions, respectively.

- CADULA supports the modelling of several versions and alternatives of design objects.

- CADULA reflects the top-down as well as the bottom-up design methodology.

- CADULA covers a wide range of CAD-applications (e.g. mechanical, electronical, and software engineering).

- CADULA can serve as a basis of a sophisticated transaction management in CAD-databases.

Acknowledgement

I like to thank Professor Peter Kandzia, University Kiel, for many discussions improving essentially the CADULA model. The very early version of the CADULA model (given in [RuSp86]) was the result of a joined work with my colleague Thomas Spindler.

References

[BhaKa87] R. Bhateja, R.H. Katz, *VALKYRIE: A Validation Subsystem of a Version Server for Computer Aided Design Data*, in: Proc. ACM/IEEE Design Automation Conf., 1987.

[BlDoMiQu85] T. Blain, M. Dohler, R. Michaelis, E. Qureshi, *Managing the Printed Circuit Board Design Process*, Proc. ACM SIGMOD Conf. on Mgmt. of Data, 1985.

[Bu88a] H.J. Bullinger (ed.), *Produktionsforum 88*, Springer, 1988.

[Bu88b] H.J. Bullinger, *CIM - Die Herausforderung der nächsten Jahre*, in: [Bu88a].

[DiLo88] K.R. Dittrich, R.A. Lorie, *Version Support for Engineering Database Systems*, IEEE TSE, Vol. 14, No. 4, 1988.

[EnGoWi87] W. End, H. Gotthardt, R. Winkelmann, *Softwareentwicklung*, Siemens Verlag, 1988.

[Ka83] R.H. Katz, *Managing the Chip Design Database*, IEEE Computer, Vol. 16, No. 12, 1983.

[KaAnCh86] R.H. Katz, M. Anwarrudin, E. Chang, *A Version Server for Computer-Aided Design Data*, in: Proc. ACM/IEEE Design Automation Conf., 1986.

[KeBe87] M.A. Ketabchi, V. Berzins, *Modeling and Managing CAD Databases*, IEEE Computer, Vol. 20, No. 2, 1987.

[Neu83] Th. Neumann, *On Representing The Design Information in a Common Database*, Proc. ACM SIGMOD Conf. on Mgmt. of Data, 1983.

[NeuHo82] Th. Neumann, C. Hornung, *Consistency and Transactions in CAD Databases*, Proc. VLDB-Conf., 1982.

[PaPe88]   St. Pardee, P. Pennino, *Software Tools Speed Circuit Board Design*, IEEE Computer, Vol. 25, No. 9, 1988.

[RuSp86]   D. Ruland, Th. Spindler, *CADULA - A Universal Model for CAD-Processes*, Universität Würzburg, Internal Report, 1986.

[Sche87]   A.W. Scheer, *CIM - Computer Integrated Manufacturing*, Springer, 1987.

[TotenH87] T. Tomiyama, P.J.W. ten Hagen, *Organization of Design Knowledge in an Intelligent CAD Environment*, Techn. Report No. CWI-CS-R8720, Center for Mathematics and Computer Science, Amsterdam, 1987.

On Hyperedge Replacement and BNLC Graph Grammars

Walter Vogler

Institut für Informatik, Technische Universität
Arcisstr. 21, D-8000 München 2

Abstract

It is shown that up to vertex labelling $BNLC$ grammars of bounded nonterminal degree generate the same languages of simple graphs as hyperedge replacement grammars. This does not hold if the vertex labelling is taken into account. Vice versa hyperedge replacement grammars generate the same languages of simple graphs as $BNLC$ grammars of bounded nonterminal degree. Furthermore the generation of loops and multiple edges by hyperedege replacement grammars is discussed.

1. Introduction

Since [PR] and [Sch] there have been made various proposals how to generalize the usual string grammars to graph grammars, see [CER, ENR, ENRR] for overviews and a rich supply of detailed papers. Unfortunately it has turned out, that for most general approaches nearly all questions of interest are undecidable. Thus in recent years research has been concentrated on finding restricted approaches, like context–free grammars, that are tractable, but still expressive. Among the most notable in this respect are two types of grammars: Hyperedge replacement (HR) grammars, where one hyperedge at a time is replaced, see [Hab, HK1, HK2, Lau2] or [BC, Cou1], and boundary node label controlled ($BNLC$) grammars, where one vertex at a time is replaced, see e.g. [RW1, RW2, RW3, Cou2]. (For a specific discussion of context-free graph grammars see [Cou2].)

In this paper we want to compare the languages generated by HR grammars with those generated by $BNLC$ grammars. The first difference between them is that HR grammars generate directed hyperedge-labelled hypergraphs which may have multiple hyperedges and loops, while $BNLC$ grammars generate undirected vertex-labelled simple graphs, i.e. graphs without multiple edges or loops. To deal with this we will restrict the comparison to HR grammars which generate directed graphs only – directed graphs can be mapped to undirected graphs in an obvious way by forgetting the directions of edges; the case of multiple edges and loops will only be discussed in a later section. Furthermore we will enrich HR grammars by introducing vertex labels in such a way that these vertex labels do not interfere with the replacement mechanism. Thus enriched HR grammars obviously generate the same graph languages as HR grammars up to the vertex labelling. Analogously we will enrich $BNLC$ grammars by introducing edge labels, again in such a way that these edge labels do not interfere with the replacement mechanism. This is very different from $B\text{-}edNCE$ grammars (see [ER, ELR]): These also belong to the 'NLC family', but their derivation mechanism is heavily influenced by the edge labels. Enriched $BNLC$ grammars obviously generate the same graph languages as $BNLC$ grammars up to the edge labels.

An (enriched) $BNLC$ grammar $\mathcal{G}$ is said to be of bounded nonterminal degree, i.e. it is an (enriched) $BNLC_{bntd}$ grammar, if there is some $k \in I\!N_0$ such that for each nonterminal graph created by $\mathcal{G}$ each nonterminal vertex has at most k neighbours. In [Lau2] it is shown that each language generated by a $BNLC_{bntd}$ grammar can be generated by an HR grammar, too. The proof given there can easily be translated to give the same result for enriched $BNLC_{bntd}$ and HR grammars. The main result of this note, given in section 4, is that vice versa each simple graph language generated by an HR grammar can be generated by an enriched $BNLC_{bntd}$ grammar, too - but only up to the vertex labelling.

In section 5 we will bring graphs with multiple edges and loops into the picture. From such a graph G one can obtain a simple graph by deleting the loops and choosing one edge from every set of multiple edges; we call the set of resulting graphs $flat(G)$. Using this notion we show that for each graph language L generated by some HR grammar, $flat(L)$ can be generated by an HR grammar, too.

Let us outline the proof of section 4 here: When comparing HR and $BNLC$ grammars it is obvious that nonterminal hyperedges must correspond to nonterminal vertices and that the sources of those hyperedges must correspond to the neighbours of those vertices. The first difficulty is that when replacing a hyperedge we might introduce terminal edges between the sources while replacing a vertex cannot create edges between its neighbours. To deal with this we give a normal form theorem for HR grammars, showing that each HR language can be generated by an HR grammar where there are no terminal edges between the sources of any right hand side.

The second difficulty is that in the HR case the context of a nonterminal edge is formed by its sources, which correspond to the sources of a right hand side replacing the nonterminal edge. In the $BNLC$ case the context of a nonterminal vertex is formed by its neighbours which are not directly present in a right hand side. Thus in the HR case edges between the context of a nonterminal edge and the vertices of a right hand side which are not sources are specified directly while the corresponding edges in the $BNLC$ case are specified production-independently by the vertex labels only. This problem is partly solved by changing the vertex labelling of an enriched HR grammar such that the sources of a nonterminal edge can also be identified by their label. For the other part we use a variant of $BNLC$ grammars as an intermediate step.

Recently, in [ER] a similar result to ours was obtained, namely that $B\text{-}edNCE_{bntd}$ grammars generate the same graph languages as HR grammars. As mentioned above in this case the replacement is controlled also by edge labels, and the connection relation is given for each rule separately. Thus the replacement mechanism is stronger (and more complex) such that the second difficulty from above does not exist. While dealing with a somewhat broader class of HR grammars and considering also hypergraph generation in some sense, a long series of normal form and other lemmas is shown, which are completely on the $BNLC$ side. Our proof is more direct and works mainly on the HR

side – and, of course, our result is different, since we deal with $BNLC$ grammars which have a simpler replacement mechanism.

2. Basic definitions

A *hypergraph* H is a tuple (V, E, s, h, ν, η), where V is the vertex and E the edge set. The function s assigns to each $e \in E$ a sequence $s(e) = v_1 \ldots v_k$ of distinct vertices, which are called the *sources* of e, k is the *rank* of e. Similarly $h = s_1 \ldots s_n$ is a sequence of distinct vertices, the *sources* of H, and n is the *rank* of H. The labelling functions $\nu : V \to A$ and $\eta : E \to B$ assign labels from some alphabets A and B to the vertices and edges. It would also be possible to encode vertex labels as the labels of edges of rank 1, see e.g. [BC], but since the derivation process of $BNLC$ grammars depends on the vertex labels, they are given explicitly for all hypergraphs and graphs here.

An edge is *incident* to each of its sources and vice versa, two vertices are *adjacent* or *neighbours* if they are sources of the same edge. The *degree* of a vertex is the number of its incident edges.

What we have defined is in fact a directed vertex- and edge-labelled hypergraph with sources. A *directed graph* is a hypergraph with empty source-sequence where for all edges e the rank of e is 1 or 2. A *loop* is an edge of rank 1. A hypergraph has no *multiple edges* if different edges have different sets of sources, i.e. multiple edges may have different labels or directions. A directed graph without loops or multiple edges is called *simple*.

An (undirected) *graph* is a tuple (V, E, s, ν, η) where V is the vertex set, E the edge set and ν and η are the vertex and the edge labelling functions as above; the source function s assigns to each $e \in E$ a two-element subset of V. The notions *incident, adjacent, neighbour, degree* and *simple* are defined as for hypergraphs.

We will always consider graphs and hypergraphs up to isomorphism. Thus we can always take a disjoint copy if the need arises. All graphs and hypergraphs are finite.

Let $H = (V, E, s, h, \nu, \eta)$ be a simple directed graph. Then its *underlying* (undirected) *graph* $und(H)$ is (V, E, s', ν, η), where for all $e \in E$ with $s(e) = v_1v_2$ we have $s'(e) = \{v_1, v_2\}$. Furthermore we have functions fel and fvl, which 'forget' the edge and the vertex labels: We have $fel(H) = (V, E, s, h, \nu, \eta')$, where $\eta'(E) = \{*\}$, and $fvl(H) = (V, E, s, h, \nu', \eta)$, where $\nu'(V) = \{*\}$ - the special symbol $*$ stands for undefined. The latter two functions are defined for graphs analogously. The three functions und, fel and fvl are extended to sets as usual by taking as image the set of images.

A *subhypergraph* of a hypergraph H is obtained by taking as edge set a subset E' of E, as vertex set a subset V' of V which contains all sources of H and of edges in E', the source-sequence h and the appropriate restrictions of s, ν and η. A *subgraph* of a graph G is obtained by taking as edge set a subset E' of E, as vertex set a subset V' of V which contains all sources of edges in E', and the appropriate restrictions of s, ν and η. A sub(hyper)graph is *induced* if the respective edge set E' consists of all those edges from E all of whose sources belong to V'.

For a hypergraph H with source-sequence $s_1 \ldots s_n$ let $\overline{H}$ be H with an additional edge with source-sequence $s_1 \ldots s_n$ and label $*$, and let $(\overline{H}, A)$ be as $\overline{H}$ but with the new edge being labelled A.

We define the union $H \cup H'$ of two hypergraphs H, H' for the case that they have the same rank n and for $i = 1, \ldots, n$ the i-th source of H has the same label as the i-th source of H': Take copies of H and H' such that for $i = 1, \ldots, n$ the i-th source of H equals the i-th source of H', their vertex sets are disjoint otherwise, and their edge sets are disjoint. Then $H \cup H' = (V_H \cup V_{H'}, E_H \cup E_{H'}, s_H \cup s_{H'}, h_H, \nu_H \cup \nu_{H'}, \eta_H \cup \eta_{H'})$. Thus $H \cup H'$ is the disjoint union of H and H' with their sources fused together. (The union of functions is defined by considering them as relations.)

3. HR and BNLC grammars

In this section we define the types of grammars we will be working with.

An *enriched hyperedge replacement (HR) grammar* $\mathcal{G} = (N, T, P, Z)$ consists of finite disjoint sets N and T of nonterminal and terminal symbols, a finite set P of productions and a start symbol $Z \in N$. Every $A \in N$ has a *rank* $\alpha(A) \in \mathbb{N}_0$. A production is a pair (A, R), where $A \in N$ and R (the right hand side) is a hypergraph of rank $\alpha(A)$, such that all vertex labels are in T, all edge labels are in $N \cup T$ and every edge with label $B \in N$ has rank $\alpha(B)$. An enriched HR grammar is an HR grammar if all vertex labels equal the special label $*$, which we assume to be contained in every set T of terminal symbols. An HR grammar is the type of grammar considered e.g. in [HK2], except that hypergraphs and edges have two sequences of sources there.

The derivation process is roughly as follows: To derive a hypergraph from some hypergraph H find an edge e of H labelled A and a production (A, R); remove the edge e, add a copy of R disjoint from H and for $i = 1, \ldots, \alpha(\eta(e))$ identify the i-th source of R with the i-th source of e. More formally:

Let $\mathcal{G} = (N, T, P, Z)$ be an enriched HR grammar, let $H = (V, E, s, h, \nu, \eta)$ be a hypergraph with $\nu(V) \subseteq T$, $\eta(E) \subseteq N \cup T$, such that for all $e \in E$ $\eta(e) \in N$ implies that e has rank $\alpha(\eta(e))$. Let $e \in E$ be an edge with $\eta(e) = A$, (A, R) be a production and R' be a copy of R disjoint from H except that the i-th source of R' equals the i-th source of e for $i = 1, \ldots, \alpha(\eta(e))$. Then H' can be derived from H in $\mathcal{G}$, denoted $H \Longrightarrow_{\mathcal{G}} H'$, if H' is (isomorphic to) the hypergraph $(V \cup V_{R'}, E \cup E_{R'} \setminus \{e\}, s \mid_{E \setminus \{e\}} \cup s_{R'}, h, \nu', \eta \mid_{E \setminus \{e\}} \cup \eta_{R'})$, where $\nu'(v)$ equals $\nu(v)$ for $v \in V$ and $\nu_{R'}(v)$ for $v \in V_{R'} \setminus V$. Thus when identifying a source of e with a source of R' the resulting vertex gets the label of the source of e. Therefore we may assume that the sources of right hand sides have label $*$.

As usual $\overset{*}{\Longrightarrow}_{\mathcal{G}}$ (also $\overset{*}{\Longrightarrow}$) denotes the reflexive, transitive closure of $\Longrightarrow_{\mathcal{G}}$. For $A \in N$ let A_0 be the hypergraph consisting of an edge e labelled A of the appropriate rank and its sources, all being labelled $*$, such that the source-sequence of e and A_0 coincide. Then we write $A \overset{*}{\Longrightarrow}_{\mathcal{G}} H$ instead of $A_0 \overset{*}{\Longrightarrow}_{\mathcal{G}} H$. The set of sentential forms of $\mathcal{G}$ is $S(\mathcal{G}) = \{H \mid Z \overset{*}{\Longrightarrow}_{\mathcal{G}} H\}$. Obviously all $H \in S(\mathcal{G})$ fulfill the condition on H required in the definition of a derivation. The language generated by $\mathcal{G}$ is $L(\mathcal{G}) = \{H \mid Z \overset{*}{\Longrightarrow}_{\mathcal{G}} H$ and $\eta(E) \subseteq T\}$. Observe that $\alpha(Z) = 0$ if $\mathcal{G}$ generates directed graphs only.

It is clear that the vertex labels and the terminal edge labels do not influence the derivation process, especially that enriched HR grammars are not stronger than HR grammars, if we forget the vertex labels:

Proposition 3.1: Let L be a set of hypergraphs. If there is an enriched HR grammar $\mathcal{G}$ with $L = L(\mathcal{G})$, then there are enriched HR grammars $\mathcal{G}'$ and $\mathcal{G}''$ such that $fvl(L) = L(\mathcal{G}')$ and $fel(L) = L(\mathcal{G}'')$. There is an HR grammar $\mathcal{G}$ with $L = L(\mathcal{G})$ if and only if there is an enriched HR grammar $\mathcal{G}'$ with $L = fvl(L(\mathcal{G}'))$.

Now we turn to $BNLC$ type grammars. An *enriched boundary node label controlled* $(BNLC)$ *grammar* $\mathcal{G} = (N, T, P, Z, C)$ consists of finite disjoint sets N and T of non-terminal and terminal symbols, a finite set P of productions, a start symbol $Z \in N$ and a connection relation C. A production is a pair (A, R) consisting of some $A \in N$ and an (undirected) simple graph R (the right hand side) such that all edge labels are in T, all vertex labels are in $N \cup T$ and no two N-labelled vertices are adjacent. The connection relation C is a subset of $(T \cup N) \times T \times T$, with the property that $(a, b, c), (a, b, d) \in C$ implies $c = d$. An enriched $BNLC$ grammar is a $BNLC$ grammar if $C \subseteq (T \cup N) \times T \times \{*\}$ and all edges of right hand sides have label $*$ - where again we assume that T always contains the special symbol $*$. A $BNLC$ grammar is the type of grammar considered e.g. in [RW1].

The derivation process is roughly as follows: To derive a graph from some graph G find a vertex x of G labelled A and a production (A, R), remove x and add a copy of R disjoint from G. Finally connect the vertices of R to the vertices of G according to the connection relation C: a vertex of R labelled a is connected to a neighbour of v in G labelled b by an edge labelled c if $(a, b, c) \in C$. More formally:

Let $\mathcal{G} = (N, T, P, Z, C)$ be an enriched $BNLC$ grammar. Let $G = (V, E, s, \nu, \eta)$ be a simple graph with $\eta(E) \subseteq T$, $\nu(V) \subseteq N \cup T$ such that all neighbours of an N-labelled vertex are T-labelled. Let x be a vertex of G labelled A, let (A, R) be a production of G and assume that R is disjoint from G. Then G derives G' in $\mathcal{G}$, denoted $G \Longrightarrow_{\mathcal{G}} G'$, if G' is the graph $(V \setminus \{x\} \cup V_R, (E_x \cup E_R) \dot{\cup} E_C, s \cup s_R \cup s_C, \nu |_{V \setminus \{x\}} \cup \nu_R, \eta |_{E_x} \cup \eta_R \cup \eta_C)$, where

$E_x = \{e \in E \mid x \text{ is not incident to } e\}$

$E_C = \{(v, w) \mid v \in V_R, w \in V, (\nu_R(v), \nu(w), a) \in C \text{ for some } a \in T\}$,

$s_C(v, w) = \{v, w\}$ for $(v, w) \in E_C$

$\eta_C(v, w) = a$ for $(v, w) \in E_C$ with $(\nu_R(v), \nu(w), a) \in C$.

As above $\overset{*}{\Longrightarrow}_{\mathcal{G}}$ or $\overset{*}{\Longrightarrow}$ is the reflexive, transitive closure of $\Longrightarrow_{\mathcal{G}}$. For $A \in N$ let A_1 be the graph consisting of a single vertex labelled A. Then we write $A \overset{*}{\Longrightarrow}_{\mathcal{G}} G$ instead of $A_1 \overset{*}{\Longrightarrow}_{\mathcal{G}} G$. The set of sentential forms of $\mathcal{G}$ is $S(\mathcal{G}) = \{G \mid Z \overset{*}{\Longrightarrow}_{\mathcal{G}} G\}$. Our definitions ensure that all $G \in S(\mathcal{G})$ are graphs labelled as required above. The language of $\mathcal{G}$ is $L(\mathcal{G}) = \{G \mid Z \Longrightarrow_{\mathcal{G}} G, \ \nu(V_G) \subseteq T\}$.

It is clear that the edge labels do not influence the derivation process, especially that enriched $BNLC$ grammars are not stronger than $BNLC$ grammars, if we forget the edge labels:

Proposition 3.2: Let L be a set of graphs. There is a $BNLC$ grammar $\mathcal{G}$ with $L = L(\mathcal{G})$ if and only if there is an enriched $BNLC$ grammar $\mathcal{G}'$ with $L = fel(L(\mathcal{G}'))$.

An (enriched) $BNLC$ grammar $\mathcal{G}$ is said to be of *bounded nonterminal degree*, i.e. it is an (enriched) $BNLC_{bntd}$ grammar, if there is some $k \in \mathbb{N}_0$ such that for each sentential form created by $\mathcal{G}$ each nonterminal vertex has at most k neighbours.

Enriched $BNLC$ grammars can be seen as special $B\text{-}edNCE$ grammars as they are used e.g. in [ER], but the latter differ mostly in the following points: First, the derivation process is heavily influenced by the edge labels, in fact it is easy to see that the connection relation in the case of $B\text{-}edNCE$ grammars can be made independent of the vertex labels, secondly there is no global connection relation for a grammar, but each production has its own connection relation, and thirdly, when the vertices of a newly added right hand side of a production are connected to the vertices of the old graph, vertices of the right hand side with the same label may be treated differently, i.e. the connection relation is not label controlled with respect to these vertices. Thus $B\text{-}edNCE$ grammars are a much stronger type of grammar than enriched $BNLC$ grammars, when we consider the derivation mechanism. Consequently our result below (4.7) on the graph generating power of $BNLC_{bntd}$ grammars is stronger than the corresponding result for $B\text{-}edNCE_{bntd}$ grammars.

We will have the need for a type of grammar, that is between enriched $BNLC$ and $B\text{-}edNCE$ grammars: In such a grammar there is no global connection relation, but each production has a separate one, which is not label controlled with respect to the vertices of the right hand side.

An enriched $P\text{-}BNLC$ grammar (P as *production oriented*) $\mathcal{G} = (N, T, P, Z)$ consists of N, T, Z as above and a finite set P of productions (A, R, C), where (A, R) is as above and C is a subset of $V_R \times T \times T$ with the property, that $(v, a, b), (v, a, c) \in C$ implies $b = c$.

The derivations of an enriched $P\text{-}BNLC$ grammar are defined similarly to those of an enriched $BNLC$ grammar. Let $G = (V, E, s, \nu, \eta)$ be a simple graph with $\eta(E) \subseteq T$, $\nu(V) \subseteq N \cup T$ such that all neighbours of an N-labelled vertex are T-labelled. Let x be a vertex of G labelled A, let (A, R, C) be a production of $\mathcal{G}$ and assume that R is disjoint from G. Then G derives G' in $\mathcal{G}$, denoted $G \Longrightarrow_{\mathcal{G}} G'$, if G' is (isomorphic to) the graph $(V \setminus \{x\} \cup V_R, (E_x \cup E_R) \dot{\cup} E_C, s \cup s_R \cup s_C, \nu |_{V \setminus \{x\}} \cup \nu_R, \eta |_{E_x} \cup \eta_R \cup \eta_C)$, where

$E_x = \{e \in E \mid x \text{ is not incident to } e\}$,

$E_C = \{(v, w) \mid v \in V_R, w \in V, (v, \nu(w), a) \in C \text{ for some } a \in T\}$,

$s_C(v, w) = \{v, w\}$ for $(v, w) \in E_C$,

$\eta_C(v, w) = a$ for $(v, w) \in E_C$ with $(v, \nu(w), a) \in C$.

The sentential forms and the language are defined analogously as above. Also an enriched $P\text{-}BNLC$ grammar of bounded nonterminal degree is defined analogously.

It is quite obvious how to transform an enriched $BNLC$ grammar into an enriched $P\text{-}BNLC$ grammar: We eliminate the global connection relation C and replace each production (A, R) by a production (A, R, C'), where $C' = \{(v, a, b) \mid \nu_R(v) = c \text{ and } (c, a, b) \in C\}$. Hence:

Proposition 3.3: For each enriched $BNLC$ grammar $\mathcal{G}$ there exists an enriched $P\text{-}BNLC$ grammar $\mathcal{G}'$ such that $L(\mathcal{G}) = L(\mathcal{G}')$.

In the next section we will show a partial converse of this proposition.

4. The comparison

In this section we will compare the generative power of enriched HR and $BNLC_{bntd}$ grammars. Since HR grammars generate hypergraphs, while $BNLC$ grammars generate graphs, we will restrict our considerations to enriched HR grammars which generate simple directed graphs only; these can be transformed to graphs by the function *und*. In [ER] also the hypergraph generating power of a $BNLC$ type grammar is considered, where hypergraphs are represented in a standard fashion as bipartite graphs.

On the one hand, we have as a slight generalization of a result in [Lau2] that enriched HR grammars can generate the same graph languages as enriched $BNLC_{bntd}$ grammars. On the other hand, our main result states that enriched $BNLC_{bntd}$ grammars can generate the same graph languages as enriched HR grammars, but only up to vertex labels. In the next section we will show that this restriction of the result cannot be avoided: There are e.g. sets of graphs, where all edge and vertex labels equal *, which can be generated by HR grammars but not by $BNLC$ grammars.

If an HR grammar generates simple directed graphs only, its sentential forms may still contain edges of rank greater than two, and this is important for the generative power of HR grammars. Such edges have nonterminal labels and it is obvious that nonterminal edges in HR grammars must correspond to nonterminal vertices in $BNLC$ grammars. Therefore we introduce functions gra_N, which transform hypergraphs to graphs such that edges with label in N are transformed to vertices with label in N.

Let $H = (V, E, s, h, \nu, \eta)$ be a hypergraph such that all edges with label not in a given set N have rank 2, and let $E = E_1 \dot\cup E_2$ with $E_1 = \{e \mid \eta(e) \in N\}$. Then the graph $gra_N(H) = (V', E', s', \nu', \eta')$ is defined by

$$V' := V \dot\cup E_1,$$

$$E' := E_2 \dot\cup \{(e, v) \mid e \in E_1, v \in V, \ v \text{ is incident to } e \text{ in } H\},$$

$$s'(e') := \begin{cases} \{v_1, v_2\} & \text{for } e' \in E_2 \text{ with } s(e') = v_1 v_2 \\ \{e, v\} & \text{for } e' = (e, v), \end{cases}$$

$$\nu'(v) := \begin{cases} \nu(v) & \text{for } v \in V \\ \eta(v) & \text{for } v \in E_1, \end{cases}$$

$$\eta'(e) := \begin{cases} \eta(e) & \text{for } e \in E_2 \\ * & \text{otherwise.} \end{cases}$$

Observe that in $gra_N(H)$ no two vertices with label in N are adjacent. Furthermore in the case that no edge has a label in N and the source sequence is empty the function gra_N simply gives the underlying graph. Thus when comparing sentential forms of HR and $BNLC$ grammars via gra_N, where N is the alphabet of nonterminal symbols, terminal simple directed graphs get immediately transformed to their underlying graphs, i.e. we get directly the comparison of the generated languages we are aiming for.

Example: Here and in the examples to come we use the following conventions for drawing hypergraphs: Edges of rank 1 or 2 are drawn as arrows in the usual fashion. Other edges are drawn as boxes with the edge label inside, such a box is connected to the incident vertices, and the connecting lines are numbered to give the ordering of the edge's sources. Also some vertices of the graph are numbered to indicate the source sequence of the graph.

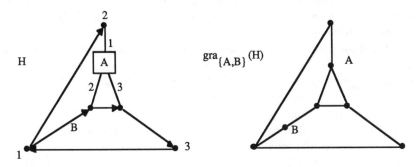

Figure 1

Figure 1 shows a graph H, where all labels not in $N = \{A, B\}$ are omitted, and $gra_N(H)$.

In the use of these functions gra_N lies a main difference to the approach of [ER]: There hypergraphs are transformed to graphs by turning every edge into a vertex. This transformation does not give for a terminal simple directed graph its underlying graph, but its underlying graph with each edge subdivided once. This is useful for the comparison of the hypergraph generating power, but the comparison of the graph generating power is obtained in an indirect way.

Our first result states that enriched $BNLC_{bntd}$ grammars can be simulated by enriched HR grammars. This result is a variation of [Lau2, Theorem 3.9], its proof is very similar and we omit it.

Theorem 4.1: For each enriched P-$BNLC_{bntd}$ grammar $\mathcal{G}$ there is an enriched HR grammar $\mathcal{G}'$ which generates simple directed graphs only such that $L(\mathcal{G}) = und(L(\mathcal{G}'))$.

Corollary 4.2: For each enriched $BNLC_{bntd}$ grammar $\mathcal{G}$ there is an enriched HR grammar $\mathcal{G}'$ which generates simple directed graphs only such that $L(\mathcal{G}) = und(L(\mathcal{G}'))$.

For each $BNLC_{bntd}$ grammar $\mathcal{G}$ there is an HR grammar $\mathcal{G}'$ which generates simple directed graphs only such that $fvl(L(\mathcal{G})) = fel(und(L(\mathcal{G}')))$.

As explained in the introduction we first show that every HR grammar can be brought into some normal form called a source-edge free grammar.

An enriched HR grammar is *source-edge free* if for any production (A, R) there is no terminal edge in R all of whose sources are sources of R.

Proposition 4.3: Let $\mathcal{G}$ be an enriched HR grammar which generates directed graphs without multiple edges only. Then there is a source-edge free HR grammar $\mathcal{G}'$ with $L(\mathcal{G}) = L(\mathcal{G}')$.

<u>Proof:</u>   The idea of the construction is to guess which terminal edges between sources of a nonterminal edge will be generated later on, and to introduce these earlier, at the same time with the nonterminal edge. For this we change the nonterminal labels such that they store the information which terminal edges between sources of a corresponding right hand side will be generated in a derivation in $\mathcal{G}$. With this information $\mathcal{G}'$ will be able to generate these edges earlier.

We will assume that all sources of right hand sides of $\mathcal{G}$ are labelled $*$. We define $\mathcal{G}'$ as (N', T, P', Z'): Let N' consist of pairs (A, H) where $A \in N$ and H is a terminally labelled hypergraph with $\alpha(A)$ vertices, all of which are sources of H and labelled $*$, H has no multiple edges and all edges are of rank 1 or 2. (Note that H is a directed graph, except that its source sequence is not empty in general.) The rank of (A, H) is $\alpha(A)$. Observe that N' is finite and that the start symbol Z is virtually unchanged, i.e. formally we have added the empty graph to it as second component to obtain Z'. The productions of $\mathcal{G}'$ are constructed from the productions of $\mathcal{G}$: Let $(A, R) \in P$ and f be a function that assigns to each nonterminal edge e of R a hypergraph $f(e)$ such that $(\eta_R(e), f(e)) \in N'$. Construct R'' as follows:

$$V_{R''} = V_R, \ E_{R''} = E_R \,\dot{\cup}\, \bigcup E_{f(e)},$$

$$s_{R''}(e') = \begin{cases} s_R(e') & \text{for } e' \in E_R \\ v_i v_j & \text{for } e' \in E_{f(e)} \text{ with } s_R(e) = v_1 \ldots v_n, \\ & h_{f(e)} = w_1 \ldots w_n, s_{f(e)}(e') = w_i w_j, \end{cases}$$

$$h_{R''} = h_R, \ \nu_{R''} = \nu_R,$$

$$\eta_{R''}(e') = \begin{cases} \eta_R(e') & \text{for } e' \in E_R \\ \eta_{f(e)}(e') & \text{for } e' \in E_{f(e)}. \end{cases}$$

If R'' has no two terminal edges with the same sets of sources, let H be the subhypergraph of R'' with the sources of R'' as vertices and the terminal edges between these sources as edges. In this case obtain R' from R'' by deleting all terminal edges between sources of R'' and add the production $((A, H), R')$ to P'.

Although (A, R) may determine several productions, P' is finite. Obviously the resulting grammar is source-edge free.

To prove the claim about $L(\mathcal{G}')$ one shows by induction on the lenght of a derivation that for any hypergraph H with all edges being terminal and without edges between the sources, and any $(A, G) \in N'$ we have: $(A, G) \overset{*}{\Longrightarrow}_{\mathcal{G}'} H$ if and only if $A \overset{*}{\Longrightarrow}_{\mathcal{G}} H \cup G$.

This induction proof is possible since the replacement in HR grammars is context-free. (Thus formally the proof depends on the context-freeness lemma of [HK], but it should be clear without spelling out the details).

Applying the above equivalence to Z gives the result, since Z has rank 0. $\square$

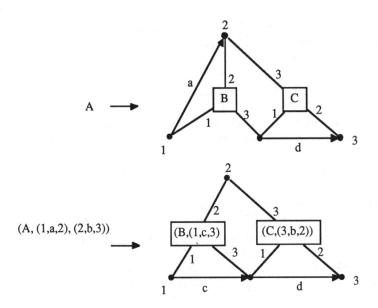

Figure 2

Example: In Figure 2 a production of $\mathcal{G}$ and a resulting production of $\mathcal{G}'$ according to the above construction are shown.

To indicate a graph H in a new edge label (A, H) we list the edges of H in the form (i, l, j), which is an edge with label l from the i-th to the j-th vertex of H. Informally speaking, the label $(A, (1, a, 2), (2, b, 3))$ means that an edge with this label 'behaves' like an edge with label A, but additionally has the 'obligation' to check for a terminal edge with label a from its first to its second source and similarly for another one with label b. The first obligation is met in our example, while the second is passed on to the C-labelled edge. Furthermore an 'early' terminal edge with label c is created, and the B-labelled edge is burdened with a corresponding obligation. In this example we have omitted all vertex labels.

Next we change the vertex labels of an enriched HR grammar in such a way that the sources of nonterminal edges of sentential forms can be distinguished by their labels. This allows a direct translation of HR grammars to P-$BNLC$ grammars. Note that such a transformation is not necessary when comparing HR and B-$edNCE$ grammars since the latter have a stronger derivation mechanism than $BNLC$ grammars. The last gap to $BNLC$ grammars will be filled on the $BNLC$ side.

A vertex labelling f is a *proper colouring* of a hypergraph H if for each edge of H all its sources have different values under f.

An enriched HR grammar $\mathcal{G} = (N, T, P, Z)$ is called *properly coloured* if for each $A \in N$ and each $i \in \{1, \dots, \alpha(A)\}$ there is a unique label $a \in T$ such that the following hold:

i) For any production (A, R) the i-th source of R is labelled a.
ii) For all $H \in S(\mathcal{G})$ the vertex labelling ν_H is a proper colouring of H, such that in H the i-th source of any edge labelled A has label a.

Lemma 4.4:   For each enriched HR grammar $\mathcal{G}$ with start symbol of rank 0 there is a properly coloured enriched HR grammar $\mathcal{G}'$ with $fvl(L(\mathcal{G}')) = fvl(L(\mathcal{G}))$. If $\mathcal{G}$ is source-edge free, then so is $\mathcal{G}'$.

<u>Proof:</u>   Let $\mathcal{G} = (N, T, P, Z)$ and $m = max\{\chi(\overline{R}) \mid (A, R) \in P\}$, where $\chi(H)$, the chromatic number, is the least n such that there is a proper colouring $f : V_H \to \{1, \ldots, n\}$ of H. Thus we have enough colours to give a proper colouring for each right hand side such that the sources get different colours. (See Section 2 for the definition of $\overline{R}$.)

We define $\mathcal{G}'$ as (N', T', P', Z'): We change the nonterminals in such a way that they contain information on the colouring of the sources of the edge they label, i.e. $N' = \{(A, f) \mid A \in N, f$ is an injection from $\{1, \ldots, \alpha(A)\}$ into $\{1, \ldots, m\}\}$. The rank of (A, f) is $\alpha(A)$. Observe that N' is finite and that the start symbol Z is virtually unchanged, i.e. formally we have added the empty function to it as second component to obtain Z'. Put $T' = T \cup \{1, \ldots, m\}$.

The productions of $\mathcal{G}'$ are constructed from those of $\mathcal{G}$. For each $(A, R) \in P$ and each proper colouring $f : V_R \to \{1, \ldots, m\}$ of R let $((A, f'), R')$ be a production of P': The function $f' : \{1, \ldots, \alpha(A)\} \to \{1, \ldots, m\}$ is defined by $f'(i) = f(v_i)$, where $h_R = v_1 \ldots v_{\alpha(A)}$. Furthermore $R' = (V_R, E_R, s_R, h_R, f, \eta)$, where the edge labelling η is defined by

$$\eta(e) = \begin{cases} \eta_R(e) & \text{for } \eta_R(e) \in T \\ (\eta_R(e), g) & \text{for } \eta_R(e) \in N, \ s_R(e) = w_1 \ldots w_{\alpha(\eta_R(e))}, \\ & g : \{1, \ldots, \alpha(\eta_R(e))\} \to \{1, \ldots, m\}, i \to f(w_i). \end{cases}$$

Again from (A, R) we obtain several productions, but P' is finite. Obviously source-edge freeness is not changed by our transformation.

An easy top down induction on the derivation length shows that $\mathcal{G}'$ is properly coloured, since all nonterminal edges store in the second component of their label the information on the labelling of their sources. A bottom up induction shows that for all $(A, f) \in N'$ and all terminal hypergraphs H' with $(A, f) \overset{*}{\Longrightarrow}_{\mathcal{G}'} H'$ there is a terminal hypergraph H with $A \overset{*}{\Longrightarrow}_{\mathcal{G}} H$ and $fvl(H') = fvl(H)$ and vice versa.

This finishes the proof. $\square$

Example: Figure 3 shows a production of $\mathcal{G}$ and a possible corrresponding production of $\mathcal{G}'$.

We have used colours red, blue and green instead of numbers, and labels (A, f) are given by A and a list of pairs $(i, f(i))$. We have omitted the original vertex and all edge labels.

Lemma 4.5:   Let $\mathcal{G}$ be a properly coloured source-edge free enriched HR grammar that generates simple directed graphs only. Then there is an enriched $P\text{-}BNLC_{bntd}$ grammar $\mathcal{G}'$ with $und(L(\mathcal{G})) = L(\mathcal{G}')$.

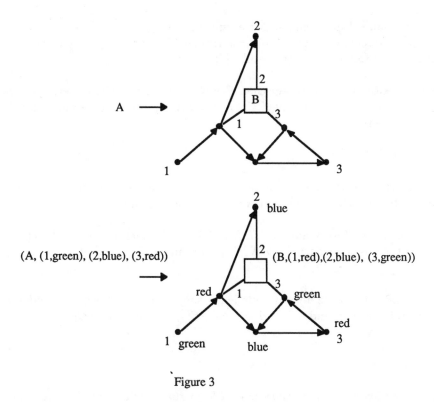

Figure 3

Proof: Let $\mathcal{G}' = (N, T, P', Z)$. We assume that in each right hand side of a production of $\mathcal{G}$ all edges with terminal label have rank 2. All other productions can be eliminated first, since they are not needed in a derivation of a directed graph. Thus gra_N is defined for all right hand sides and all sentential forms of $\mathcal{G}$.

From each production $(A, R) \in P$ we will construct a production of $\mathcal{G}'$. When we derive some hypergraph H' from a sentential form H of $\mathcal{G}$ by applying (A, R) to some edge e of H, then e is replaced by (a copy of) R via fusing of equally labelled sources of e and R since $\mathcal{G}$ is properly coloured. Now consider $gra_N(H)$ and $gra_N(R)$. If we replace e in $gra_N(H)$ (e is a vertex of $gra_N(H)$!) by $gra_N(R)$ via fusing the sources of e and R (which are vertices of $gra_N(H)$ and $gra_N(R)$), we obviously get $gra_N(H')$. But we can also use the $P\text{-}BNLC$ mechanism to derive $gra_N(H')$: Let R' be the subgraph of $gra_N(R)$ induced by those vertices which are not sources of R. Replace e in $gra_N(H)$ by R'; to get $gra_N(H')$ from the resulting graph only some edges have to be added: These are edges not between sources of e, since $\mathcal{G}$ is source-edge free, but they are all incident with one source of e and one vertex of R'. The sources of e in H are the neighbours of e in $gra_N(H)$, and they are distinguished by their labels. Hence the missing edges can be added according to the connection relation $C = \{(v, a, b) \mid v \in V_{R'}, \text{ there is a source } w \text{ of } R \text{ with } \nu_R(w) = a \text{ and an edge of } gra_N(R) \text{ with sources } v \text{ and } w \text{ and label } b\}$. For each $(A, R) \in P$ let the corresponding (A, R', C) be in P'.

One shows by induction on the derivation length that $H \in S(\mathcal{G})$ if and only if $gra_N(H) \in S(\mathcal{G}')$. Since all $H \in L(\mathcal{G})$ are terminally labelled and have no sources we have $und(H) = gra_N(H)$. Therefore $und(L(\mathcal{G})) = L(\mathcal{G}')$.

$\mathcal{G}'$ is of bounded nonterminal degree since the neighbours of nonterminal vertices in $G \in S(\mathcal{G}')$ correspond to sources of nonterminal edges and obviously there is a bound on their numbers for $\mathcal{G}$. Hence the result follows. □

Proposition 4.6: Let $\mathcal{G}$ be an enriched P-$BNLC$ grammar. Then there is an enriched $BNLC$ grammar $\mathcal{G}'$ such that $fvl(L(\mathcal{G})) = fvl(L(\mathcal{G}'))$. If $\mathcal{G}$ is of bounded nonterminal degree then so is $\mathcal{G}'$.

Proof: omitted □

Putting together the above results we get the main theorem and a corollary where a graph is called *unlabelled* if all vertex and edge labels equal *.

Theorem 4.7: For every enriched HR grammar $\mathcal{G}$ which generates directed simple graphs only there is some enriched $BNLC_{bntd}$ grammar $\mathcal{G}'$ such that $fvl(und(L(\mathcal{G}))) = fvl(L(\mathcal{G}'))$.

Corollary 4.8: Let L be a set of unlabelled simple graphs. Then $L = und(L(\mathcal{G}))$ for some HR grammar $\mathcal{G}$ if and only if $L = fvl(L(\mathcal{G}'))$ for some $BNLC_{bntd}$ grammar $\mathcal{G}'$.

5. Multiple edges and vertex labels

In this section we want to show how HR languages with multiple edges and loops can be compared with $BNLC$ languages. We will also give a result on sets of unlabelled graphs generated by $BNLC$ grammars which shows that $BNLC$ grammars cannot generate all unlabelled graph languages which can be generated by HR grammars. Due to lack of space proofs had to be omitted.

Note that one could easily add loops to $BNLC$ generated graphs by allowing loops on the right hand side of $BNLC$ productions. But it is not clear how to treat multiple edges, especially if the sets of multiple edges are unbounded in size for a graph language. Thus what we want to do is flattening out all multiple edges and loops in hypergraphs generated by HR grammars.

For a hypergraph H let $flat(H)$ be the set of all maximal subhypergraphs H' of H without multiple edges or loops. Since subhypergraphs are not necessarily induced, this simply means that we delete all loops and from every set of multiple edges all but one to get some $H' \in flat(H)$. As usual $flat$ is generalized to sets of hypergraphs by taking as image the union of the images of the elements.

We have that flattening transforms an HR language of directed graphs into an HR language. This result is also useful when proving a set of simple directed graphs to be an HR language: When exhibiting a suitable HR grammar one does not have to care about unwanted multiple edges.

Theorem 5.1: For every enriched HR grammar $\mathcal{G}$ that generates directed graphs only there is an enriched HR grammar $\mathcal{G}'$ with $flat(L(\mathcal{G})) = L(\mathcal{G}')$.

Corollary 5.2: Let L be a set of unlabelled simple graphs. Then $L = und(flat(L(\mathcal{G})))$ for some HR grammar $\mathcal{G}$ if and only if $L = fvl(L(\mathcal{G}'))$ for some $BNLC_{bntd}$ grammar $\mathcal{G}'$.

The following theorem gives an insight into the structure of unlabelled graphs generated by $P\text{-}BNLC$ grammars. It strengthens for these grammars a result of [EMR] on NLC grammars. We need some definitions first:

A *cycle* of length n of a simple graph is a sequence $v_0, \ldots, v_{n-1}$ of distinct vertices with v_i adjacent to v_{i+1} for $i = 0, \ldots, n-1$ and indices taken modulo n. A cycle is *chordless* if v_i adjacent to v_j implies $i = j + 1 \pmod{n}$ or $j = i + 1 \pmod{n}$.

Theorem 5.3: Let L be a set of unlabelled graphs generated by some $P\text{-}BNLC$ grammar $\mathcal{G}$. Then there is some $k \in I\!N_0$ such that for all $G \in L$ all chordless cycles of G have length less or equal k.

It is easy to see that there is an HR grammar which generates all unlabelled directed cycles. By the last theorem the corresponding set of underlying graphs cannot be generated by a $P\text{-}BNLC$ or a $BNLC$ grammar. On the other hand, there is a $BNLC$ grammar which generates all unlabelled complete graphs (i.e. simple graphs where any two vertices are adjacent). A corresponding set of directed graphs cannot be generated by an HR grammar, see [HK2]. Thus we have that the graph generating power of HR and $BNLC$ grammars are incomparable, if we take the vertex labelling into account.

Corollary 5.4: There exists some HR grammar $\mathcal{G}$ generating unlabelled simple directed graphs only such that for no $BNLC$ grammar $\mathcal{G}'$ we have $L(\mathcal{G}') = und(L(\mathcal{G}))$.

There exists some $BNLC$ grammar $\mathcal{G}$ generating unlabelled graphs only such that for no HR grammar $\mathcal{G}'$ generating simple directed graphs we have $L(\mathcal{G}) = und(L(\mathcal{G}'))$.

6. Concluding remarks

We have shown that graph languages generated by $BNLC_{bntd}$ grammars can be generated by enriched HR grammars, and that graph languages generated by HR grammars can be generated by enriched $BNLC_{bntd}$ grammars up to the vertex labelling. This allows to transfer results obtained for one model, like the polynomial recogizability for certain $BNLC$ languages [RW1], to the other model. We refer the reader to the discussion in [ER] and only mention an example here that could not be treated there. If we call the maximal rank of an edge appearing in some HR grammar $\mathcal{G}$ the order of $\mathcal{G}$, then the class of languages of unlabelled graphs generated by HR grammars of order r is properly contained in the respective class for $r + 1$ [HK2], thus we get a proper hierarchy of classes of HR languages. This result carries over to $BNLC_{bntd}$ languages where the order of a $BNLC$ grammar is the maximal degree of a nonterminal vertex in some sentential form: The constructions of 4.1 and 4.5 transform $P\text{-}BNLC$ grammars of order r to HR grammars of order r and vice versa. The transformations of 4.3, 4.4 and 4.6 obviously preserve order. Thus we get a proper hierarchy of classes of

$BNLC_{bntd}$ languages. A corresponding result could not be obtained in [ER], since one of the transformations used there does not preserve order. This shows the advantage of our more direct transformation.

One might interpret our result as saying that HR grammars are just a special case of $BNLC$ grammars and that consequently it is enough to study the latter. But there are several reasons why this is not so:

Restrictions are not necessarily bad, e.g. known polynomial recognition algorithms for $BNLC$ languages only work for connected graphs of bounded degree anyway [RW1]. HR languages are necessarily of bounded tree-width, thus they are algorithmically more accessible, see [RS], [ACP], [Lau1], [ALS]. HR grammars are context-free in the sense of [Cou2], while $BNLC$ grammars are only confluent.

The order of HR grammars is defined statically, i.e. it can be determined by simply looking at the productions, while the order of a $BNLC_{bntd}$ grammar depends on the sentential forms.

HR grammars generate hypergraphs, which might be useful in applications e.g. to Petri nets, where the transitions might be seen as hyperedges, a hyperedge replacement as a transition refinement. (For this point see [ER] where hypergraphs are also treated with $BNLC$-type grammars.) They may also generate multiple edges, which can be useful when specifying networks with edge weight.

As we have seen in Section 5, $BNLC$ grammars can generate all unlabelled HR graph languages only if we ignore the vertex labels. Hence HR grammars are not really a special type of $BNLC$ grammars.

We conclude with an open problem: Our method and the method of [Lau2] allow to mimick the derivation process of one type of grammar with the other type. It would be interesting to have a correspondence not on the grammar level, but on the language level. While it is clear that every HR graph language and thus every (enriched) $BNLC_{bntd}$ language is of bounded tree-width [RS], and hence every HR graph language is an (enriched) $BNLC$ language of bounded tree-width, is it also true that every $BNLC$ language of bounded tree-width is an HR language?

References

[ACP]   Arnborg, S.; Corneil, D.G.; Proskurowski, A.: Complexity of Finding Embeddings in a k-Tree. SIAM J. Alg. and Discr. Methods 8 (1987) 277 – 284

[ALS]   Arnborg, S.; Lagergreen, J.; Seese, D.: Problems Easy for Tree-Decomposable Graphs. In: T.Lepistö et al.(eds.): Automata, Languages and Programming. Proc. 15th ICALP, Tampere, 1988, Springer, Lect. Notes Comp. Sci. 317 (1988) 38–51

[BC]   Bauderon, M.; Courcelle, B.: Graph Expressions and Graph Rewritings. Math. Systems Theory 20 (1987) 83–127

[CER]   Claus, V.; Ehrig, H.; Rozenberg, G. (eds.): Graph-Grammars and Their Application to Computer Science and Biology. Springer, 1979, LNCS 73

[Cou1]   Courcelle, B.: The Monadic Second-Order Theory of Graphs I: Recognizable Sets of Finite Graphs. To appear in Information and Computation

[Cou2]   Courcelle, B.: An Axiomatic Definition of Context-Free Rewriting and its Application to NLC Graph Grammars. Theor. Comp. Sci. 55 (1987)

[EMR]   Ehrenfeucht, A.; Main, M.G.; Rozenberg, G.: Restrictions on NLC Grammars. Theor. Comp. Sci. 31 (1984) 211 – 223

[ENR]   Ehrig, H.; Nagl, M.; Rozenberg, G. (eds.): Graph Grammars and Their Application to Computer Science. Springer, 1983, Lect. Notes Comp. Sci. 153

[ENRR]   Ehrig, H.; Nagl, M.; Rozenberg, G.; Rosenfeld, A. (eds.): Graph-Grammars and Their Application to Computer Science. Springer, 1987, Lect. Notes Comp. Sci. 291

[ELR]   Engelfriet, J.; Leih, G.; Rozenberg, G.: Apex Graph Grammars. In [ENR] 167 – 185

[ER]   Engelfriet, J.; Rozenberg, G.: A Comparison of Boundary Graph Grammars and Context-Free Hypergraph Grammars. Dept. of Comp. Sci., Leiden University, 1988

[Hab]   Habel, A.: Hyperedge Replacement: Grammars and Languages. Diss. Uni. Bremen, FB Mathem./Inform., 1988

[HK1]   Habel, A.; Kreowski, H.-J.: Characteristics of Graph Languages Generated by Edge Replacement. Uni. Bremen, FB Mathem./Inform. Report No. 3/85, 1985; also: Theor. Comp. Sci. 51 (1987) 81 –115

[HK2]   Habel, A.; Kreowski, H.-J.: Some Structural Aspects of Hypergraph Languages Generated by Hyperedge Replacement. Techn. Report Nr. 12, TU Berlin, Uni. Bremen, 1985; extended Abstract in: F.J. Brandenburg et al. (eds.): STACS '87, Proc. of the 4th Annual Symposium on Theor. Aspects of Comp. Sci., Passau, 1987, Lect. Notes Comp. Sci. 247 (1987) 207 – 219

[Lau1]   Lautemann, C.: Efficient Algorithms on Graphs Represented by Decomposition Trees, Uni. Bremen, FB Mathem./Inform., Bericht 6/87, 1987; also in: M. Dauchet; M. Nivat (eds.): CAAP '88, Lect. Notes Comp. Sci. 299 (1988) 28 – 39

[Lau2]   Lautemann, C.: Efficient Algorithms on Context-Free Languages. In: T. Lepistö et al.(eds.): Automata, Languages and Programming. Proc. 15th ICALP, Tampere, 1988, Springer, Lect. Notes Comp. Sci. 317 (1988) 362–378

[PR]   Pfaltz, J.; Rosenfeld, A.: Web Grammars. Proc. Int. Joint Conf. Art. Intell. Washington, 1969, 609 – 619

[RS]   Robertson, N.; Seymour, P.D.: Graph Minors. II. Algorithmic Aspects of Tree Width. J. Algorithms 7 (1986) 309–322

[RW1]   Rozenberg, G.; Welzl, E.: Boundary NLC Graph Grammars – Basic Definitions, Normal Forms and Complexity. Inf. and Control 69 (1986) 136 – 167

[RW2]   Rozenberg, G.; Welzl, E.: Combinatorial Properties of Boundary NLC Graph Languages. Discr. Appl. Math. 16 (1987) 59 – 73

[RW3]   Rozenberg, G.; Welzl, E.: Graph Theoretic Closure Properties of the Family of Boundary NLC Graph Languages. Acta Informatica 23 (1986) 289 – 309

[Sch]   Schneider, H.J.: Chomsky-Systems for Partial Orderings (in German). Technical Report / MMD-3-3, University of Erlangen, 1970

GRAPH REWRITING SYSTEMS WITH PRIORITIES

Michel BILLAUD, Pierre LAFON

Yves METIVIER, Eric SOPENA

LABORATOIRE BORDELAIS DE RECHERCHE EN INFORMATIQUE
Université de Bordeaux
U.A. 726 du C.N.R.S.
351, Cours de la Libération
33405 TALENCE - FRANCE

Abstract : *In this paper, we develop a new theory of attribute graph rewriting systems with priorities. This theory provides a very general tool for describing algorithms from classical graph theory, and for algorithms implemented on networks of communicating processors and distributed systems. Moreover, this theory gives an algebraic model which allows us to mathematically prove properties of distributed algorithms.*

Mailing Address :   *Yves METIVIER, ENSERB, 351 cours de la Libération*
                       33405 TALENCE (FRANCE)
Electronic Mail :   *metivier@geocub-prog.fr*

SECTION 0. INTRODUCTION

In this paper, we introduce a new formalism to describe distributed algorithms. This formalism is based on a special kind of graph rewriting systems working on networks (viewed as graphs) of communicating processors. The main objectives of this work is to provide a general tool allowing to prove some properties of distributed algorithms such as termination and correctness [5,10]. We also show that we can use it to express some classical algorithms of graph theory [1,3].

As usual (see e.g. [8,9]), we will represent a network of communicating processors as a graph, whose vertices stand for processors and whose edges stand for communication channels. By labeling the graph components (vertices and edges) we can simulate processor or channel states, channels contents, ... By this way, a distributed algorithm will be described by an initial graph (the initial state of the network) and a set of "calculation rules" expressed as graph rewriting rules.

Due to the initial aims of our work, the graph rewriting systems we use are not very powerful, in the sense that they never modify the underlying structure of the graph they work on, and can not be used as a graph generating device. For this reason, we do not intend to rely our graph rewriting systems to the field of graph grammar theory.

In order to increase the expressive power of our model, we use the classical concept of rewriting with priorities [2,7], which allows us to simulate the different kinds of control statements necessary to express distributed algorithms. This priority concept is intended to solve strictly local rewriting conflicts and is used whenever we need to translate statements such as *"for each neighbour of processor x ..."*, *"if processor x has no neighbour in state s ..."*, ...

To handle the possibility for a given processor to realize some local computations (on some local variables) we introduce the concept of attributed rewriting which only provides a simplification of the expression of graph rewriting systems whenever the attribute domains are finite (which is obviously always the case when we work on computers).

Our paper is organized as follows : in the first section, we recall some basic definitions concerning binary relations, graphs and labeled graphs. The second section is devoted to the definition of graph rewriting systems with priorities (PGRSs for short). In the third section we give some examples of PGRSs devised to solve some classical graph problems. The last section is devoted to the concept of attribute graph rewriting systems with priorities (APGRSs), for which we give formal definitions and an illustrating example.

The reader will find all proofs and more examples in [4].

SECTION 1. BASIC DEFINITIONS

In this section, we review some basic definitions and general notations concerning binary relations, graphs and labeled graphs which we shall use throughout this paper.

1.1 BINARY RELATIONS

(1) Let X be a set. $\mathcal{P}(X)$ is the powerset of X; $\#X$ is the cardinality of X. Let $\longrightarrow$ be a binary relation on X; $\overset{*}{\longrightarrow}$ denotes the reflexive-transitive closure of $\longrightarrow$. For a given binary relation $\longrightarrow$ we let $\Delta(x) = \{ y \ / \ x \longrightarrow y \}$. Let $(x_i)_{0 \leq i \leq n}$ be a sequence of elements of X such that :
$$x = x_0 \longrightarrow x_1 \longrightarrow \ldots \longrightarrow x_n = y,$$
we say that n is the *length* of this sequence, and we write $x \overset{n}{\longrightarrow} y$.

(2) An element x of X is said to be *irreducible* with respect to relation $\longrightarrow$ if $\Delta(x)$ is empty ; we say that x is a $\longrightarrow$-*normal form*. We let :
$$Irred(x) = \{ y \ / \ x \overset{*}{\longrightarrow} y \text{ and } y \text{ is irreducible} \}.$$

(3) We say that the relation $\longrightarrow$ is :
i) *noetherian* if there is no infinite sequence of the form
$$x_1 \longrightarrow x_2 \longrightarrow \ldots \longrightarrow x_n \longrightarrow \ldots, \quad \text{then } \overset{*}{\longrightarrow} \text{ is } well\text{-}founded,$$
ii) *compatible with an order* $>$ if for any elements x and y, we have :
$$x \longrightarrow y \implies x > y.$$

1.2 Graphs and labeled graphs

We recall here some (classical) graph theoretical definitions (see e.g. [1,3]).

Definition 1.1 : A *graph* G consists in a finite set of *vertices* V, a finite set of *edges* E, and a mapping *Ends* from E to V×V, assigning to each edge e two, not necessarily distinct, vertices of V (the extremities of edge e). The graph G will be denoted by G = (V, E, Ends).

Definition 1.2 : A *directed graph* G, or simply a *d-graph*, consists in a finite set of vertices V, a finite set of (directed) *edges* E, and two mappings s and t from E to V×V, assigning to each edge e its source and target nodes respectively. The graph G will be denoted by G = (V, E, s, t).

Definition 1.3 : Let G = (V, E, s, t) be a d-graph ; we define the *underlying* (undirected) *graph of* G, as Und(G) = (V, E, Ends) with :
$$\forall\ e \in E, \quad Ends(e) = \{\ s(e),\ t(e)\ \}.$$

As our rewriting systems are based on edge- or node-labels, we now introduce the concept of labeled graph. Let $C = (C_E, C_V)$ be a pair of disjoint sets of labels ; C_V (resp. C_E) stands for node (resp. edge) labels.

Definition 1.4 : A *labeled graph over* C, or simply a *C-graph*, consists in a graph G = (V, E, Ends) and a labeling function which is a pair of mappings $l = (l_V, l_E)$ such that :
$$l_V : V \longrightarrow C_V \text{ is the node-labeling function,}$$
$$l_E : E \longrightarrow C_E \text{ is the edge-labeling function.}$$
Similarly, a labeled directed graph, or simply a *C-d-graph*, is defined as a directed graph with a labeling function l defined as above.

If x stands for a component of a graph G (i.e. a vertex or an edge) we will denote by l(x) its label (which stands for $l_V(x)$ if x is a vertex and $l_E(x)$ otherwise). If c is a node- (resp. an edge-) label, $|G|_c$ is the number of c-labeled nodes (resp. edges) in the graph G. For any graph (or d-, C-, C-d-graph) G, we will denote its components by V_G, E_G, $Ends_G$, s_G, t_G or l_G.

We now introduce the concepts of subgraph, partial labeling functions and graph morphisms we shall use in the next section. All these definitions will be given for C-d-graphs as they can easily be transposed to the other kinds of graphs we deal with.

Definition 1.5 : Let G and H be two C-d-graphs, we say that G is a *subgraph* of H, denoted by G < H, if :
   i) $V_G \subset V_H$ and $E_G \subset E_H$,

ii) s_G and t_G are respectively the restrictions of s_H and t_H to E_G,

iii) l_{V_G} (resp. l_{E_G}) is the restriction of l_{V_H} (resp. l_{E_H}) to V_G (resp. E_G).

Definition 1.6 : Let G be a C-d-graph and $\lambda = (\lambda_V, \lambda_E)$ a partial labeling function of G (i.e. λ_V is a mapping from $X \subseteq V$ to C_V and λ_E a mapping from $Y \subseteq E$ to C_E) ; we define the C-d-graph $H = \lambda G$ as follows :

   i) $V_H = V_G$, $E_H = E_G$, $s_H = s_G$, $t_H = t_G$,

   ii) $\forall\ v \in V_H$, if $v \in X$ then $l_{V_H}(v) = \lambda_V(v)$ else $l_{V_H}(v) = l_{V_G}(v)$,

   iii) $\forall\ e \in E_H$, if $e \in Y$ then $l_{E_H}(e) = \lambda_E(e)$ else $l_{E_H}(e) = l_{E_G}(e)$.

Definition 1.7 : Let G and H be two C-d-graphs ; a *graph morphism* from G to H is a pair $m = (m_V, m_E)$ of mappings $m_V : V_G \longrightarrow V_H$, $m_E : E_G \longrightarrow E_H$, such that :

   i) $\forall\ e \in E_G$, $s_H(m_E(e)) = m_V(s_G(e))$,

   ii) $\forall\ e \in E_G$, $t_H(m_E(e)) = m_V(t_G(e))$,

   iii) $l_{V_G} = l_{V_H} \circ m_V$ and $l_{E_G} = l_{E_H} \circ m_E$.

If m_V and m_E are injective (resp. surjective) mappings, we say that m is injective (resp. surjective). If m is injective and surjective we say that m is an isomorphism and we write $H \cong G$.

We shall denote by mG the subgraph of H defined by $mG = (m_V V, m_E E, s', t', l')$ where s', t' and l' stand for the adhoc restrictions of s_H, t_H and l_H respectively.

SECTION 2 : GRAPH REWRITING SYSTEMS WITH PRIORITIES (PGRSs)

In this section, we introduce the main concepts of our graph rewriting model. We deal with a special kind of graph rewriting systems in which the rewriting rules do not modify the underlying structure of the rewritten graphs but only the coloring of their components (vertices and edges).

Let $C = (C_V, C_E)$ be a color alphabet as defined in the previous section.

Definition 2.1 : A *rewriting rule* over C consists in a connected C-d-graph D and a partial labeling function of D denoted by λ. We shall write $r = (D, \lambda)$.

Note that such a rewriting rule can be more classically viewed as a pair of graphs (the left-hand-side and the right-hand-side of the rule) given by $(D, \lambda D)$.

Definition 2.2 : Let G be a C-d-graph and $r = (D, \lambda)$ a rewriting rule. We say that the graph G is *reducible by using rule* r if there exists an injective morphism μ from D to G. We call the subgraph μD of G an *occurrence* of D in G.

Definition 2.3 : A *rewriting step* is defined by a 4-tuple (G, r, μ, H) where :
   i) $r = (D, \lambda)$ is a rewriting rule,
  ii) G is a C-d-graph reducible by using rule r (with corresponding injective morphism μ),
 iii) H is a C-d-graph obtained from graph G by relabeling vertices and edges of occurrence μD by function λ.

    We will say that G is rewritten into H and denote it by $G \xrightarrow{r,\mu} H$, or simply $G \xrightarrow{r} H$.

Example 2.4 : Figure 2.1 shows a rewriting rule $r = (D, \lambda)$ and two graphs G and H such that $G \xrightarrow{r} H$. The color alphabet is given by $C_V = \{a, b, c\}$ and $C_E = \{x, y, z\}$.

Definition 2.5 : A *rewriting chain* is a sequence $(G_1, r_1, \mu_1, \ldots, G_{n-1}, r_{n-1}, \mu_n, G_n)$ such that : $\forall\ i,\ 1 \le i < n,\quad (G_i, r_i, \mu_i, G_{i+1})$ is a rewriting step.

Definition 2.6 : A *graph rewriting system with priorities* (a PGRS for short) is defined by a triple $(C, P, >)$ where :
   i) $C = (C_V, C_E)$ is the color alphabet,
  ii) P is a finite set of rewriting rules over C,
 iii) $>$ is a partial order on the rules of P.

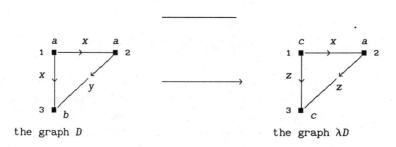

the graph D the graph λD

(a) the rewriting rule $r = (D, \lambda)$

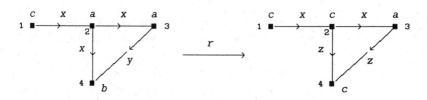

(b) a rewriting step $G \xrightarrow{r} H$

- *figure 2.1* -

    The concept of correctness, introduced below, enables us to precise the effect of priorities within a PGRS.

Definition 2.7 : Let $\mathbb{R} = (C, P, >)$ be a PGRS, $r = (D, \lambda)$ a rule of P and G a C-d-graph. A rewriting step (G, r, μ, H) is said to be *correct* with respect to $\mathbb{R}$ if and only if :

    for any rule $r' = (D', \lambda')$ of P such that $r' > r$, there exists no occurrence $\mu' D'$ of r' which intersects the occurrence μD of r.

For such a rewriting step, we will write $G \xrightarrow[\mathbb{R}]{r} H$.

    We can easily extend this notion of correctness to rewriting chains in the following way :

    a rewriting chain is said to be correct with respect to $\mathbb{R}$ if and only if all of its rewriting steps are correct. We shall then write $G_1 \xrightarrow[\mathbb{R}]{*} G_n$.

Example 2.8 : Let r and G be the rewriting rule and the C-d-graph of example 2.4, and $r' = (D', \lambda')$ the rewriting rule shown by figure 2.2.

Let us consider the PGRS $\mathbb{R} = (C, \{r, r'\}, >)$ defined by $r' > r$ (rule r' has a greater priority than rule r) ; then :

    $G \xrightarrow{r} H$ is not correct with respect to $\mathbb{R}$,

    $G \xrightarrow{r'} H'$ is correct with respect to $\mathbb{R}$,

where H' denotes the graph obtained from graph G by applying rule r'.

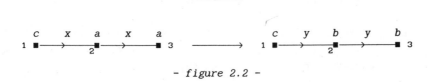

- figure 2.2 -

    Note that the effect of the priorities is purely local (when there is a conflict between two intersecting occurrences): which means that if two rules r and r' such that $r' > r$ can be applied in two *distinct* parts of a graph G, one can freely choose to apply r or r'.

    This property of Church-Rosser type is precised by the following lemma :

Lemma 2.9 : *Let $\mathbb{R} = (C, P, >)$ be a PGRS, and G a C-d-graph ; let (G, r_1, μ_1, H_1) and (G, r_2, μ_2, H_2) be two rewriting steps correct with respect to $\mathbb{R}$ and such that occurrences $\mu_1 D_1$ and $\mu_2 D_2$ does not intersect. Then :*

    i) $(H_1, r_2, \mu_2, H_{12})$ and $(H_2, r_1, \mu_1, H_{21})$ are both correct with respect to $\mathbb{R}$,

    ii) moreover, we have $H_{12} = H_{21}$.

    Hence, one can view the behaviour of a PGRS on a given graph G in the following way :

1. Sequential behaviour :

    Consider all the occurrences in graph G which can be *correctly* rewritten, choose one of them and apply the corresponding rewriting rule. Repeat this process while possible. If no more rule can be applied, then the computation is terminated.

2. *Distributed behaviour* :

Consider all the occurrences in graph G which can be *correctly* rewritten, choose one or several not intersecting ones of them and apply the corresponding rewriting rule(s). In this case, two (or more) rewriting steps may be applied at the same time. Repeat this process while possible. If no more rule can be applied, then the computation is terminated.

This behaviour is clearly equivalent to one (or more) sequential ones.

SECTION 3. EXAMPLES

In this section we use PGRSs to describe some graph algorithms. We first present two techniques for the computation of a directed spanning tree in a connected graph (or a network of processors [9]). The first one is rather similar to the (sequential) Tremaux algorithm, whereas the second one is a distributed computation.

3.1 COMPUTING A SPANNING TREE IN A CONNECTED GRAPH

This example demonstrates the use of a PGRS for the computation of a spanning tree in a non-directed connected graph. This PGRS acts in the very same way as the classical Tremaux algorithm [3].

Example 3.1 : To solve this problem, we use a PGRS $\mathbb{R}_1 = (C, >, P)$ defined by :
1. $C = (C_V, C_E)$ with :

- $C_V = \{ A, M, N, F \}$ for vertices, where A stands for *active*, M for *marked*, N for *not yet visited* and F for *finished*.
- $C_E = \{ t, f \}$ for edges, where t indicates that the edge belongs to the computed tree.

2. $P = \{ r_1, r_2 \}$, with :

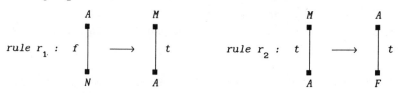

3. $>$ is defined by $r_1 > r_2$.

Proposition 3.2 : *i) relation $\xrightarrow[\mathbb{R}_1]{}$ is noetherian,*
ii) Let G be a connected graph with n vertices such that :

- each edge is labeled with f,
- exactly one vertex r (the root) is labeled with A, the other ones with N,
then:
    a) every graph G' in Irred(G) satisfies the following properties :
        - the root r is labeled with A, all other vertices with F,
          - the t-edges of G' constitute a spanning tree for the underlying
        graph.
    b) $G' \in Irred(G)$ iff $G \xrightarrow[\mathbb{R}_1]{2n-2} G'$.

Sketch of the proof : One can easily check that the quantity $(|G|_N, |G|_M)$ induces a noetherian order compatible with $\mathbb{R}_1$. As this quantity is always positive, we obtain a termination argument.

The correctness of $\mathbb{R}_1$ is deduced from the following invariants :
(I1) the t-labeled edges constitute a tree ; a vertex is in the tree iff it is not labeled with N,
(I2) there is one and only one A-labeled vertex ; the M-labeled vertices constitute a chain in the tree from the root to this vertex,
(I3) a F-labaled vertex has no neighbour with label N. □

3.2 COMPUTING A DIRECTED SPANNING TREE

A slight modification in the preceding PGRS allows the computation of a directed spanning tree. The trick is the following : we use a modulo-3 numbering for successive layers in the tree.

For any i in $\{0,1,2\}$, let $s(i) = (i+1) \mod 3$.

Example 3.3 : We consider the new PGRS $\mathbb{R}_2 = (C, >, P)$ defined by :

1. the colour set $C = (C_V, C_E)$ is given by :
  - $C_V = \{ A, F, M \} \times \{ 0, 1, 2 \} \quad \cup \quad \{ N \}$,
  - $C_E = \{ f, t \}$,

2. $P = \{ r_1^i, r_2^i \}$ for any $i \in \{0,1,2\}$ with :

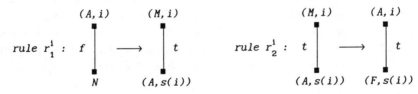

3. relation < is defined by : $\forall \ i,j \in \{0,1,2\} \quad r_1^i > r_2^i$.

Proposition 3.4 : i) relation $\xrightarrow[\mathbb{R}_2]{}$ is noetherian,

ii) Let G be a connected graph with n vertices such that :

- each edge is labeled with f,
- exactly one vertex r (the root) is labeled with (A,0), the other ones with N,

then :

a) every graph G' in Irred(G) satisfies the following properties :
- the root r is labeled with (A,0), and all other vertices with (F,i), i ∈ {0,1,2},
- the t-edges of G' constitute a spanning tree for the underlying graph, directed from the root to leaves by successive indexes.

b) G' ∈ Irred(G) iff $G \xrightarrow[\mathbb{R}_2]{2n-2} G'$.

Remark 3.5 : We can also present this technique under the form of a "cartesian product" between the Spanning Tree Construction PGRS above and the simple Tree Directing PGRS below :

1. $C = (C_V, C_E)$ with :
   - $C_V = \{ -, 0, 1, 2 \}$ (- stands for *not numbered*),
   - $C_E = \{ t \}$.

2. $P = \{ r_i, i = 0,1,2 \}$ is given by :

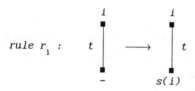

$$\text{rule } r_i :$$

which obviously provides an orientation on a tree (here t-edges) where :
   i) the root is initially numbered 0,
   ii) all other vertices are initially not numbered (labeled with -).

Nota : In the rest of the paper we will omit indices and edge labels, and simply draw arrows on those edges which belong to the directed tree.

Remark 3.6 : an important property of both these PGRSs is the local termination detection property : one easily notices that the computation stops as soon as the root has label A and all of its neighbours have label F. This property will be fundamental when we attempt to solve problems on networks (see e.g. [8,9,10]) : the node (i.e. the processor) which initiates the computation often has to know whether the computation has terminated or not.

3.3 DISTRIBUTED COMPUTATION OF A SPANNING TREE

In this section we first present a very simple PGRS for the distributed computation of a directed spanning tree in a connected graph. As this PGRS does not possess the local termination detection property (LTDP), we also present a more sophisticated PGRS which fills this requirement.

Example 3.7 : We consider here the simple PGRS $\mathbb{R}_3 = (C, >, P)$ defined by :

1. $C = (C_V, C_E)$ is given by $C_V = \{ A, N \}$ and $C_E = \{ t, f \}$ with the same meaning as in example 3.1.

2. $P = \{ r \}$ is given below :

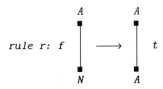

$$\text{rule } r: \quad f \quad \longrightarrow \quad t$$

Proposition 3.8 : *i) relation $\xrightarrow[\mathbb{R}_3]{}$ is noetherian,*

ii) Let G be a connected graph with n vertices such that :
- *each edge is labeled with f,*
- *exactly one vertex r (the root) is labeled with A, the other ones with N,*

then :
    a) every graph G' in Irred(G) satisfies the following properties :
- *all vertices are labeled with A,*
- *the t-edges of G' constitute a spanning tree for the underlying graph,*

    b) $G' \in Irred(G)$ iff $G \xrightarrow[\mathbb{R}_3]{n-1} G'$.

Example 3.9 : PGRS with local termination detection
    Here we show how to achieve the LTD requirement by mean of "termination messages" emitted by vertices when a subtree has been explored.
    We consider the PGRS $\mathbb{R}_4 = (C, >, P)$ defined by :

1. $C = (C_V, C_E)$ is given by :
- $C_V = \{ A, N, M, R, F \}$ where A stands for *active*, N for *not_reached*, M for *marked*, R for *ready_to_finish* and F for *finished*,
- $C_E = \{ \, - \, , \, \rightarrow \, \}$ (as we are building a directed tree we use the same trick as in example 3.3 and we use arrows instead of t-labeled edges and modulo 3 numbering of successive layers).

2. $P = \{ r_1, r_2, r_3, r_4, r'_4, r_5 \}$ is given below :

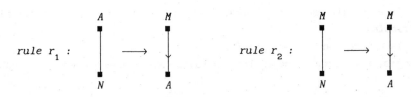

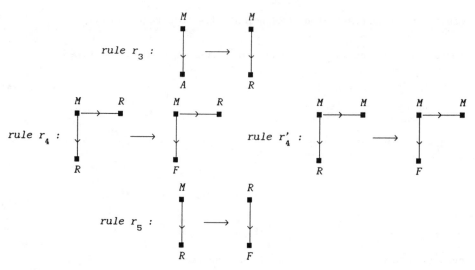

rule r_3 :

rule r_4 :

rule r'_4 :

rule r_5 :

3. relation > is defined by : $\{\ r_1,\ r_2\ \}\ >\ \{\ r_3,\ r_4,\ r'_4\ \}\ >\ \{\ r_5\ \}$.

Proposition 3.10 : *i) relation $\xrightarrow[R_4]{}$ is noetherian,*

ii) Let G be a connected graph with n (n>1) vertices such that :
- *each edge is labeled with f,*
- *exactly one vertex r (the root) is labeled with A, the other ones with N,*

then every graph G' in Irred(G) satisfies the following properties :
- *a) the root r is labeled with R, all other vertices with F,*
- *b) the t-edges of G' constitute a spanning tree for the underlying graph.*

SECTION 4. ATTRIBUTE GRAPH REWRITING SYSTEMS

In this section, we introduce the concept of attribute graph rewriting systems (APGRSs for short) which allows us to do some computations by means of semantic rules associated with each rewriting rule of a given PGRS.

Definitions will be given for C-graphs as they can easily be transposed in a natural way to the other kinds of graphs.

Definition 4.1 : An attribute graph rewriting system (with priorities) is given by $(C, P, >, A, Att, Dom, Sem)$ where :
- *i) $(C, P, >)$ is a PGRS,*
- *ii) A is a finite set of symbols called attribute names or simply attributes,*
- *iii) Att is a function from C into $\mathcal{P}(A)$,*
- *iv) Dom is a mapping from A into domains which associates with each*

attribute a its domain of values $Dom(a)$,

v) for each rewriting rule $r = (D,\lambda)$ of P, $Sem(r)$ is the set of semantic rules of r, which is defined as follows :

Let $D = <V,E,Ends,l>$; an *attribute of* r is a pair (a,x) with $x \in V \cup E$, and $a \in Att(\lambda(x)) \cup Att(l(x))$.

A *semantic rule* α in $Sem(r)$ has the following form :

$$(a_0,x) = f ((a_1,x_1), ..., (a_n,x_n))$$

with $x \in V \cup E$, $a_0 \in Att(\lambda(x))$, and for i, $1 \le i \le k$, $x_i \in V \cup E$, $a_i \in Att(l(x))$ and f is a mapping from $Dom(a_1) \times ... \times Dom(a_n)$ into $Dom(a_0)$.

Remark 4.2 : Semantic rules added to a PGRS are always well defined, provided the initialization of attributes has been done correctly. This is obvious because the semantic rules do not introduce circular dependencies between attributes.

Example 4.3 : In this example we use the PGRS of 3.9 to solve the well-known problem [9] of assigning a unique identifier to each vertex. We will use the *Dewey identifiers*. With each label A or M, we associate an integer attribute a, which denotes the first free value to be assigned to a son of the labeled vertex. We use the following integer attributes : $i(A)$, $i(M)$, $i(R)$, $i(F)$, $a(A)$ and $a(M)$. Attribute i stands for *identifier* of a vertex ; a for the first free value to be assigned to a son of the labeled vertex.

We obtain the following APGRS :

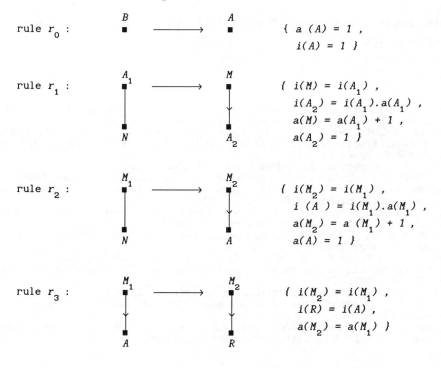

rule r_0 : $B \longrightarrow A$ $\{ a (A) = 1 ,$
$i(A) = 1 \}$

rule r_1 : $A_1 \longrightarrow M$ $\{ i(M) = i(A_1) ,$
$N \quad\quad A_2$ $i(A_2) = i(A_1).a(A_1) ,$
$a(M) = a(A_1) + 1 ,$
$a(A_2) = 1 \}$

rule r_2 : $M_1 \longrightarrow M_2$ $\{ i(M_2) = i(M_1) ,$
$N \quad\quad A$ $i (A) = i(M_1).a(M_1) ,$
$a(M_2) = a (M_1) + 1 ,$
$a(A) = 1 \}$

rule r_3 : $M_1 \longrightarrow M_2$ $\{ i(M_2) = i(M_1) ,$
$A \quad\quad R$ $i(R) = i(A) ,$
$a(M_2) = a(M_1) \}$

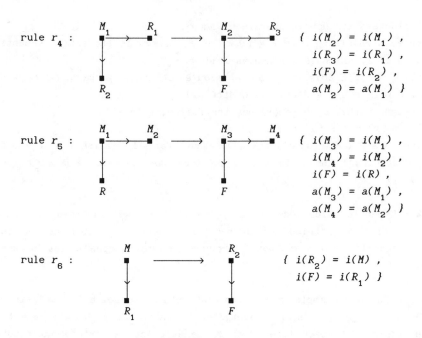

rule r_4 :   ... { $i(M_2) = i(M_1)$,
$i(R_3) = i(R_1)$,
$i(F) = i(R_2)$,
$a(M_2) = a(M_1)$ }

rule r_5 :   ... { $i(M_3) = i(M_1)$,
$i(M_4) = i(M_2)$,
$i(F) = i(R)$,
$a(M_3) = a(M_1)$,
$a(M_4) = a(M_2)$ }

rule r_6 :   ... { $i(R_2) = i(M)$,
$i(F) = i(R_1)$ }

Lemma 4.4 : *Let G be a connected graph (with at least 2 vertices) with each vertex labeled N except one of them which has label B.*

Then any graph G' in Irred(G) satisfies the following property :

let x and y be two any distinct vertices of G', then their associated identifiers (given by attribute i) are distinct.

Acknowledgements : We are very indebted to R. Cori and B. Vauquelin who have inspired this work.

REFERENCES

[1] Aho, A.O., Hopcroft, J.E., and Ullman, J.D., *The Design and Analysis of Computer Algorithms*, Addison-Wesley, Reading, Mass., 1974.
[2] Baeten, J.C.M., Bergstra, J.A., Klop, J.W., *Term Rewriting Systems with Priorities*, Rewriting Techniques and Applications, Lecture Notes in Comp. Science 256, pp 83-94, 1987.
[3] Berge, C., *Graphes et Hypergraphes*, Dunod, Paris, 1970.
[4] Billaud, M., Lafon, P., Metivier, Y., Sopena, E., *Graph Rewriting Systems with Priorities : Definitions and Applications*, Technical Report n°8909, LABRI, Université Bordeaux I, 1989.
[5] Dijkstra, E.W., and Scholten, C.S, *Termination Detection for Diffusing Computations*, Information Processing Letters. 11, pp 1-4, 1980.
[6] Hoare, C.A.R., *Communicating Sequential Processes*, Comm. ACM, Vol 21 n°8 pp 666-676, Aug. 1978.
[7] Mohan, C.K., *Priority Rewriting : Semantics, Confluence and Conditionnals*, Rewriting Techniques and Applications, Lecture Notes in Comp. Science 256, pp 278-291, 1987.
[8] Raynal, M., *Algorithmes Distribués et Protocoles*, Eyrolles 1985.
[9] Raynal, M., *Systèmes Répartis et Réseaux : Concepts Outils et Algorithmes*, Eyrolles 1987.
[10] Topor, R.W., *Termination Detection for Distributed Computations*, Information Processing Letters 18 (1984) pp 33-36.

Filtering Hyperedge–Replacement Languages Through Compatible Properties

Annegret Habel, Hans-Jörg Kreowski
Universität Bremen
Fachbereich Mathematik and Informatik
D–2800 Bremen

Abstract

In this paper, we demonstrate that the set of all hypergraphs of a hyperedge–replacement language that satisfy a so–called compatible property can again be generated by a hyperedge–replacement grammar. Because many familiar graph–theoretic properties like connectedness, k–colorability, planarity, the existence of Eulerian and Hamiltonian paths and cycles, etc. are compatible and compatibility is closed under Boolean operations, the members of a hyperedge–replacement language can be filtered through the sieve of various desirable and undesirable properties easily on the syntactic level of the generative grammar rules.

1. Introduction

Imagine the following situation: You have to deal with a set of items given – say – by a generative device, yet you know of additional requirements and objections. For instance, the programs of a programming language specified by the context–free syntax rules are assumed to satisfy additional context conditions. Similarly, a grammar infered from a set of positive examples may be assumed to avoid a set of negative examples in addition. In such a case, it would be nice if the generative devive could be adapted such that exactly those items of the original set are generated that fulfil the requirements and avoid the objections. A classical result of this kind is that the intersection of a context–free language and a regular set is context–free (see, e.g., [Sa 73]) or, more general, that the intersection of an equationally defined set and a recognizable set is equational again (see, e.g., [MW 67], [Co 88]). Within the area of graph grammars, the reader can find various results of this type. Rozenberg and Welzl show in [RW 86] that the family of BNLC graph languages is closed under many graph–theoretic properties. Courcelle shows in [Co 87] that the class of so–called context–free sets of graphs is closed under properties definable by monadic 2nd–order formulae. Lengauer and Wanke show in [LW 88] that a certain type of hyperedge–replacement languages is closed under so–called finite properties (these are, roughly speaking, graph properties for which only a finite number of graphs exist that behave essentially different with respect to the insertion into a bigger context).

In this paper, we present a further result of this type based on the concept of compatibility as introduced in [HKV 87]. In Section 4, we demonstrate that the set of all hypergraphs of a hyperedge–replacement language that satisfy a property compatible with the derivation process in a certain way can again be generated by a hyperedge–replacement grammar. The adapted grammar is effectively constructed from the originally given one.

In Section 5, we discuss consequences of the main result in three directions. First, because many familiar graph–theoretic properties like connectedness, k–colorability, planarity, the existence of Eulerian and Hamiltonian paths and cycles, etc. are known to be compatible (cf. [HKV 87+89, Ha 89]) and compatibility is closed under Boolean operations, the members of a hyperedge–replacement language can be filtered through the sieve of various desirable and undesirable properties easily on the syntactic level of the generative grammar rules. Second, because the Emptiness Problem and the Finiteness Problem are decidable for hyperedge–replacement languages, we can apply these results to the intersections of hyperedge–replacement languages with the sets of all hypergraphs satisfying a compatible property $PROP$. In this way, the following questions turn out to be decidable for all languages L generated by hyperedge–replacement grammars.

(1) Does a hypergraph of L exist satisfying $PROP$?
(2) Do all hypergraphs of L satisfy $PROP$?
(3) Does at most a finite number of hypergraphs in L satisfy $PROP$?
(4) Do all hypergraphs of L except perhaps for a finite number of them satisfy $PROP$?

The reader may note that the first two decidability results are obtained in a different way in [HKV 87]. The proofs of these two results in this paper follow the line of argumentation of Rozenberg and Welzl [RW 86], Courcelle [Co 87], Lengauer and Wanke [LW 88]. Third, we get criteria at hand for deducing that certain languages cannot be generated by hyperedge–replacement grammars and that certain properties cannot be compatible. An example of the former is the set of 5–colorable graphs, an example of the latter is the regularity of a hypergraph (i.e., equal node degree for all nodes).

Section 2 and 3 recall the basic notions and results for hypergraphs and hyperedge–replacement. The paper is closed by a short discussion.

2. Preliminaries

This section provides the basic notions on hypergraphs as far as needed in the paper. The key construction is the replacement of some hyperedges of a hypergraph by hypergraphs yielding an expanded hypergraph.

2.1 Definition (hypergraphs)

1. Let C be an arbitrary, but fixed alphabet, called a set of *labels* (or *colors*).
2. A *hypergraph* over C is a system (V, E, s, t, l) where V is a finite set of *nodes* (or *vertices*), E is a finite set of *hyperedges*, $s : E \to V^*$ and $t : E \to V^*$ [1] are two mappings assigning a sequence of *sources* $s(e)$ and a sequence of *targets* $t(e)$ to each $e \in E$, and $l : E \to C$ is a mapping *labeling* each hyperedge.
3. A hyperedge $e \in E$ of a hypergraph (V, E, s, t, l) is called an (m, n)–*edge* for some $m, n \in \mathbb{N}$ if $|s(e)| = m$ and $|t(e)| = n$. [2] The pair (m, n) is the *type* of e, denoted by $type(e)$. e is said to be *well–formed* if its sources and targets are pairwise distinct.
4. A *multi–pointed hypergraph* over C is a system $H = (V, E, s, t, l, begin, end)$ where the first five components define a hypergraph over C and $begin, end \in V^*$. Components of H are denoted by $V_H, E_H, s_H, t_H, l_H, begin_H, end_H$, respectively. The set of all multi–pointed hypergraphs over C is denoted by $\mathcal{H}_C$. A set $L \subseteq \mathcal{H}_C$ is called a *hypergraph language* over C.

---

[1] For a set A, A^* denotes the set of all words over A, including the empty word λ.
[2] For a word $w \in A^*$, $|w|$ denotes its length.

5. $H \in \mathcal{H}_C$ is said to be an (m,n)–hypergraph for some $m, n \in I\!N$ if $|begin_H| = m$ and $|end_H| = n$. The pair (m,n) is the type of H, denoted by $type(H)$. H is said to be well–formed if all hyperedges are well–formed and the begin–nodes and end–nodes of H are pairwise distinct.

6. Let $H \in \mathcal{H}_C$, $begin_H = begin_1 \ldots begin_m$ and $end_H = end_1 \ldots end_n$ with $begin_i, end_j \in V_H$ for $i = 1, \ldots, m$ and $j = 1, \ldots, n$. Then $EXT_H = \{begin_i | i = 1, \ldots, m\} \cup \{end_j | j = 1, \ldots, n\}$ denotes the set of external nodes of H. Moreover, $INT_H = V_H - EXT_H$ denotes the set of internal nodes of H.

Remark: There is a 1–1–correspondence between hypergraphs and $(0,0)$–hypergraphs so that hypergraphs may be seen as special cases of multi–pointed hypergraphs.

2.2 Definition (special hypergraphs)

1. A multi–pointed hypergraph H is said to be a singleton if $|E_H| = 1$ and $|V_H - EXT_H| = 0$. $e(H)$ refers to the only hyperedge of H and $l(H)$ refers to its label.

2. A singleton H is said to be a handle if $s_H(e) = begin_H$ and $t_H(e) = end_H$. If $l_H(e) = A$ and $type(e) = (m,n)$ for some $m, n \in I\!N$, then H is called an (m,n)–handle induced by A.

Remark: Given $H \in \mathcal{H}_C$, each hyperedge $e \in E_H$ induces a handle $e^\bullet$ by restricting the mappings s_H, t_H, and l_H to the set $\{e\}$, restricting the set of nodes to those ones occurring in $s_H(e)$ and $t_H(e)$, and choosing $begin_{e^\bullet} = s_H(e)$ and $end_{e^\bullet} = t_H(e)$.

2.3 Definition (subhypergraphs and isomorphic hypergraphs)

1. Let $H, H' \in \mathcal{H}_C$. Then H is called a (weak) subhypergraph of H', denoted by $H \subseteq H'$, if $V_H \subseteq V_{H'}$, $E_H \subseteq E_{H'}$, and $s_H(e) = s_{H'}(e)$, $t_H(e) = t_{H'}(e)$, $l_H(e) = l_{H'}(e)$ for all $e \in E_H$ [where nothing is assumed on the relation of the distinguished nodes].

2. Let $H, H' \in \mathcal{H}_C$ and $i_V : V_H \to V_{H'}$, $i_E : E_H \to E_{H'}$ be bijective mappings. Then $i = (i_V, i_E) : H \to H'$ is called an isomorphism from H to H' if $i_V^*(s_H(e)) = s_{H'}(i_E(e))$, $i_V^*(t_H(e)) = t_{H'}(i_E(e))$, $l_H(e)) = l_{H'}(i_E(e))$ for all $e \in E_H$ as well as $i_V^*(begin_H) = begin_{H'}$, $i_V^*(end_H) = end_{H'}$ [3]. H and H' are said to be isomorphic, denoted by $H \cong H'$, if there is an isomorphism from H to H'.

Now we are ready to introduce how hypergraphs may substitute hyperedges. An (m,n)–edge can be replaced by an (m,n)–hypergraph in two steps:

(1) Remove the hyperedge,

(2) add the hypergraph except the external nodes and hand over each tentacle of a hyperedge (of the replacing hypergraph) which grips to an external node to the corresponding source or target node of the replaced hyperedge.

Moreover, an arbitrary number of hyperedges can be replaced simultaneously in this way.

2.4 Definition (hyperedge replacement)

Let $H \in \mathcal{H}_C$ be a multi–pointed hypergraph, $B \subseteq E_H$, and $repl : B \to \mathcal{H}_C$ a mapping with $type(repl(b)) = type(b)$ for all $b \in B$. Then the replacement of B in H through $repl$ yields the multi–pointed hypergraph X given by

- $V_X = V_H + \sum_{b \in B}(V_{repl(b)} - EXT_{repl(b)})$, [4]
- $E_X = (E_H - B) + \sum_{b \in B} E_{repl(b)}$,
- each hyperedge of $E_H - B$ keeps its sources and targets,
- each hyperedge of $E_{repl(b)}$ (for all $b \in B$) keeps its internal sources and targets

---

[3] For a mapping $f : A \to B$, the free symbolwise extension $f^* : A^* \to B^*$ is defined by $f^*(a_1 \ldots a_k) = f(a_1) \ldots f(a_k)$ for all $k \in I\!N$ and $a_i \in A$ $(i = 1, \ldots, k)$.

[4] The sum symbols $+$ and $\sum$ denote the disjoint union of sets; the symbol $-$ denotes the set–theoretic difference.

and the external ones are handed over to the corresponding sources and targets of b, i.e.,

$$s_X(e) = h^*(s_{repl(b)}(e)) \text{ and } t_X(e) = h^*(t_{repl(b)}(e)) \text{ for all } b \in B \text{ and } e \in E_{repl(b)}$$

where $h : V_{repl(b)} \to V_X$ is defined by $h(v) = v$ for $v \in V_{repl(b)} - EXT_{repl(b)}$,

$h(b_i) = s_i$ $(i = 1,\ldots,m)$ for $begin_{repl(b)} = b_1 \ldots b_m$ and $s_H(b) = s_1 \ldots s_m$,

$h(e_j) = t_j$ $(j = 1,\ldots,n)$ for $end_{repl(b)} = e_1 \ldots e_n$ and $t_H(b) = t_1 \ldots t_n$,

- each hyperedge keeps its label,
- $begin_X = begin_H$ and $end_X = end_H$.

The resulting multi–pointed hypergraph X is denoted by $REPLACE(H, repl)$.

Remark: The construction above is meaningful and determines (up to isomorphism) a unique hypergraph X if h is a mapping. This is automatically fulfilled whenever the $begin$–nodes and end–nodes of each replacing hypergraph are pairwise distinct. If one wants to avoid such a restriction, one has to require that the following application condition is satisfied for each $b \in B$: If $begin_{repl(b)} = x_1 \ldots x_m$ and $end_{repl(b)} = x_{m+1} \ldots x_{m+n}$ as well as $s_H(b) = y_1 \ldots y_m$ and $t_H(b) = y_{m+1} \ldots y_{m+n}$, then, for $i,j = 1,\ldots,m+n$, $x_i = x_j$ implies $y_i = y_j$.

3. Hyperedge–Replacement Grammars and Languages

In this section we give a short summary of the basic notions of hyperedge–replacement grammars and languages. Details and examples can be found in [HK 87].

3.1 Definition (productions and derivations)

1. Let $N \subseteq C$. A *production* (over N) is an ordered pair $p = (A, R)$ with $A \in N$ and $R \in \mathcal{H}_C$. A is called *left–hand side* of p and is denoted by $lhs(p)$, R is called *right–hand side* and is denoted by $rhs(p)$. The *type* of p, denoted by $type(p)$, is given by the type of R.

2. Let $H \in \mathcal{H}_C$, $B \subseteq E_H$, and P be a set of productions. A mapping $prod : B \to P$ is called a *production base* in H if $l_H(b) = lhs(prod(b))$ and $type(b) = type(rhs(prod(b)))$ for all $b \in B$.

3. Let $H, H' \in \mathcal{H}_C$ and $prod : B \to P$ be a production base in H. Then H *directly derives* H' *through* $prod$ if H' is isomorphic to $REPLACE(H, repl)$ where $repl : B \to \mathcal{H}_C$ is given by $repl(b) = rhs(prod(b))$ for all $b \in B$. We write $H \underset{P}{\Longrightarrow} H'$ or $H \Longrightarrow H'$ in this case.

4. A sequence of direct derivations $H_0 \Longrightarrow H_1 \Longrightarrow \ldots \Longrightarrow H_k$ is called a *derivation* from H_0 to H_k (of length k). Additionally, in the case $H \cong H'$, we speak of a *derivation* from H to H' of length 0. A derivation from H to H' is shortly denoted by $H \underset{P}{\overset{*}{\Longrightarrow}} H'$ or $H \overset{*}{\Longrightarrow} H'$. If the length of the derivation should be stressed, we write $H \underset{P}{\overset{k}{\Longrightarrow}} H'$ or $H \overset{k}{\Longrightarrow} H'$.

5. A direct derivation through $prod : \emptyset \to P$ is called a *dummy*. [5] A derivation is said to be *valid* if at least one of its steps is not a dummy.

Remarks: 1. The application of a production $p = (A, R)$ of type (m, n) to a multi–pointed hypergraph H requires the following two steps only:

(1) Choose a hyperedge e of type (m, n) with label A.

(2) Replace the hyperedge e in H by R.

2. On the one hand, the definition of a direct derivation includes the case that no hyperedge is replaced. This dummy step derives a hypergraph isomorphic to the initial one. On the other hand, it includes the case that all hyperedges are replaced in one step. Moreover, whenever some hyperedges can be replaced in parallel, they can be replaced one after the other leading to the same derived hypergraph.

---

[5] A production base $prod : B \to P$ in H may be *empty*, i.e., $B = \emptyset$. In this case $H \Longrightarrow H'$ through $prod$ implies $H \cong H'$, and there is always a trivial direct derivation $H \Longrightarrow H$ through $prod$.

Using the introduced concepts of productions and derivations hyperedge–replacement grammars and languages can be introduced in a straightforward way.

3.2 Definition (hyperedge–replacement grammars and languages)

1. A *hyperedge–replacement grammar* is a system $HRG = (N, T, P, Z)$ where $N \subseteq C$ is a set of *nonterminals*, $T \subseteq C$ is a set of *terminals*, P is a finite set of *productions* over N, and $Z \in \mathcal{H}_C$ is the *axiom*. The class of all hyperedge–replacement grammars is denoted by $\mathcal{HRG}$.

2. HRG is said to be *typed* if there is a mapping $ltype : N \cup T \to I\!N \times I\!N$ such that, for each production $(A, R) \in P$, $ltype(A) = type(R)$ and $ltype(l_R(e)) = type(e)$ for all $e \in E_R$ and $ltype(l_Z(e)) = type(e)$ for all $e \in E_Z$. HRG is said to be *well–formed* if the right–hand sides of the productions are well–formed and all hyperedges in Z are well–formed.

3. The *hypergraph language* $L(HRG)$ *generated by* HRG consists of all hypergraphs which can be derived from Z applying productions of P and which are terminally labeled:

$$L(HRG) = \{H \in \mathcal{H}_T | Z \xRightarrow[P]{*} H\}.$$

4. A hypergraph language $L \subseteq \mathcal{H}_C$ is said to be a *hyperedge–replacement language* if there is a hyperedge–replacement grammar HRG with $L(HRG) = L$.

Remark: Without effecting the generative power, we will assume in the following that N and T are finite, $N \cap T = \emptyset$, and Z is a handle with $l(Z) \in N$. Furthermore, we will assume that the hyperedge–replacement grammars considered in this paper are typed and well–formed.

The results presented in the following sections are mainly based on some fundamental aspects of hyperedge–replacement derivations. Roughly speaking, hyperedge–replacement derivations cannot interfere with each other as long as they handle different hyperedges. On the one hand, a collection of derivations of the form $e^{\bullet} \xRightarrow{*} H(e)$ for $e \in E_R$ can be simultaneously embedded into R leading to a single derivation $R \Longrightarrow H$. On the other hand, restricting a derivation $R \Longrightarrow H$ to the handle $e^{\bullet}$ induced by the hyperedge $e \in E_R$ one obtains a so–called "restricted" derivation $e^{\bullet} \xRightarrow{*} H(e)$ where $H(e) \subseteq H$. Finally, restricting a derivation to the handles induced by the hyperedges, and subsequently embedding them again returns the original derivation. In other words, hyperedge–replacement derivations can be distributed to the handles of the hyperedges without losing information. We state and use this result in the following recursive version concerning terminal hypergraphs which are derivable from handles.

3.3 Fact

Let $HRG = (N, T, P, Z)$ be a typed and well–formed hyperedge-replacement grammar, $A \in N \cup T$, and $H \in \mathcal{H}_T$. Then there is a derivation $A^{\bullet} \Longrightarrow R \xRightarrow{k} H$ for some $k \geq 0$ [6] if and only if $A^{\bullet} \Longrightarrow R$ and, for each $e \in E_R$, there is a derivation $l_R(e)^{\bullet} \xRightarrow{k} H(e)$ with $H(e) \subseteq H$ such that $H \cong REPLACE(R, repl)$ with $repl(e) = H(e)$ for $e \in E_R$.

Remarks: 1. The derivation $l_R(e)^{\bullet} \xRightarrow{k} H(e)$ may be valid or not. In the first case, it has the same form as the original derivation, but it is shorter as the original one. In the latter case, $H(e)$ is isomorphic to $e^{\bullet}$ (resp. $l_R(e)^{\bullet}$) and hence a terminal handle.

2. Given a derivation $R \xRightarrow{k} H$, the derivation $l_R(e)^{\bullet} \xRightarrow{k} H(e)$ for each $e \in E_R$ is called the *fibre* of e and — the other way round — the given derivation is the *joint embedding* of its fibres.

---

[6] For a symbol $A \in N \cup T$ with $ltype(A) = (m, n)$, $A^{\bullet}$ denotes an (m, n)–handle induced by A. Note that (m, n)–handles induced by a symbol A are isomorphic.

4. Filtering Through Compatibility

In this section, the central question is the following:

> "If L is a hyperedge–replacement language and $PROP_0$ a graph–theoretic property, is the set of all hypergraphs from L satisfying $PROP_0$ again a hyperedge–replacement language?"

We demonstrate that the class of hyperedge–replacement languages behaves nicely in the sense that for all compatible properties the resulting languages are hyperedge–replacement languages. In particular, the above question gets an affirmative answer, if the property $PROP_0$ is: being connected, planar, cycle–free, Eulerian, Hamiltonian, resp. k–colorable.

Moreover, we show that the filtering mechanisms fail in general if the compatible property is replaced by a hyperedge–replacement language. More exactly speaking, the intersection of two hyperedge–replacement languages is not a hyperedge–replacement language, in general.

We start by recalling the notion of compatibility. Roughly speaking, a property $PROP_0$ is compatible with the derivation process of hyperedge–replacement grammars if it can be tested for each hypergraph H and each derivation of the form $A^\bullet \Longrightarrow R \overset{*}{\Longrightarrow} H$ by testing the property (or related properties) for the hypergraphs derived from the hyperedges of R and composing the results of the tests to the result for H.

4.1 Definition (compatible predicates)

1. Let $\mathcal{C}$ be a class of hyperedge–replacement grammars, I a finite index set, $PROP$ a decidable predicate[7] defined on pairs (H, i) with $H \in \mathcal{H}_{\mathcal{C}}$ and $i \in I$, and $PROP'$ a decidable predicate on triples $(R, assign, i)$ with $R \in \mathcal{H}_{\mathcal{C}}$, a mapping $assign : E_R \to I$, and $i \in I$. Then $PROP$ is called $\mathcal{C}, PROP'$-compatible if, for all $HRG = (N, T, P, Z) \in \mathcal{C}$ and all derivations $A^\bullet \Longrightarrow R \overset{*}{\Longrightarrow} H$ with $A \in N$ and $H \in \mathcal{H}_T$, and for all $i \in I$, $PROP(H, i)$ holds if and only if there is a mapping $assign : E_R \to I$ such that $PROP'(R, assign, i)$ holds and $PROP(H(e), assign(e))$ holds for all $e \in E_R$.

2. A predicate $PROP_0$ on $\mathcal{H}_{\mathcal{C}}$ is called $\mathcal{C}$–compatible if predicates $PROP$ and $PROP'$ and an index i_0 exist such that $PROP_0 = PROP(-, i_0)$[8] and $PROP$ is $\mathcal{C}, PROP'$-compatible.

Remarks: 1. Intuitively, a property is compatible if it can be tested for a large hypergraph with a long fibre by checking the smaller components of the corresponding shorter fibres.

2. Examples of compatible properties are: connectedness, planarity, cycle–freeness, existence of Hamiltonian and Eulerian paths and cycles, k–colorability for $k \geq 0$ (see [HKV 87] and [Ha 89]).

3. If $PROP$ is $\mathcal{C}, PROP'$–compatible and $\mathcal{C}' \subseteq \mathcal{C}$, then $PROP$ is $\mathcal{C}', PROP'$-compatible, too. In particular, if $PROP_0 = PROP(-, i_0)$ for some $i_0 \in I$, then $\mathcal{C}, PROP'$-compatibility of $PROP$ implies $\mathcal{C}'$-compatibility of $PROP_0$.

4. If $PROP$ is $\mathcal{C}_1, PROP'$–compatible and $\mathcal{C}_2, PROP'$–compatible for some $\mathcal{C}_1, \mathcal{C}_2 \subseteq \mathcal{HRG}$, then $PROP$ is $\mathcal{C}_1 \cup \mathcal{C}_2, PROP'$-compatible, too.

$\mathcal{C}$–compatible properties are closed under Boolean operations (cf. [Ha 89]).

4.2 Fact (closure under Boolean operations)

Let $PROP_1$, $PROP_2$ be $\mathcal{C}$–compatible for some class $\mathcal{C} \subseteq \mathcal{HRG}$. Let, for $H \in \mathcal{H}_{\mathcal{C}}$,

$(PROP_1 \wedge PROP_2)(H)$ if and only if $PROP_1(H) \wedge PROP_2(H)$,

$(PROP_1 \vee PROP_2)(H)$ if and only if $PROP_1(H) \vee PROP_2(H)$,

$(\neg PROP_1)(H)$ if and only if $\neg PROP_1(H)$.

Then $(PROP_1 \wedge PROP_2)$, $(PROP_1 \vee PROP_2)$, and $(\neg PROP_1)$ are $\mathcal{C}$–compatible.

---

[7] We assume that all considered predicates are *closed under isomorphisms*, i.e., if a predicate Φ holds for $H \in \mathcal{H}_{\mathcal{C}}$ and $H \cong H'$, then Φ holds for H', too.

[8] For $i \in I$, $PROP(-, i)$ denotes the unary predicate defined by $PROP(-, i)(H) = PROP(H, i)$ for all $H \in \mathcal{H}_{\mathcal{C}}$.

Filtering a hyperedge–replacement language L through a compatible property $PROP_0$ yields the language $L \cap L(PROP_0)$ of all hypergraphs from L that satisfy $PROP_0$. We are going to show now that $L \cap L(PROP_0)$ is a hyperedge–replacement language again. For this purpose, a hyperedge–replacement grammar generating L is transformed in such a way that the resulting grammar generates $L \cap L(PROP_0)$. The transformation works, roughly speaking, as follows. Let $PROP$ be a $C, PROP'$–compatible property with index set I. Then the nonterminals are indexed by elements from I where A_i for a nonterminal A and $i \in I$ is going to derive all hypergraphs derivable from A that satisfy $PROP(-, i)$. Moreover, each old production (A, R) leads to a new production (A_i, R_{assign}) for $i \in I$ and $assign : E_R \to I$ (where R_{assign} is obtained from R by indexing the label of each nonterminal hyperedge e by $assign(e)$) whenever $assign$ is a successful candidate due to the compatibility of $PROP$. We need some more notation to formulate the main result of this section.

4.3 Notations

1. For a property $PROP_0$, $L(PROP_0)$ denotes the set of all hypergraphs satisfying $PROP_0$.
2. For a class $\mathcal{L}$ of hypergraph languages and a hypergraph property $PROP_0$, let
$$\mathcal{L} \cap PROP_0 = \{L \cap L(PROP_0) | L \in \mathcal{L}\} \quad \text{and} \quad \mathcal{L} - PROP_0 = \{L - L(PROP_0) | L \in \mathcal{L}\}.$$
3. Conversely, for a hypergraph language L, the induced property $PROP_L$ is defined as follows: $PROP_L(H)$ holds if and only if $H \in L$.
4. $\mathcal{HRG}$ denotes the class of all hyperedge–replacement grammars.
5. For $\mathcal{C} \subseteq \mathcal{HRG}$, let $\mathcal{L}(\mathcal{C}) = \{L(HRG) | HRG \in \mathcal{C}\}$.

4.4 Theorem

Let $\mathcal{C} \subseteq \mathcal{HRG}$ and $PROP_0$ be a $\mathcal{C}$–compatible property. Then

$$\mathcal{L}(\mathcal{C}) \cap PROP_0 \subseteq \mathcal{L}(\mathcal{HRG}).$$

Proof: Let $PROP_0$ be $\mathcal{C}$–compatible and $PROP$ and $PROP'$ be the corresponding predicates over the index set I such that $PROP$ is $\mathcal{C}, PROP'$–compatible and $PROP_0 = PROP(-, i_0)$ for some $i_0 \in I$. Let $HRG = (N, T, P, Z) \in \mathcal{C}$. Then, we construct a hyperedge–replacement grammar $HRG' = (N', T, P', Z')$ with $L(HRG') = L(HRG) \cap L(PROP_0)$ as follows. Let

- $N' = N \times I$, the elements of which are written A_i rather than (A, i),
- $P' = \{(A_i, R_{assign}) | (A, R) \in P, \; assign : E_R \to I, \; i \in I, \; PROP'(R, assign, i),$
$$PROP(l_R(e)^\bullet, assign(e)) \text{ for all terminal } e \in E_R\}$$
where R_{assign}, for $R \in \mathcal{H}_{\mathcal{C}}$ and $assign : E_R \to I$, is obtained from R by relabeling the nonterminal hyperedges according to $assign$, i.e., $R_{assign} = (V_R, E_R, s_R, t_R, l, begin_R, end_R)$ with $l(e) = l_R(e)_{assign(e)}$ if $l_R(e) \in N$ and $l(e) = l_R(e)$ otherwise,
- $Z' = Z_{assign}$ where $assign(e(Z)) = i_0$, this means that the label of the only hyperedge of Z is indexed by i_0.

To show $L(HRG') = L(HRG) \cap L(PROP_0)$, we prove Claim 1 and Claim 2.

Claim 1: For all $A \in N$ and all $i \in I$, if $A^\bullet \overset{*}{\Longrightarrow} H \in \mathcal{H}_T$ is a derivation in HRG and $PROP(H, i)$ holds, then there is a derivation $A_i^\bullet \overset{*}{\Longrightarrow} H$ in HRG'.

Proof of Claim 1 (by induction on the number of steps in the derivation):
Let $A^\bullet \overset{1}{\Longrightarrow} H \in \mathcal{H}_T$ be a one–step derivation in HRG and $PROP(H, i)$ be satisfied. Then we can assume that $(A, H) \in P$ and $H = REPLACE(H, repl)$ with $repl(e) = l_H(e)^\bullet$ for $e \in E_H$. By $\mathcal{C}, PROP'$–compatibility of $PROP$, there is a mapping $assign : E_H \to I$ such that

$PROP'(H, assign, i)$ and $PROP(l_H(e)^{\bullet}, assign(e))$ hold for $e \in E_H$. Hence, $(A_i, H_{assign}) \in P'$. Moreover, $H_{assign} = H$. Thus, there is a derivation $A_i^{\bullet} \overset{1}{\Longrightarrow} H$ in HRG'.

Suppose now that the statement is true for k–step derivations. Let $A^{\bullet} \overset{k+1}{\Longrightarrow} H \in \mathcal{H}_T$ be a $(k+1)$–step derivation in HRG and $PROP(H, i)$ be satisfied. Then we can assume that the derivation is of the form $A^{\bullet} \Longrightarrow R \overset{k}{\Longrightarrow} H$ for some $(A, R) \in P$. Consider now the fibres $l_R(e)^{\bullet} \overset{k}{\Longrightarrow} H(e)$ of $e \in E_R$. Since $PROP(H, i)$ holds and $PROP$ is $\mathcal{C}, PROP'$–compatible, there is a mapping $assign : E_R \to I$ such that $PROP'(R, assign, i)$ and $PROP(H(e), assign(e))$ hold for $e \in E_R$. By construction of P', $(A_i, R_{assign}) \in P'$. Moreover, by inductive hypothesis, for $e \in E_R$ with $l_R(e) \in N$, there is a derivation $l_R(e)^{\bullet}_{assign(e)} \overset{\bullet}{\Longrightarrow} H(e)$ in HRG'. Embedding of these derivations into R_{assign} yields a derivation $R_{assign} \overset{\bullet}{\Longrightarrow} H'$. Since R_{assign} and R differ only in the labels of the nonterminal hyperedges and $H = REPLACE(R, repl)$ with $repl(e) = H(e)$ for $e \in E_R$, $H' = REPLACE(R_{assign}, repl)$ equals H. Thus, there is a derivation $A_i^{\bullet} \Longrightarrow R_{assign} \overset{\bullet}{\Longrightarrow} H$ in HRG'.

Claim 2: For all $A \in N$ and all $i \in I$, if $A_i^{\bullet} \overset{\bullet}{\Longrightarrow} H \in \mathcal{H}_T$ is a derivation in HRG', then there is a derivation $A^{\bullet} \overset{\bullet}{\Longrightarrow} H$ in HRG, and $PROP(H, i)$ holds.

Proof of Claim 2 (by induction on the number of steps in the derivation):

Let $A_i^{\bullet} \overset{1}{\Longrightarrow} H \in \mathcal{H}_T$ be a one–step derivation in HRG'. Then we can assume that $(A_i, H) \in P'$ and $H = REPLACE(H, repl)$ with $repl(e) = l_H(e)^{\bullet}$ for $e \in E_H$. By construction of P', there are a production $(A, H) \in P$ and a mapping $assign : E_H \to I$ such that $PROP'(H, assign, i)$ and $PROP(l_H(e)^{\bullet}, assign(e))$ hold for $e \in E_R$. Thus, there is a derivation $A^{\bullet} \overset{1}{\Longrightarrow} H$ in HRG and $PROP(H, i)$ is satisfied.

Suppose that the statement is true for k–step derivations. Let $A_i^{\bullet} \overset{k+1}{\Longrightarrow} H \in \mathcal{H}_T$ be a $(k+1)$–step derivation in HRG'. Then we can assume that the derivation decomposes into $A_i^{\bullet} \Longrightarrow R_{assign} \overset{k}{\Longrightarrow} H$ for some $(A_i, R_{assign}) \in P'$. Hence $PROP'(R, assign, i)$ and, for terminal hyperedges $e \in E_R$, $PROP(l_R(e)^{\bullet}, assign(e))$ hold. For nonterminal hyperedges $e \in E_{R_{assign}}$, the fibres of the derivation $R_{assign} \overset{k}{\Longrightarrow} H$ are of the form $l_R(e)^{\bullet}_{assign(e)} \overset{k}{\Longrightarrow} H(e)$. By inductive hypothesis, for each nonterminal $e \in E_R$, there is a derivation $l_R(e)^{\bullet} \overset{\bullet}{\Longrightarrow} H(e)$ in HRG and $PROP(H(e), assign(e))$ holds. Embedding the derivations into R yields a derivation $R \overset{\bullet}{\Longrightarrow} H$ in HRG. Thus, there is a derivation $A^{\bullet} \Longrightarrow R \overset{\bullet}{\Longrightarrow} H$ in HRG. Moreover, by $\mathcal{C}, PROP'$–compatibility of $PROP$, $PROP(H, i)$ holds. This completes the inductive step.

By Claim 1 and 2, for $l(Z) \in N$ and $i_0 \in I$, there is a derivation $l(Z)^{\bullet}_{i_0} \overset{\bullet}{\Longrightarrow} H$ in HRG' if and only if there is a derivation $l(Z)^{\bullet} \overset{\bullet}{\Longrightarrow} H \in \mathcal{H}_T$ in HRG and $PROP(H, i_0)$ is satisfied. This completes the proof of the theorem. $\square$

Similar to the string case, it turns out that the class of all hyperedge–replacement languages is not closed under intersection. But note that the usual counterexample, i.e., the intersection $L = \{a^n b^n c^n | n \geq 1\}$ of the context–free languages $\{a^m b^m c^n | m, n \geq 1\}$ and $L = \{a^m b^n c^n | m, n \geq 1\}$, does not work because L can be generated by a hyperedge–replacement grammar (see [HK 87]). Nevertheless, a more sophisticated counterexample works.

4.5 Theorem

$$\mathcal{L}(\mathcal{HRG}) \cap \mathcal{L}(\mathcal{HRG}) \not\subseteq \mathcal{L}(\mathcal{HRG}).$$

Proof: Let

$$L_1 = \{(a^{n_1} b^{n_1} a^{n_2} b^{n_2} \ldots a^{n_k} b^{n_k})^{\bullet} | k \geq 1, n_1, \ldots, n_k \geq 1\},$$
$$L_2 = \{(a^{n_1} b^{n_2} a^{n_2} b^{n_3} \ldots a^{n_k} b^{n_{k+1}})^{\bullet} | k \geq 1, n_1, \ldots, n_k \geq 1\}.$$

where, for a string $w = x_1 x_2 x_3 \ldots x_n$, $w^{\bullet}$ denotes the graph $\xrightarrow{x_1} \xrightarrow{x_2} \xrightarrow{x_3} \ldots \xrightarrow{x_n}$.

Clearly, L_1 and L_2 are hyperedge–replacement languages. Moreover,

$$L_1 \cap L_2 = \{((a^n b^n)^k)^{\bullet} | k, n \geq 1\}.$$

By the Pumping Lemma for hyperedge–replacement languages (see [HK 87]), it can be shown that $L = L_1 \cap L_2 \notin \mathcal{L}(\mathcal{HRG})$. $\square$

5. Some Consequences

In this section, we show various corollaries of the main theorem.

First, it is pointed out that compatible properties cannot only be used as additional requirements to accept members of hyperedge–replacement languages, but also as objections to refuse some of them. Moreover, compatible requirements can be combined conjunctively and disjunctively such that the filtering can be done through a set of requirements and objections (including alternatives) simultaneously. This follows easily from Theorem 4.4 in combination with the Fact 4.2.

5.1 Corollary

Let $PROP_1$, $PROP_2$ be $\mathcal{C}$–compatible for some $\mathcal{C} \subseteq \mathcal{HRG}$. Then
(1) $\mathcal{L}(\mathcal{C}) - PROP_1 \subseteq \mathcal{L}(\mathcal{HRG})$,
(2) $\mathcal{L}(\mathcal{C}) \cap (PROP_1 \wedge PROP_2) \subseteq \mathcal{L}(\mathcal{HRG})$,
(3) $\mathcal{L}(\mathcal{C}) \cap (PROP_1 \vee PROP_2) \subseteq \mathcal{L}(\mathcal{HRG})$.

It is not difficult to see that the Emptiness Problem for hyperedge–replacement grammars is decidable. If this is applied to grammars generating the languages $L(HRG) \cap L(PROP_0)$ and $L(HRG) \cap L(\neg PROP_0)$, where HRG is some hyperedge–replacement grammar and $PROP_0$ is some compatible property, one gets the decidability of the existential and universal questions,
(1) Does a hypergraph of $L(HRG)$ exist satisfying $PROP_0$?,
(2) Do all hypergraphs of $L(HRG)$ satisfy $PROP_0$?,
that are investigated directly in [HKV 87]. Note that the line of argumentation presented here follows Rozenberg and Welzl [RW 86] and Courcelle [Co 87].

5.2 Fact (Emptiness Problem)

For hyperedge–replacement grammars, there is an algorithm for deciding whether the language of a hyperedge–replacement grammar is empty or not.

Remark: Note that the decidability of the Emptiness Problem can be proved by adapting the well-known ideas for proving the decidability of the Emptiness Problem for context–free string grammars to hyperedge–replacement grammars.

5.3 Corollary

Let $PROP_0$ be $\mathcal{C}$–compatible with respect to some class $\mathcal{C}$ of hyperedge–replacement grammars. Then, for all $HRG \in \mathcal{C}$, it is decidable whether
(1) $PROP_0$ holds for some $H \in L(HRG)$,
(2) $PROP_0$ holds for all $H \in L(HRG)$.

Proof: Let $HRG \in \mathcal{C}$ and HRG_{PROP_0}, $HRG_{\neg PROP_0}$ be the hyperedge–replacement grammars generating $L(HRG) \cap L(PROP_0)$ and $L(HRG) \cap L(\neg PROP_0)$, respectively, due to Fact 4.2 and Theorem 4.4. Then there is a hypergraph $H \in L(HRG)$ satisfying $PROP_0$ if and only if $L(HRG_{PROP_0})$ is non–empty, and all hypergraphs in $L(HRG)$ satisfy $PROP_0$ if and only if $L(HRG_{\neg PROP_0})$ is empty. Therefore, the decidability of the Emptiness Problem implies the decidability of the problems (1) and (2). $\qquad\square$

It is also not difficult to see that the Finiteness Problem is decidable for hyperedge–replacement grammars. If this is applied to the grammars generating the languages $L(HRG) \cap L(PROP_0)$ and $L(HRG) \cap L(\neg PROP_0)$ for some hyperedge–replacement grammar HRG and some compatible property $PROP_0$, one gets the decidability of the following questions.

(1) Does at most a finite number of hypergraphs in $L(HRG)$ satisfy $PROP_0$?

(2) Do all hypergraphs of $L(HRG)$ except perhaps for a finite number of them satisfy $PROP_0$?

5.4 Fact (Finiteness Problem)

For hyperedge–replacement grammars, there is an algorithm determining whether a given hyperedge–replacement grammar generates a finite or infinite number of (non–isomorphic) hypergraphs.

5.5 Corollary

Let $PROP_0$ be $\mathcal{C}$–compatible for some $\mathcal{C} \subseteq \mathcal{HRG}$. Then, for all $HRG \in \mathcal{C}$, it is decidable whether

(1) $PROP_0$ holds for at most a finite number of $H \in L(HRG)$,

(2) $PROP_0$ holds for all $H \in L(HRG)$ except perhaps a finite number.

Proof: Let $HRG \in \mathcal{C}$ and HRG_{PROP_0}, $HRG_{\neg PROP_0}$ be the hyperedge–replacement grammars generating $L(HRG) \cap L(PROP_0)$ and $L(HRG) \cap L(\neg PROP_0)$, respectively, due to Fact 4.2 and Theorem 4.4. Then, $PROP_0$ holds for at most a finite number of $H \in L(HRG)$ if and only if $L(HRG_{PROP_0})$ is finite, and $PROP_0$ holds for all $H \in L(HRG)$ except perhaps a finite number if and only if $L(HRG_{\neg PROP_0})$ is finite. Therefore, the decidability of the Finiteness Problem implies the decidability of the problems (1) and (2). $\qquad\square$

In addition to the problems outlined above, one may investigate questions of the following type:

(1) If $PROP_1$ and $PROP_2$ are graph–theoretic properties and L is a hyperedge–replacement language, is the set of all hypergraphs of L satisfying $PROP_1$ included in the one satisfying $PROP_2$?

(2) If L is a hyperedge–replacement language and all hypergraphs of L satisfy $PROP_1$, do all hypergraphs of L satisfy $PROP_2$, too?

Both types of questions turn out to be decidable if the properties $PROP_1$ and $PROP_2$ are compatible.

5.6 Corollary

Let $PROP_1$ and $PROP_2$ be $\mathcal{C}$–compatible with respect to some class $\mathcal{C}$ of hyperedge–replacement grammars. Then, for all $HRG \in \mathcal{C}$, it is decidable whether

(1) $L(PROP_1) \cap L(HRG) \subseteq L(PROP_2)$,

(2) $L(HRG) \subseteq L(PROP_1)$ implies $L(HRG) \subseteq L(PROP_2)$.

Proof: Let $HRG \in \mathcal{C}$ and HRG_{PROP_1} be the hyperedge–replacement grammar generating $L(HRG) \cap L(PROP_1)$ due to Theorem 4.4. Then

$$L(PROP_1) \cap L(HRG) \subseteq L(PROP_2) \iff L(HRG_{PROP_1}) \subseteq L(PROP_2)$$
$$\iff \forall H \in L(HRG_{PROP_1}) : PROP_2(H)$$

Let $\bar{\mathcal{C}} \subseteq \mathcal{HRG}$ consist of all hyperedge–replacement grammars one gets by transforming the grammars of $\mathcal{C}$ using the transformation in 4.4 with respect to $PROP_1$. Then it is easy to show that $PROP_2$ (being $\mathcal{C}$–compatible) is $\bar{\mathcal{C}}$–compatible, too. Hence, due to Corollary 5.3, the universal question (and, with this, the first stated problem) turns out to be decidable. Moreover,

$L(HRG) \subseteq L(PROP_1)$ implies $L(HRG) \subseteq L(PROP_2)$

$$\Longleftrightarrow \quad L(HRG) \not\subseteq L(PROP_1) \quad \text{or} \quad L(HRG) \subseteq L(PROP_2)$$

$$\Longleftrightarrow \quad \exists H \in L(HRG) : \neg PROP_1(H) \quad \text{or} \quad \forall H \in L(HRG) : PROP_2(H).$$

Due to Fact 4.2 and Corollary 5.3, the existential as well as the universal question (and, therefore, the second stated problem) are decidable. $\quad\square$

Theorem 4.4 can be used to show that specific languages cannot be generated by a hyperedge–replacement grammar and that specific properties cannot be compatible in the sense of Def. 4.1. Simple reformulations of Theorem 4.4 make this transparent.

5.7 Corollary

Let $PROP_0$ be $\mathcal{C}$–compatible for some $\mathcal{C} \subseteq \mathcal{HRG}$ and L be a hypergraph language. Then, $L \cap L(PROP_0) \notin \mathcal{L}(\mathcal{HRG})$ implies $L \notin \mathcal{L}(\mathcal{C})$.

Proof: Suppose $L \in \mathcal{L}(\mathcal{C})$. Then, by Theorem 4.4, $L \cap L(PROP_0) \in \mathcal{L}(\mathcal{HRG})$, a contradiction. $\quad\square$

5.8 Example

Let L be the set of all 5–colorable graphs and $PLANAR$ be the predicate on $\mathcal{H}_C$ given by $PLANAR(H)$ if and only if H is planar. Since every planar graph is 5–colorable, $L \cap L(PLANAR)$ consists of all planar graphs. On the one hand, $PLANAR$ is $\mathcal{ERG}$–compatible where $\mathcal{ERG}$ denotes the class of all edge–replacement grammars (see [HKV 87]). On the other hand, the set of all planar graphs cannot be generated by an edge–replacement grammar. Therefore, $L \notin \mathcal{L}(\mathcal{ERG})$.

5.9 Corollary

Let $PROP_0$ be a predicate on $\mathcal{H}_C$ and $HRG \in \mathcal{HRG}$. Then, $L(HRG) \cap L(PROP_0) \notin \mathcal{L}(\mathcal{HRG})$ implies that $PROP_0$ is not $\{HRG\}$–compatible. Moreover, $PROP_0$ is not $\mathcal{C}$–compatible for all $\mathcal{C}$ with $HRG \in \mathcal{C}$.

Proof: Suppose $PROP_0$ is $\mathcal{C}$–compatible. Then, by Theorem 4.4, $L(HRG) \cap L(PROP_0) \in \mathcal{L}(\mathcal{HRG})$, a contradiction. Thus, $PROP_0$ is not $\mathcal{C}$–compatible. $\quad\square$

5.10 Examples

1. Let $SQUARE$ be the predicate on $\mathcal{H}_C$ given by $SQUARE(H)$ if and only if H has a square number of nodes. Moreover, let HRG be a hyperedge–replacement grammar which generates the set of totally disconnected hypergraphs (with at least one node). Then $L(HRG) \cap L(SQUARE)$ consists of all totally disconnected hypergraphs with a square number of nodes. This set cannot be generated by a hyperedge–replacement grammar because the growth with respect to the number of nodes is not sublinear (see [HK 87], Corollary 4.5). Therefore, $L(HRG) \cap L(SQUARE) \notin \mathcal{L}(\mathcal{HRG})$ and $SQUARE$ is not $\{HRG\}$–compatible.

2. By the same argumentation, $SQUARE$ is not $\{HRG\}$–compatible where HRG is a grammar that generates, e.g., the set of all stars, wheels, ladders, outerplanar graphs, respectively.

3. Let $REGULAR$ be the predicate on $\mathcal{H}_C$ given by $REGULAR(H)$ if and only if H is regular, i.e., there is a natural number $k \in I\!N$ such that all nodes in H are of degree k. Let HRG be a hyperedge–replacement grammar which generates the set of trees (with at least one node) the leaves of which are equipped with loops. Then $L(HRG) \cap L(REGULAR)$ consists of all "regular" tree structures (with at least one node) where all nodes have the same odd degree. By the Pumping Lemma for hyperedge–replacement languages (see [HK 87], Theorem 3.4), it can be shown that $L(HRG) \cap L(REGULAR) \notin \mathcal{L}(\mathcal{HRG})$. Thus, $REGULAR$ is not $\{HRG\}$–compatible.

Furthermore, Corollary 5.9 can be combined with Theorem 4.5 leading to the observation that a hyperedge–replacement language does not define a compatible property in general.

5.11 Corollary

Let $HRG_1, HRG_2 \in \mathcal{HRG}$ with $L(HRG_1) \cap L(HRG_2) \notin \mathcal{L}(\mathcal{HRG})$. Then, $PROP_{L(HRG_2)}$ is not $\{HRG_1\}$–compatible. Moreover, $PROP_{L(HRG_2)}$ is not $\mathcal{C}$–compatible for all $\mathcal{C}$ with $HRG_1 \in \mathcal{C}$.

Proof: By assumption, $L(HRG_1) \cap PROP_{L(HRG_2)} = L(HRG_1) \cap L(HRG_2) \notin \mathcal{L}(\mathcal{HRG})$. Therefore, Corollary 5.11 immediately follows from Corollary 5.9. $\qquad\square$

Finally, we will show that the undecidability of the Post Correspondence Problem can be used to prove non–compatibility.

Let $2 - PATHS$ be the predicate on $(1,1)$–graphs given by $2 - PATHS(G)$ if and only if G has two edge–disjoint paths, $p_1 = e_1, \ldots, e_m$ and $p_2 = e'_1, \ldots, e'_n$, between $begin_G$ and end_G which are equally labeled, i.e., $l_G(e_1) \ldots l_G(e_m) = l_G(e'_1) \ldots l_G(e'_n)$.

5.12 Theorem (non–compatibility)

$2 - PATHS$ is not $\mathcal{HRG}$–compatible.

Proof: Let $U = u_1, \ldots, u_n$ and $V = v_1, \ldots, v_n$ be two lists of words over a finite alphabet T. Let $HRG(U, V)$ be the hyperedge–replacement grammar

$$HRG(U, V) = \{S, A\} \cap \{u_1, v_1 \ldots, u_n, v_n\}, T, P, Z)$$

where Z is given by

and P consists of the following productions:

$$u_i ::= (u_i)^\bullet, \quad v_i ::= (v_i)^\bullet \qquad \text{for } i = 1, \ldots, n$$

Then the instance of PCP has a solution if and only if there exists a graph $H \in L(HRG(U, V))$ satisfying $2 - PATHS$.

Assume $2-PATHS$ is C–compatible where $C = \{HRG(U,V)|(U,V)$ is an instance of PCP$\}$. Then, for all instance (U,V) of PCP, we can construct a grammar $HRG(U,V)'$ generating the graphs in $L(HRG(U,V)$ satisfying $2-PATHS$. Finally, we may test whether $L(HRG(U,V)')$ is non–empty. If so, then (U,V) has a solution; otherwise, (U,V) does not have a solution. Therefore, there is an algorithm for PCP, which does not exist. Hence $2-PATHS$ is not C–compatible and, therefore, not $\mathcal{HRG}$–compatible. $\square$

6. Discussion

In the present paper, we have continued the investigation of hyperedge–replacement grammars and languages focussed on compatibility (cf. [HKV 87+89, Ha 89]). We have demonstrated that the family of hyperedge–replacement languages is closed under compatible properties. Filtering all hypergraphs from such a language that satisfy such a property does not lead out of the family. In this respect, our hypergraph languages behave quite similar with respect to compatible properties as the languages studied by Courcelle [Co 87] and Lengauer and Wanke [LW 88] with respect to definable and finite properties, respectively, and the family of BNLC languages investigated by Rozenberg and Welzl [RW 86] with respect to several explicit graph–theoretic properties.

Although hyperedge replacement has turned out to be mathematically promising, many problem areas are not yet sufficiently studied, among them one finds the following.
(1) While the known results for definable, finite, and compatible properties are quite similar, the relationship between the three notions is not fully revealed.
(2) An obvious technical difference between compatibility on the one hand and finiteness as well as definability on the other hand is that compatibility is defined relative to an arbitrary class C of hyperedge–replacement grammars while the other two notions are introduced on the level of graphs and graph operations independent of a certain class of grammars. We don't know yet whether this is an essential difference. Are there "interesting" properties that are C–compatible for some C, but neither definable, nor finite, nor $\mathcal{HRG}$–compatible where $\mathcal{HRG}$ is the class of all hyperedge–replacement grammars?
(3) The notion of compatibility, that leads to very nice structural and decidability results for hyperedge–replacement grammars and languages, concerns and deals with hypergraphs as singletons. What about problems and properties that relate several hypergraphs with each other like the isomorphism problem?
(4) The fact that compatible properties are closed under Boolean operations provides a way to combine additional requirements on a language, but not an awfully exciting one. What about other more sophisticated ways of combining properties like the questions at the end of Section 5 (cf. also the discussion of Rozenberg and Welzl [RW 86])?

References

[Co 87]   B.Courcelle: On Context–Free Sets of Graphs and Their Monadic Second–Order Theory, Lect. Not. Comp. Sci. 291, 133-146, 1987

[Co 88]   B.Courcelle: The Monadic Second–Order Logic of Graphs I: Recognizable Sets of Finite Graphs, revised version of "Recognizability and Second Order Definability for Sets of Finite Graphs" (Research Report 8634), Report 8837, Bordeaux I University, 1988

[Ha 89]   A. Habel: Graph–Theoretic Properties Compatible with Graph Derivations, Proc. Graph–Theoretic Concepts in Computer Science 1988 (WG'88), Lect. Not. Comp. Sci. 344, 11-29, 1989

[HK 87]   A. Habel, H.-J. Kreowski: Some Structural Aspects of Hypergraph Languages Generated by Hyperedge Replacement, Proc. STACS'87, Lect. Not. Comp. Sci. 247, 207-219, 1987

[HKV 87] A. Habel, H.-J. Kreowski, W. Vogler: Metatheorems for Decision Problems on Hyperedge Replacement Graph Languages, to appear in Acta Informatica, short version with the title "Compatible Graph Properties are Decidable for Hyperedge Replacement Graph Languages" in: Bull. EATCS 33, 55-62, 1987

[HKV 89] A. Habel, H.-J. Kreowski, W. Vogler: Decidable Boundedness Problems for Hyperedge–Replacement Graph Grammars, Proc. Joint Conference on Theory and Practice of Software Development (TAPSOFT'89), Lect. Not. Comp. Sci. 351, 275-289, 1989

[LW 88]   T. Lengauer, E. Wanke: Efficient Analysis of Graph Properties on Context–Free Graph Languages, Proc. ICALP'88, Lect. Not. Comp. Sci. 317, 379-393, 1988

[MW 67]   J. Mezei, J. B. Wright: Algebraic Automata and Context–Free Sets, Information and Control 11, 3-29, 1967

[RW 86]   G. Rozenberg, E. Welzl: Graph Theoretic Closure Properties of the Family of Boundary NLC Graph Languages, Acta Informatica 23, 289-309, 1986

[Sa 73]   A. Salomaa: Formal Languages, Academic Press, New York, 1973

Describing Distributed Systems by Categorical Graph Grammars

H.J. Schneider, Erlangen

Abstract

The structure of an asynchronous system of processes is described by a
labeled hypergraph. It represents both the past and the present of the system. The set of all possible traces is defined by a hypergraph grammar. In
the graph, actions and process states are represented by hyperedges. Each
hyperedge is connected to some event nodes, some of which are considered to
be predecessors of the edge, whereas others are successor nodes. This induces a partial ordering of the hyperedges. Some aspects of the Ada rendez-
vous are used as an example and translated into hypergraph productions.

1. The hypergraph model of an asynchronous system

Graph-theoretic structures are an obvious means to describe systems of
asynchronous processes. We are interested in generating the set of all
possible traces the system of processes may run through. The trace of a
sequential process is a linear sequence of actions. The trace of a system
of asynchronous processes, however, is a partially ordered set of actions.
We represent it by a labeled hypergraph the hyperedges of which denote actions. The partial order is induced by connecting an edge to some predecessor nodes and to some successor nodes. Furthermore, it may be connected
to channel nodes.

Generation of traces is modeled by a graph grammar. Contextfree productions describe stand-alone evolution of processes, whereas synchronizing
operations require more general productions. The synchronization mechanism
defines how to build these productions out of the contextfree ones. We use
the categorical graph-grammar approach. Its main advantage is that we can
use a lot of theorems concerning concurrency as well as amalgamation of
productions. We refer to the work of Ehrig, Kreowski and others [4, 6, 7].
We assume the reader to be familiar with this approach to such an extent as
it is presented by Ehrig [2].

This work was partly supported by Deutsche Forschungsgemeinschaft (SFB 182
- Project B1).

We start with giving an example. We consider a system consisting of two producers and one consumer. These components communicate with one another by using a common buffer. The start graph is given in Fig.1; b_init, p_init and c_init denote the initial states of the buffer, the producer and the consumer respectively. The consumer as well as both producers are connected to the buffer. In this case, the channel nodes indicate logical connections. (A different solution is possible: If you want to study visibility problems, you may use a common channel node connected to all processes.) Fig.2 is an example of a system graph after some computation steps have been performed. This graph can be derived from the start graph by applying the productions that we present in a later section.

In our figures, hyperedges are denoted by rectangulars, whereas lines represent channel nodes and arcs correspond to event nodes. This is unusual from a graph-theoretic point of view, but we decided in favor of it because the diagrams look like the well-known flow-charts. The hypergraph represents the past and the present of the system. A computation corresponds to a finite or infinite sequence of derivable hypergraphs. Of course, a derivation step must not change the past. This can be realized by dividing the set of edge labels into terminal and nonterminal labels. The nonterminal ones identify computation states within a process, i.e., they indicate the edges that allow further derivation steps. In Fig.2, the labels at the bottom (p_prod, b_wait, c_init) are nonterminal edge labels, indicating that the consumer process is in its initial state, both producers have generated new items and the buffer is waiting for a rendez-vous. These hyperedges may be replaced by applying productions, whereas the hyperedges that belong to the past are labeled by actions. They are considered to be terminal edges and must not be replaced.

Each hyperedge is connected to some nodes. We distinguish between event nodes (arcs) and channel nodes (lines). An event node may be either a predecessor or a successor to some hyperedge. We indicate its meaning by the direction of the arc. Most of the hyperedges in our example have exactly one predecessor and one successor node, but there are some exceptions. The hyperedge labeled by "accept insert item" has two predecessors and one successor, whereas, e.g., "b_wait" is without a successor. (The hyperedges at the top of the figure have predecessors in a task creating the process sys-

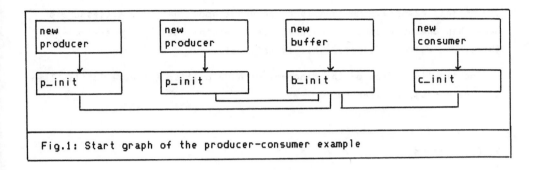

Fig.1: Start graph of the producer-consumer example

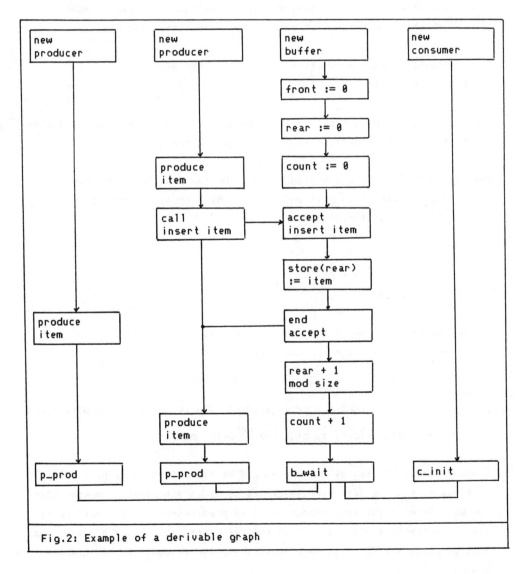

Fig.2: Example of a derivable graph

tem of this example.) An example of an event node that is a successor to more than one hyperedge is the predecessor of the last "produce item" in the second column. Therefore, it is split of.

For simplification of the diagram, we have omitted the node labels. They will be discussed later on.

2. Definitions

As we have already mentioned, we use the categorical graph-grammar approach, which is meaningful in each category that has pushouts.

Definition 1: A graph production is a pair of morphisms
$$p = ('B \longleftarrow K \longrightarrow B').$$

The graphs 'B and B' are called the left-hand side and the right-hand side of the production respectively. K is the gluing graph.

Definition 2: Given a graph production p. We say that G' is derivable from 'G with respect to p if and only if there exists a graph D such that both parts of the diagram

$$
\begin{array}{ccccc}
'B & \longleftarrow & K & \longrightarrow & B' \\
\downarrow & & \downarrow & & \downarrow \\
'G & \longleftarrow & D & \longrightarrow & G'
\end{array}
$$

are pushouts.

Now, we have to define a suitable category, i.e., objects describing distributed systems and morphisms reflecting the relationships between them.

Definition 3: A hypergraph H is a quintuple $H = (E, V, s, t, c)$ with E and V being sets and $s, t, c: E \longrightarrow V^*$ being mappings.

E and V denote the set of hyperedges and vertices (nodes) respectively. The mappings s, t and c determine sequences of nodes a hyperedge is connected to. We interpret the source nodes $s(e)$ as predecessors, the target nodes $t(e)$ as successors and the $c(e)$ as channel nodes. (Usual definitions of a hypergraph do not distinguish between these three sequences and concatenate them.)

Definition 4: A hypergraph morphism

$$f: (E_G, V_G, s_G, t_G, c_G) \longrightarrow (E_H, V_H, s_H, t_H, c_H)$$

is a pair $f = (f_E: E_G \longrightarrow E_H, f_V: V_G \longrightarrow V_H)$ of set morphisms such that

$$f_V^* s_G = s_H f_E \quad \& \quad f_V^* t_G = t_H f_E \quad \& \quad f_V^* c_G = c_H f_E$$

where $f_V^*: V_G^* \longrightarrow V_H^*$ is the extension of f_V to sequences of nodes.

Considering s, t and c component by component, we can use all the well-known constructions of the category GRAPH in the category HYPERGRAPH, too.

Definition 5: Let L_E and L_V be two sets of labels. An (L_E, L_V)-labeled

hypergraph is a septuple

$$H = (E, V, s, t, c, l_E, l_V)$$

where

(a) (E, V, s, t, c) is a hypergraph,

(b) $l_E: E \longrightarrow L_E$, $l_V: V \longrightarrow L_V$ are mappings.

The elements of L_V and L_E are called node labels and edge labels respect-ively. At the first glance, the type

$$(|s(e)|, |t(e)|, |c(e)|)$$

of an edge e would be expected to be determined by its label (cf. Degano/ Montanari [1]). In the next section, however, there is an interesting counterexample: In Fig. 6b, the edge e labeled by the call statement has the type

$$|s(e)| = 1, \quad |t(e)| = 1, \quad |c(e)| = 1$$

on the left-hand side of the production and

$$|s(e)| = 1, \quad |t(e)| = 2, \quad |c(e)| = 0$$

on the right-hand side. Thus, we have decided not to bind the type of an edge to its label. (Please note that this terminal edge is not replaced by the production; only its embedding into the context is changed.)

Definition 6: The state of a process system is an (L_E, L_V)-labeled hyper-

graph where the elements of L_E are statements and those of L_V are as-

sertions.

Strictly speaking, each hyperedge is labeled by a statement scheme. Substituting arguments for the variables allows us to construct a family of states with the same structure.

Parisi-Presicce, Ehrig and Montanari have extended the algebraic approach to graph grammars to deal with variables by imposing a simple structure on the set of labels [8]. They use a reflexive and transitive binary relation on the set of labels. Graph morphisms must preserve this structure. The following definition is a special case of this approach:

Definition 7: A state morphism is a hypergraph morphism

$$h: \quad B \longrightarrow G$$

together with a substitution σ satisfying:

(a) $(\forall v \epsilon V_B)(1_{VG}(h_V(v))$ implies $\sigma(1_{VB}(v)))$

(b) $(\forall e \epsilon E_B)(1_{EG}(h_E(e)) = \sigma(1_{EB}(e)))$

The edges are labeled by statements, which may include variables. Condition (b) means that the subgraph of G that is to be replaced by a production need not be equal to the left-hand side B, but it may also be obtained by substituting other terms or objects for these variables.

Similarly, condition (a) requires that each assertion occurring at a node on the left-hand side of the production must be a consequence of the corresponding assertion in the host graph. Thus, the node labeling in the productions can be restricted to the relevant conditions. In Fig.5, we shall give a production replacing a hyperedge labeled by "b_wait". It may only be applied if the predecessor node in the host graph G is labeled by an assertion that implies that the buffer is not full.

Corollary: The set of states together with the state morphisms is a category STATE which has all finite colimits.

The proof is straight-forward.

In this paper, the labeling of the channel nodes is simple: two processes are connected with one another or they are not. Therefore, all channel nodes are implicitly labeled by "true". This is no longer possible if we take into consideration special restrictions, e.g., access rights or physical characteristics of channels. Discussing these aspects would blow up this paper too much.

3. A grammar describing some aspects of the Ada rendez-vous

As we have already mentioned, we use some aspects of the Ada rendez-vous to illustrate our concept. First, we describe the buffer process separately. We use the well-known Ada version of the bounded buffer (Fig.3). The productions fall under two heads: local productions and synchronization productions. Fig.4 shows the hypergraph productions that do not require synchronization. We call them contextfree because the left-hand side of each production contains only one hyperedge. We have depicted only the left-hand sides and the right-hand sides of these productions; we have omitted the gluing graphs. It is well-known that the gluing graph can be restricted to some fringe nodes [3]; furthermore, the mappings from the gluing graph into both sides are injections in our example. Therefore, it is sufficient to number the gluing nodes in such a way that the correspondence becomes obvious. In our figures, 'i and i' denote the nodes onto which gluing node i, which is not explicitly given, is mapped by the left-hand and the right-hand side morphism, respectively.

Fig.5 describes the part of the buffer process which realizes synchronization. Application of such a production requires that a corresponding production is used in the buddy process. Furthermore, new event nodes are generated; they are used to establish the ordering of actions that are part of different processes. In our figure, these event nodes are denoted by the additional arcs.

Remark: At this point, we ought to discuss generating post-conditions, but we leave this to a more detailed technical report. Here, we only refer to the categorical treatment of assertions that was given by Wagner [9].

Discussing synchronizing operations with respect to only one process is not the plain truth. Such a derivation step depends on what is happening in the other process or what has happened just before. This means that the left-hand side of a synchronizing production contains more than one hyperedge. There are two possibilities:

(a) We consider such a production being composed of two contextfree productions that correspond to one another; the way they are fitted together is determined by the semantics of the programming language. Applying the production changes the state of both processes. This solution is a synchronous interpretation (cf. Degano/Montanari [1]).

```
task buffer is
     entry insert(item: in element);
     entry remove(item: out element);
end;

task body buffer is
     size:      constant integer := 100;
     store:     array (1..size) of element;
     rear,
     front:     integer range 1 .. size := 1;
     count:     integer range 0 .. size := 0;
begin
     loop
         select
             when count < size =>
                 accept insert(item: in element) do
                         store(rear) := item;
                 end;
                 rear := (rear mod size) + 1;
                 count := count + 1;
         or
             when count > 0 =>
                 accept remove(item: out element) do
                         item := store(front);
                 end;
                 front := (front mod size) + 1;
                 count := count - 1;
         end select;
     end loop;
end buffer;
```

Fig.3: Ada-version of the bounded buffer

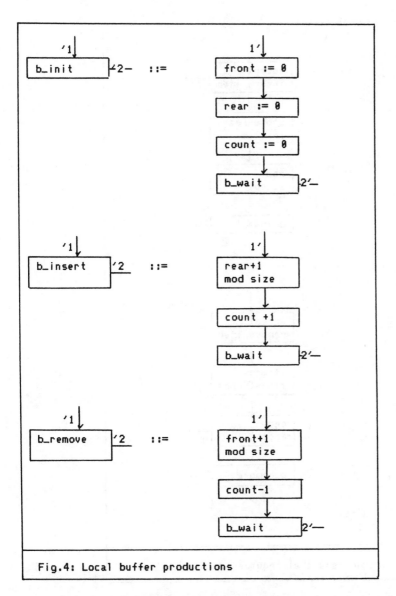

Fig.4: Local buffer productions

(b) The second possibility is a context-sensitive production in the narrow
    sense, i.e., only one of the hyperedges is replaced when applying the
    production; but this is only possible if the other hyperedge exists.
    Each derivation step, therefore, changes the state of only one process.
    This solution is also suited for describing asynchronous and quasi-
    synchronous communication.

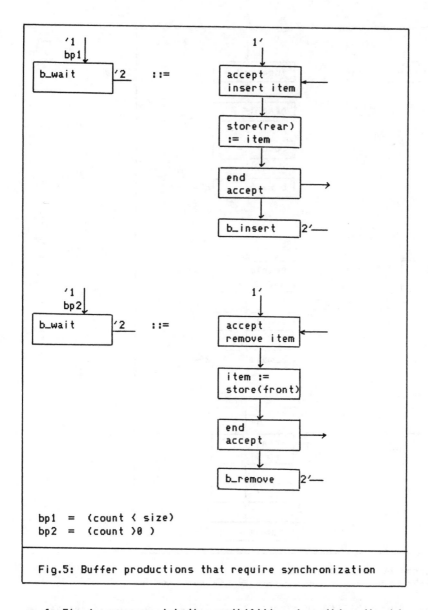

bp1 = (count < size)
bp2 = (count >0)

Fig.5: Buffer productions that require synchronization

In Fig.6, we present both possibilities describing the Ada rendez-vous.
In order to keep our example simple, we have reduced the producer and the
consumer process to small infinite loops. The producer alternates between
producing an item and calling the insert entry of the buffer. The consumer
process is organized analogously. Solution (a) has two major disadvantages:

- The resulting production depends on the actual processing state of the
  producer process, i.e., it contains hyperedges labeled by local computation states (p_init, p_prod). Therefore, we cannot describe synchronization without knowing the details of the client process.

- Although we connect the call edge to the accept edge by an event node, this solution suggests that the call statement is delayed until the corresponding accept statement can be performed: Both edges are generated in the same derivation step. Ada semantics, however, specifies that a calling task can issue an entry call statement before a corresponding accept statement is reached [10, no. 9.5(12)].

Solution (b) avoids this and assumes the hyperedge labeled by the call statement to be already generated in a previous derivation step. In accordance with the definition of Ada, a derivable hypergraph may then contain more than one unmatched call statement. Two graph-theoretic consequences arise from solution (b):

- In the producer process, we have to connect not only the nonterminal hyperedge, but also the hyperedge labeled by the call statement to the channel node (Fig.7). This ensures that the synchronizing production can later draw an arc from the call to the accept statement.

- After having applied the production of Fig.7, we may continue to derive the producer process without waiting for the rendez-vous. Thus, hypergraphs may be derived that do not correspond to real situations. If we insist on such a correspondence, we have to add the property of being blocked or not blocked to the assertions that we use to label event nodes. (All predecessor nodes on the left-hand sides of productions are assumed to be labeled by "not blocked" unless the contrary is stated explicitly.)

Fig.8 presents three hypergraph retracts illustrating the second solution. Our figures show in accordance with the Ada semantics that the entry call statement is always issued before the accept statement can be executed. But this does not exclude that the rendez-vous point may be first reached by the called task because reaching the rendez-vous point corresponds to generating a hyperedge labeled by "b_wait".

4. Related work

In 1986, Jackel presented a graph-grammar based approach to describe Ada concurrency [5]. The idea of this approach is representing the abstract syntax of the program by a program graph and emulating the operations by applying graph-productions to a runtime graph. The program graph depicts

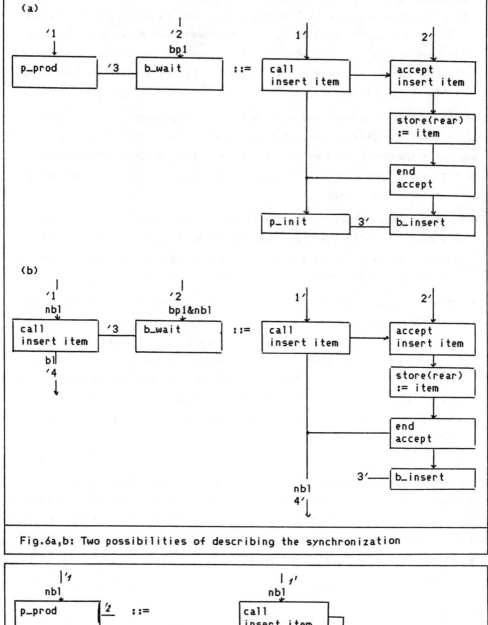

Fig.6a,b: Two possibilities of describing the synchronization

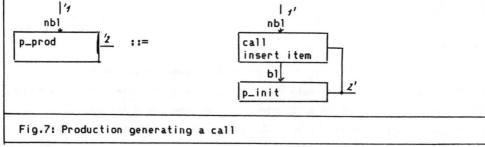

Fig.7: Production generating a call

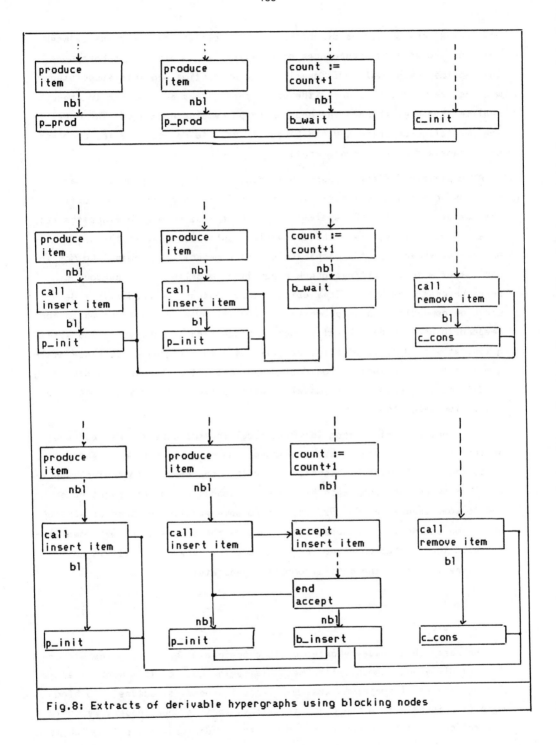

Fig.8: Extracts of derivable hypergraphs using blocking nodes

the contextfree structure of the program as well as the relations between defining and applied occurrences of entities. It is a proper subgraph of the runtime graph that additionally contains runtime segments associated with active units. Of course, the graph-productions are applied to these runtime segments, whereas the program graph remains unchanged. The model is implementation-oriented and can be easily used to demonstrate the problems of storage and process management.

A completely different approach to describing nets of processes was given by Degano and Montanari [1]. Whereas Jackel's graph grammar defines the set of all states the system can reach, a Pisa grammar defines the set of all computations, i.e., the derivable graphs also contain the history. Another difference is the mechanism of graph generation. Degano and Montanari start with contextfree graph productions modeling the behaviour of every process separately. They obtain context-sensitive productions describing communication of processes by combining the productions which correspond to one another. Finally, such a production is applied to the system graph; at each derivation step, exactly one event is generated. Since amalgamation of productions is controlled by lumping together actions occurring in different processes, this model is well-suited to study synchronous communication mechanisms.

Our approach differs from the Pisa model in some aspects. We represent actions and process states by hyperedges, too. But we consider events to be nodes that are labeled by assertions and that can be used to control applicability of productions. Each hyperedge is connected to some event nodes and to some channel nodes. By considering some event nodes predecessors of an action and some other nodes successors of the same action, we immediately embed the partial ordering in the grammatical frame-work. It is not necessary to define the partial ordering separately.

5. Conclusion

The research we have reported on is concerned with studying the structure of different communication mechanisms. The goal of this work is analyzing classes of morphisms that map logical process structure into hardware structures with given characteristics. In this short paper, we could present only the basic concepts and how the rendez-vous concept, which is typical of the remote procedure call, can be described. Another interesting, but not surprising result is that the monitor concept leads to more complicated grammatical structures.

Finally, I wish to think B. Hindel, C. Schiedermeier, P. Wilke and the referees for their helpful comments.

References:

[1] P. Degano/U. Montanari: "A model for distributed systems based on graph rewriting", J. Assoc. Comput. Mach. 34, 2 (1987), p.411-449

[2] H. Ehrig: "Introduction to the algebraic theory of graph grammars (a survey)", Lect. Notes in Computer Science 73 (1979), p.1-69

[3] H. Ehrig: "Tutorial introduction to the algebraic approach of graph-grammars", Lect. Notes Computer Science 291 (1987), p.3-14

[4] H. Ehrig et al.: "Distributed parallelism of graph transformations", Lect. Notes Computer Science 314 (1988), p.1-19

[5] M. Jackel: "ADA-concurrency specified by graph grammars", Lect. Notes Computer Science 291 (1987), p.262-279

[6] H.J. Kreowski: "Is parallelism already concurrency? Part 1: Derivations in graph grammars", Lect. Notes in Computer Science 291 (1987), p.343-360

[7] H.J. Kreowski/A. Wilharm: "Is parallelism already concurrency? Part 2: Non sequential processes in graph grammars", Lect. Notes in Computer Science 291 (1987), p.361-377

[8] F. Parisi-Presicce et al.: "Graph rewriting with unification and composition", Lect. Notes in Computer Science 291 (1987), p.496-514

[9] E.G. Wagner: "A categorical treatment of pre- and post-conditions", Theoretical Computer Science 53 (1987), p.3-24

[10] "The programming language Ada", Lect. Notes in Computer Science 155 (1983)

Author's address:

Prof. Dr. H.J. Schneider
Computer Science Department
University of Erlangen-Nuernberg
Martensstr.3, D-8520 Erlangen

E-mail: schneider@informatik.uni-erlangen.de

A parser for context free plex grammars

H. Bunke and B. Haller

Institut fuer Informatik und Angewandte Mathematik

Laenggassstr. 51, CH - 3012 Bern, Switzerland

Abstract

Plex grammars according to [13], generating two-dimensional plex structures, are a generalization of string grammars. In this paper we describe a parser for context free plex grammars. The parser is an extension of Earley's algorithm, which was originally developed for context free string grammars. Our parser is able to recognize not only complete structures generated by a plex grammar but also partial ones. The algorithm has been implemented and tested on a number of examples. The time complexity of the parser is exponential in general, but there exist subclasses of plex languages for which the parser has a polynomial time complexity.

Key words: formal languages, parsing, Earley's algorithm, two-dimensional structures, plex grammars, plex grammar parser.

1   Introduction

Formal grammars operating on two-dimensional structures rather than strings of symbols have been a subject of research from both a theoretical and practical point of view for many years. Although quite a number of papers on various types of graph grammars and graph rewriting systems have been published [1,2,3], only little attention has been paid to the parsing problem. Franck was the first who had developed a parser for graph languages [4]. For reasons of efficiency, he restricted his considerations to a special subclass of graph languages which are parsable in linear time. Later on, the work of Franck was extended by Kaul [5,6]. Another early paper on parsing of graph languages is [7]. In the beginning of the 1980's Fu and his collaborators studied special subclasses of graph languages and developed corresponding parsers [8,9]. Similar classes of graphs have been studied by

Brandenburg [10]. Recently, a parser for a subclass of node label controlled (NLC) graph grammars has been proposed [11]. The relationships between hyperedge replacement systems and NLC graph grammars have been studied in another recent paper [12] which also introduces a polynomial time parsing algorithm for a special subclass of graphs generated by hyperedge replacement systems.

In this paper we introduce a parser for context free plex grammars according to [13]. Although plex grammars are known for quite a long time, only one parser for plex grammars has been reported in the literature to our knowledge [14]. This parser works top-down with backtracking and is restricted to context free plex grammars without left-recursions. In this paper we describe another parsing algorithm for context free plex grammars without any restriction. In contrast with the parser by Lin and Fu [14], our approach is based on a generalization of Earley's algorithm for context free string grammars [15]. It is able to recognize not only complete structures generated by a plex grammar but also partial ones. This ability makes our algorithm interesting for applications in pattern recognition and scene analysis, where, due to occlusions and other types of distortions, parts of an input structure may be missing. Other potential applications of our parser include syntactic analysis of two-dimensional, i.e. graphical or visual, programming languages [16]. An application of plex grammars to the analysis of three–dimensional scenes is described in [17].

Recently, so–called hyperedge replacement systems have been introduced and studied by a number of researchers [18,19]. As we will explain in the next section, they are identical to plex grammars according to [13]. So all results obtained for hyperedge replacement systems can be applied to plex grammars. Vice versa, the parser described in this paper is applicable to hyperedge replacement systems as well.

2 Preliminaries

The basic constituents of plex grammars and languages are napes (n-attaching point entities). As a generalization of a symbol occurring in a string, a nape may have any number of points where it can be connected to one or more other napes. Thus, particular two-dimensional structures called plex structures are obtained. In this paper, we will restrict our considerations to context free plex grammars, i.e. plex grammars with only one nape in the left-hand side of any production. The following basic definitions are included in this paper for the purpose of self-containedness. For more details the reader is referred to [13].

Definition 1 *A* **plex structure** *is a triple* $\Pi = (\mathcal{X}, \Gamma, \Delta)$ *where*

- $\mathcal{X} = (\chi_1, \ldots, \chi_{n_\mathcal{X}})$ *is a list of napes; each nape* χ_i *has* $n_{\chi_i} \geq 0$ *connection points*

- $\Gamma = (\gamma_1, \ldots, \gamma_{n_\Gamma})$ *is a list of internal connections*

- $\Delta = (\delta_1, \ldots, \delta_{n_\Delta})$ *is a list of external connections*

such that each connection point of each nape belongs to exactly one internal or external connection.

We write $a(\chi_i, j) = \gamma_k$ iff the connection point j of the nape χ_i belongs to the connection γ_k.

Definition 2 *A context free plex grammar is a 4-tuple* $G = (V_T, V_N, P, S)$ *where*

- V_T *is a finite, nonempty set of napes, called the set of terminal napes; each terminal nape has a nonempty set of connection points*

- V_N *is a finite, nonempty set of napes, called the set of nonterminal napes where* $V_T \cap V_N = \emptyset; V = V_N \cup V_T$

- P *is a finite set of context free productions*

- $S \in V_N$ *is the starting nape; this nape is the only one with an empty set of connection points.*

Definition 3 *A context free production is an entity* $A\Delta \rightarrow X\Gamma\Delta$ *where*

- (X, Γ, Δ) *is a plex structure according to Definition 1*

- A *is a single nonterminal nape and* Δ *is the list of its connections*

The components $A\Delta$ and $X\Gamma\Delta$ are called the left-hand and right-hand side of the production, respectively. Notice that the list Δ of external connections in the right-hand side is identical to the list of (external) connections of the single nonterminal nape A in the left-hand side. The left-hand side can be interpreted as a plex structure consisting of only one (nonterminal) nape with an empty list of internal connections.

The derivation of a context free plex grammar proceeds by replacing the nonterminal nape in the left-hand side of a production by the plex structure in the right-hand side. Such a replacement usually occurs in a larger context. The list Δ plays the role of the embedding transformation [1,2,3]. That is, the role of the connections of the nonterminal in the left-hand side is taken over by the (identical) external connections of the right-hand side, which are both given by Δ. The language L(G) generated by a plex grammar G consists of the set of all terminal plex structures which can be derived from the starting nape by successive application of productions.

Comparing plex grammars with hyperedge replacement systems [18] we notice that napes are identical to hyperedges, and plex structures are identical to multi–pointed hypergraphs. Furthermore, there is a one–to–one correspondence between context free productions of a plex grammar and productions of a hyperedge replacement system. Finally, any plex grammar can be considered as a hyperedge replacement system and vice versa. As a consequence, the class of languages generated by context free plex grammars is identical to the class of languages generated by hyperedge replacement systems. In this paper, we always use the terminology developed for plex grammars.

As an example, consider the two nonterminal napes PROG and P and the four terminal napes START, HALT, FUNCT, and PRED in Fig. 1. For instance, PRED has three connection points 1, 2, and 3, while P has only two connection points 1 and 2. In this paper we represent nonterminals by double-bordered boxes while terminals are represented by graphical symbols with a single border line.

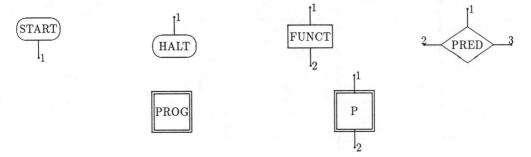

Figure 1: Napes of a sample plex grammar, terminals above and nonterminals below

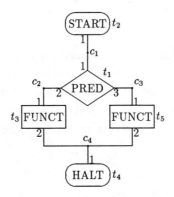

Figure 2: A sample plex structure

PROG

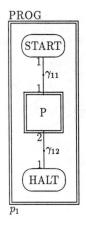

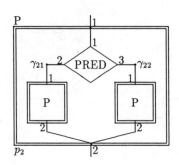

p_1 p_2 p_3

Figure 3: Productions of the sample plex grammar

A graphical representation of a plex structure is given in Fig. 2. Its description according to Def. 1 is $\Pi = (\mathcal{X}, \Delta, \Gamma)$ with

$\mathcal{X} = (t_1 = \mathrm{PRED}, t_2 = \mathrm{START}, t_3 = \mathrm{FUNCT}, t_4 = \mathrm{HALT}, t_5 = \mathrm{FUNCT})$

$\Gamma = (c_1, \ldots, c_4)$

$\Delta = ()$

The connections are:

$a(t_1, 1) = c_1,$ $\qquad a(t_1, 2) = c_2,$ $\qquad a(t_1, 3) = c_3,$

$a(t_2, 1) = c_1,$ $\qquad a(t_3, 1) = c_2,$ $\qquad a(t_3, 2) = c_4,$

$a(t_4, 1) = c_4,$ $\qquad a(t_5, 1) = c_3,$ $\qquad a(t_5, 2) = c_4.$

Three productions of a context free plex grammar are shown in Fig. 3. The graphical representation for each production is such that the outermost box corresponds to the nonterminal in the left-hand side while the right-hand side is drawn inside this box. For example, production p_2 replaces the nonterminal nape P by a plex structure consisting of one terminal and two nonterminal napes, two internal connections, and two external connections. According to Def. 3, the productions p_1, p_2 and p_3 can be written as

- p_1: $A_1() \rightarrow (\chi_{11}, \chi_{12}, \chi_{13})(\gamma_{11}, \gamma_{12})()$
    $= \mathrm{PROG}() \rightarrow (\mathrm{START}, \mathrm{P}, \mathrm{HALT})(\gamma_{11}, \gamma_{12})()$

    connection points:

    $a(\chi_{11}, 1) = \gamma_{11},$ $\qquad a(\chi_{12}, 1) = \gamma_{11},$ $\qquad a(\chi_{12}, 2) = \gamma_{12},$ $\qquad a(\chi_{13}, 1) = \gamma_{12}$

- p_2: $A_2(\delta_{21}, \delta_{22}) \rightarrow (\chi_{21}, \chi_{22}, \chi_{23})(\gamma_{21}, \gamma_{22})(\delta_{21}, \delta_{22})$
    $= \mathrm{P}(\delta_{21}, \delta_{22}) \rightarrow (\mathrm{PRED}, \mathrm{P}, \mathrm{P})(\gamma_{21}, \gamma_{22})(\delta_{21}, \delta_{22})$

$$a(\chi_{21},1) = \delta_{21}, \qquad a(\chi_{21},2) = \gamma_{21}, \qquad a(\chi_{21},3) = \gamma_{22}, \qquad a(\chi_{22},1) = \gamma_{21},$$
$$a(\chi_{22},2) = \delta_{22}, \qquad a(\chi_{23},1) = \gamma_{22}, \qquad a(\chi_{23},2) = \delta_{22}$$

- $p_3 : A_3(\delta_{31},\delta_{32}) \rightarrow (\chi_{31})()(\delta_{31},\delta_{32}) = P(\delta_{31},\delta_{32}) \rightarrow (\text{FUNCT})()(\delta_{31},\delta_{32})$

$$a(\chi_{31},1) = \delta_{31}, \qquad a(\chi_{31},2) = \delta_{32}$$

If we use the nonterminal PROG as starting symbol, then the productions shown in Fig. 3 generate a language which represents a special subclass of flowchart diagrams. One element of this language is given in Fig. 2. (A more complex plex grammar generating the class of all well structured flowchart diagrams is described in [20].)

There are some slight changes in the definitions of this section as compared to [13]. These changes have been made for the purpose of notational convenience only. From a theoretical point of view, the above definitions are equivalent to the original ones. For the parsing algorithm described in the following sections of this paper we assume that any plex structure to be analysed is connected and that the starting symbol doesn't have external connections. It is furthermore supposed that any internal connection consists of at least two connection points and that not more than one connection point of the same nape participates in any internal or external connection. Notice, however, that these restrictions aren't essential and could be overcome by slight modifications to the parsing algorithm described in the following section.

3 The Parsing Algorithm

Our algorithm is based on the same idea as Earley's parser [15]. I.e., we sequentially read one terminal nape after the other from the plex structure to be parsed and construct a list for each terminal. In the list belonging to the terminal read in the j-th step we keep entries describing all possibilities to derive a nonterminal into a plex structure which is compatible with all the terminals read up to step j. The construction of the lists is done by three different subroutines called scanner, completer, and predictor.

An entry in a list consists of a production and additional information concerning the degree to which this production has been recognized. Consider the list I_j and the production $p : A\Delta \rightarrow X\Gamma\Delta$. If p is contained in I_j then the additional information consists of

- a mapping between the terminals in the right-hand side X and the terminals $t_1, \ldots, t_j$ read by the parser so far. This mapping contains the information which of the terminals $t_1, \ldots, t_j$ in the input plex structure can be recognized as instances of the terminals in X. It also indicates whether or not the right-hand side of p has been completely recognized.

- a mapping similar to the above one for external and internal connections.

- all productions which contain, in their right-hand side, the nonterminal A in a way consistent with the terminals $t_1, \ldots, t_j$ read as far.

- for each nonterminal nape A' in the right-hand side $\mathcal{X}$ all productions $p' : A'\Delta' \rightarrow \mathcal{X}'\Gamma'\Delta'$ which derive A' into a plex structure $\mathcal{X}'$ compatible with $t_1, \ldots, t_j$.

- a pointer to production p'' in list I_{j-1} if the entry concerning the production p in I_j has been generated from p'' in I_{j-1} (see also steps 2c and 2d of the parsing algorithm).

In the rest of this paper we will use only the word "production" in order to refer to an entry in any of the lists $I_j; j = 1, \ldots, n$. However, the reader should keep in mind that for each production, i.e. each entry, the additional information described above is always present in I_j, too. Two productions in a list are identical if and only if they are identical with respect to their left-hand and right-hand side, and their additional information.

Algorithm: parser for context free plex grammars

Input: a context free plex grammar and a terminal plex structure with terminal napes $t_1, t_2, \ldots, t_n$ and internal connections

Output: a list I_j for each terminal nape $t_j; j = 1, \ldots, n$

1. **Construction of the list I_1:**

   (a) Select a terminal nape t_1.

   (b) *Scanner:* Add each production which contains t_1 in its right-hand side to I_1.

   /* After completion of the scanner, I_1 shows all possibilities to derive t_1, and perhaps additional terminal napes, in one derivation step */

   (c) *Completer:* If there is a production p_1 contained in I_1 which has the nonterminal N in its left-hand side, and if there exists another production p_2 with the same nonterminal N in its right-hand side, then add p_2 to I_1. Repeat this step as long as new entries are generated.

   /* After completion of the completer, I_1 shows all possibilities to derive t_1, and perhaps additional terminal napes, in $n \geq 1$ derivation steps from a nonterminal. */

   (d) *Predictor:* If there is a production p in I_1 and if the right-hand side of p contains a nonterminal N which shares a connection with t_1, then add any production to I_1 which replaces the nonterminal N. Repeat this step as long as new entries are generated.

   /* The predictor prepares the processing of the next terminal t_2 by the scanner. It enters all those productions to I_1 which derive, in $n \geq 1$ steps, a terminal in the immediate neighborhood of t_1. I.e. the predictor predicts all terminals which can share a common connection with t_1. */

2. **Construction of the list I_j based on I_{j-1} for $j = 2, \ldots, n$:**

(a) Select a new terminal nape t_j which has at least one connection in common with the connection points of the terminal napes $t_1, \ldots, t_{j-1}$.

(b) *Scanner:* Select a production p from I_{j-1} which contains t_j in its right-hand side in such a way that it is compatible with $t_1, \ldots, t_{j-1}$. Modify it (i.e., update its additional information) in order to take into account that t_j has been read, and add it to I_j. Repeat these operations as long as new entries can be added to I_j.

/* The productions in I_{j-1} which possibly fulfill this condition have been added to I_{j-1} by the predictor. After completion of the scanner, I_j shows all possibilities to derive t_j, and perhaps additional terminal napes, in one derivation step */

(c) *Completer:* If there is a production p in I_j with a nonterminal N in its left-hand side, select the corresponding production p' in I_{j-1} from which p in I_j has been generated. (This production p' also has the nonterminal N in its left-hand side.) Take each production p'' in I_{j-1} which contains this nonterminal in its right-hand side, update it by the information about t_j, and add it to I_j. Repeat this procedure for all productions which are originally in I_j, and for all productions which are added to I_j by the completer.

/* After completion of the completer, I_j shows all possibilities to derive t_j, and perhaps additional terminal napes, in $n \geq 1$ derivation steps form a nonterminal. */

(d) *Predictor:* If there is a production p_1 in I_j which contains a nonterminal nape N in its right-hand side, consider all productions p_2 which contain N in their left-hand side, and distinguish between the following cases:

   i. p_2 is contained in I_j: do nothing;

   ii. p_2 isn't contained in I_j, but in I_{j-1}: in this case update p_2 by information about t_j and add it to I_j;

   iii. p_2 is contained neither in I_{j-1}, nor in I_j: if the nonterminal N shares a common connection point with t_j then add p_2 to I_j, else do nothing.

Repeat this procedure for all productions which are originally in I_j, and for all productions which are added to I_j by the predictor.

/* The predictor updates all information which is relevant with respect to t_j and transfers it from I_{j-1} to I_j (case ii). It also prepares the processing of the next terminal t_{j+1} by the scanner (case iii). After completion of the predictor, I_j shows all possibilities to derive one or more of the terminals $t_1, \ldots, t_j$ and / or one or more of the terminals in the immediate neighborhood of $t_1, \ldots, t_j$, and perhaps additional terminals. Thus, it predicts the neighborhood of $t_1, \ldots, t_j$, i.e. all terminals which share a common connection with at least one of the terminals $t_1, \ldots, t_j$. */

End of Algorithm

The behaviour of the above parsing algorithm can be characterized by the following theorem.

Theorem: Let G be a plex grammar and Π a terminal plex structure consisting of napes $t_1, t_2, \ldots, t_n$ and internal connections $c_1, c_2, \ldots, c_m$.

1. The plex structure Π is generated by G, i.e. $\Pi \in L(G)$, if and only if the list I_n constructed by the parsing algorithm contains an instance of a production p such that

   (a) the left–hand side of p is the starting nape

   (b) all terminals and all connections of Π either directly occur in the right–hand side of p, or can be derived by means of productions from nonterminals occurring in the right–hand side of p

   (c) the right–hand side of p has been completely recognized

2. The plex structure Π is a substructure of another plex structure $\Pi' \in L(G)$, i.e. $\Pi \subseteq \Pi'$, if the conditions (a) and (b) above are fulfilled. □

The proof of this theorem can be done by induction on the number of terminal napes in Π. All details are given in [20]. Each of the conditions (a) – (c) in the theorem can be easily checked using the additional information which is associated with the productions in the list I_n. Similar to Earley's parser for string grammars it is possible to reconstruct, from the entries in the list I_n, the sequence of derivation steps leading from starting nape to the plex structure to be parsed. This sequence is equivalent to the derivation tree. A detailled description of the recovery of the derivation tree is given in [20]. The second part of the theorem says that our parser is able to recognize not only complete structures but also partial ones.

4 An Example

Consider the productions in Fig. 3 and the terminal plex structure in Fig. 2. The construction of the list I_1 proceeds as follows:

1. The terminal t_1 is read. (The order of the terminals in the example has been arbitrarily assumed. The parser can process any plex structure independent of the order of the terminals.)

2. The scanner adds p_2 to the (empty) list I_1 and adds the following information:

   - t_1 is recognized as terminal PRED in the right-hand side of p_2

   - c_1 is recognized as external connection 1 in Δ; external connection 2 in Δ is not recognized; c_2 and c_3 are recognized as internal connections γ_{21} and γ_{22}, respectively.

   (This instance of p_2 in I_1 will be referred to as $p_{2,1}$ in the following.)

3. The completer adds one instance of p_1 and two instances of p_2 to I_1. (These instances will be referred to as $p_{1,1}, p_{2,2}$, and $p_{2,3}$ in the following.) They describe how the nonterminal P in the left-hand side of $p_{2,1}$ can be derived by means of other productions. Notice that the completer generates two instances of p_2 since the nonterminal P in the left-hand side of $p_{2,1}$ may correspond to either the nonterminal in the left or in the right branch of the right-hand side of p_2.

The completer adds one more instance of p_1 and two more instances of p_2 (being referred to as $p_{1,2}, p_{2,4}$, and $p_{2,5}$ in the following), which describe how the nonterminal P in the left-hand sides of $p_{2,2}$ and $p_{2,3}$ can be derived by means of other productions. (Notice that $p_{2,4}$ and $p_{2,5}$ are different from $p_{2,2}$ and $p_{2,3}$ because the connection point 1 of the nonterminal P is unknown.) By recursion, the productions $p_{1,2}, p_{2,4}$, and $p_{2,5}$ describe how the nonterminal P in the left-hand sides of $p_{2,4}$ and $p_{2,5}$ can be derived.

4. The predictor adds two instances for each of the productions p_2 and p_3. (They will be referred to as $p_{2,6}, p_{2,7}, p_{3,1}$, and $p_{3,2}$ in the following.) They are for predicting the neighborhood shared by the terminal t_1 at c_2 and c_3, respectively.

Now the list I_1 is complete and we proceed by constructing I_2.

1. The terminal t_2 is read.

2. The scanner selects $p_{1,1}$ from I_1 and adds to it the information that the terminal START in the right-hand side represents the terminal t_2 in the structure. This updated instance of p_1 is added to the (empty) list I_2 and will be referred to as $p_{1,3}$ in the following.

3. The completer selects $p_{1,3}$ from I_2 and gets to $p_{1,1}$ from which $p_{1,3}$ has been generated by the scanner. Because the nonterminal PROG doesn't occur on the right-hand side of any production, the list of such productions in $p_{1,1}$ is empty. So the completer adds no productions to I_2.

4. The predictor selects $p_{1,3}$ from I_2 and finds that the nonterminal P in the right-hand side is described by the production p_2, i.e. by its instance $p_{2,1}$, which is contained in list I_1 (case ii). So the predictor updates $p_{2,1}$ by the information that the nonterminal P in its left-hand side is described by $p_{1,3}$, and adds it to I_2 as instance $p_{2,8}$ of p_2. Furthermore, the predictor selects $p_{2,6}, p_{2,7}, p_{3,1}$ and $p_{3,2}$, updates them, and adds them to I_2 for predicting the neighborhood of t_1.

Now the list I_2 is complete and the parser continues by constructing I_3. The detailled steps which follow are omitted. After the last terminal t_5 has been read and the list I_5 has been completely constructed, I_5 contains an instance of p_1, an instance of p_2 for replacing the nonterminal P in the right-hand side of p_1, and two instances of p_3 for replacing the two nonterminals in the right-hand side of p_2. There are pointers between these productions such that the derivation of the plex structure

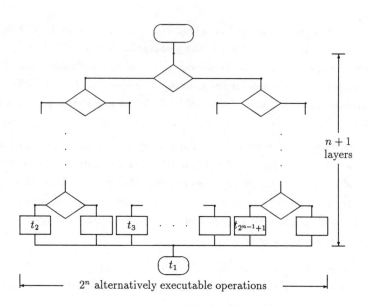

Figure 4: A plex structure with exponential parsing time and space complexity

given in Fig. 2 can be uniquely reconstructed. Each of the conditions (a) – (c) in the first part of the theorem of the last section is fulfilled. So we can conclude that the plex structure shown in Fig. 2 is generated by the grammar given in Fig. 3.

5   Computational Complexity and Experimental Results

In the worst case, the parser described in section 3 has an exponential time and space complexity. As an example, consider the flowchart shown in Fig. 4. This flowchart can be generated by the productions given in Fig. 3. There are n successive layers of conditions each having two exits in Fig. 4. In layer $n+1$ we observe 2^n terminals FUNCT. Let $m = 2^{n-1}+1$ and assume that the parser reads the terminals in order $t_1, t_2, \ldots, t_m$ (see Fig. 4). In I_1 the parser enters instances of p_1, p_2 and p_3. (Notice that the terminal PRED in the right-hand side of p_2 corresponds to the terminal in the topmost layer in Fig. 4.) After reading t_2, the parser can conclude that t_2 belongs to the nonterminal P in either the left or ther right branch of the right-hand side of p_2 in I_1. Since there isn't any further information which case is the correct one, both alternatives are entered in I_2. The same happens independently to all the terminals $t_3, t_4, \ldots, t_m$. For example, in list I_3 the parser has to take into account four possibilities: both t_1 and t_2 in the left branch; both t_1 and t_2 in the right branch; t_1 in the left and t_2 in the right branch; t_1 in the right and t_2 in the left branch. Therefore, the number of entries in the list I_j exponentially grows with $j = 1, \ldots, m$. So both the space and time complexity are exponential.

The exponential time complexity of our parser is not surprising because of the equivalence of plex grammars and hyperedge replacement systems, which are known to be able to generate graph languages with an NP–complete membership problem [12,21]. Notice, that there are subclasses of graph languages generated by hyperedge replacement systems which are parsable in polynomial time [12]. It is an open question at the moment if the parser described in this paper works in polynomial time on this subclass.

However, there is another subclass of plex languages where our parser as described in section 3 — without any modifications — works in polynomial time. It is the class of context free string languages which can be considered as a special subclass of plex structures if we interpret each symbol of an alphabet as a nape with two external connection points. Earley's parser for string grammars [15] has a time complexity of $O(n^2)$ if the underlying grammar is unambiguous and $O(n^3)$ otherwise, where n gives the length of an input string. A simple analysis shows that our parser, which is an extension of Earley's algorithm, has a polynomial time complexity of $O(n^5)$ for plex languages representing string languages. Let $I_{E,j}$ and $I_{P,j}$ denote the lists generated by Earley's parser and our parsing algorithm, respectively, when the j-th terminal is read. The entries in list $I_{P,j}$ correspond to a subset of all the entries which are generated in $I_{E,1}, \ldots, I_{E,j}$. The number of entries in each list $I_{E,i}$ grows linearly with $i = 1, \ldots, n$ (see [15]). So the number of entries in $I_{P,j}$ is of order $O(j^2)$. Notice that the length of each entry in $I_{P,j}$ is linearly dependant on j. Since our parser has to keep only the lists I_{j-1} and I_j in step j, the space complexity is $O(n^3)$. For each new entry in $I_{P,j}$ the parser has to check if an identical entry is already contained in $I_{P,j}$. Since this check can be done in constant time, we get a time complexity of $O(j^2) \cdot O(j^2) = O(j^4)$ for each list $I_{P,j}$. Summing over $j = 1, \ldots, n$ a time complexity of $O(n^5)$ results.

The parser described in section 3 has been fully implemented in C under UNIX on a Nixdorf Targon (a 4 MIPS machine, approximately). The length of the program code is about 6000 lines, including a number of input and output routines. As executable code, the program takes about 60 Kbyte. We did a number of experiments primarily concerned with the computation time needed by the parser. From those experiments we found out that the actual runtime of the program depends on a number of factors like the order in which the terminals are read, and particular characteristics of the underlying grammar.

An example is shown in Fig. 5. In this example, we used an extended version of the grammar shown in Fig. 3, generating the set of well structured flowchart diagrams [20]. As heuristics for determining the order in which the terminals are read we used two rules. First, we prefer that terminal which has the least number of "unknown" and the maximum number of "known" connection points. A connection point is called unknown if it is not connected to any connection point of a terminal already read, otherwise it is a known connection point. If there are ties, we use a second rule and prefer that terminal which has the minimum number of neighboring terminals not yet inspected and the maximum number of neighboring terminals already read. Informally speaking, this strategy keeps

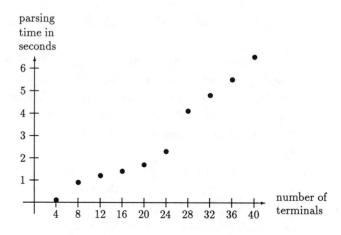

Figure 5: Parsing time for simple plex structures

the structure inspected during the course of the analysis in a compact form. The strategy has not been developed particularly for the flowchart grammar and has proven useful for other applications, too. The planarity of the graphs to be parsed doesn't have any direct influence on the parsing time in this example.

As one can conclude from Fig. 5, there is a subexponential growth of the computation time in the considered range. Generally, we strongly believe that there are many applications where it is possible to find suitable heuristics such that the computation time actually needed in the average case is better than exponential.

6 Conclusions

An algorithm for parsing plex structures generated by context free plex grammars is proposed in this paper. It follows the same strategy as Earley's parser does for context free string languages. For the string case, a number of different versions of Earley's parser have been developed like an error correcting parser [22] or parsers which can handle probabilistic information, either in the productions [23] or in the input string [24]. Such extensions potentially have a number of interesting applications, primarily in the domain of pattern recognition and image analysis. The clear logical structure of the plex grammar parser described in this paper and its similarities with Earley's algorithm seem to be a promising basis for similar extensions. They would certainly further enhance the applicability of our proposed parser to problems in pattern recognition and image analysis.

Perhaps the most critical problem with our parser is its high computational complexity. As mentioned in section 5, however, there seems to be a potential for incorporating heuristics for speeding up the parsing time actually needed. Such heuristics may be put into the grammar itself or into the order in which the terminals of the plex structures under consideration are read. Furthermore, as already discussed in section 5, there exist subclasses of plex structures with a lower time complexity. A detailed investigation of the properties of such subclasses will be an interesting topic for further research, from both a practical and theoretical point of view.

Acknowledgements

We would like to thank H.–J. Kreowski, A. Habel and W. Vogel for interesting discussions which led to an improvement of this paper. Thanks are due also to the anonymous referees for their valuable comments. A. Ueltschi provided help in typing the paper. The implementation of the parser was done on a Nixdorf Targon which was made available to us by Nixdorf AG, Switzerland. We gratefully acknowledge their generous support.

References

[1] Claus, V. / Ehrig, H. / Rozenberg, G. (eds.): *Graph-grammars and their application to computer science and biology*, Proc. 1st Int. Workshop, Lecture Notes in Comp. Sci. 73, Springer Verlag, 1979.

[2] Ehrig, H. / Nagl, M. / Rozenberg, G. (eds.): *Graph-grammars and their application to computer science*, Proc. 2nd Int. Workshop, Lecture Notes in Comp. Sci. 153, Springer Verlag, 1982.

[3] Ehrig, H. / Nagl, M. / Rozenberg, G. / Rosenfeld, A. (eds.): *Graph-grammars and their application to computer science*, Proc. 3rd Int. Workshop, Lecture Notes in Comp. Sci. 291, Springer Verlag, 1987.

[4] Franck, R.: *A class of linearly parsable graph grammars*, Acta Informatica 10, 1978, 175 – 201.

[5] Kaul, M.: *Syntaxanalyse von Graphen bei Präzedenz–Graph–Grammatiken*, Techn. Report MIP–8610, University of Passau, FRG, 1986.

[6] Kaul, M.: *Computing the minimum error distance of graphs in $O(n^3)$ time with precedence graph grammars*, in Ferrate, G. / Paulidis, T. / Sanfeliu, A. / Bunke, H.: *Syntactic and Structural Pattern Recognition*, Springer Verlag, NATO ASI Series, 1988, 69 – 83.

[7] Della Vigna, P. / Ghezzi, C.: *Context–free graph grammars*, Information and Control 37, 1978, 207 – 233.

[8] Sanfeliu, A. / Fu, K.S.: *Tree graph grammars for pattern recognition*, in [2], 349 – 368.

[9] Shi, Q.–Y. / Fu, K.S.: *Parsing and translation of (attributed) expansive graph languages for scene analysis*, IEEE Trans. PAMI-5, 1983, 472 – 485.

[10] Brandenburg, F.J.: *On partially ordered graph grammars*, in [3], 99 – 111.

[11] Flasinski, M.: *Parsing of edNLC–grammars for scene analysis*, Pattern Recognition 21, 1988, 623 – 629.

[12] Lautemann, C.: *Efficient algorithms on context–free graph languages*, in Lepistö, T. / Salomaa, A. (eds.): *Automata, Languages and Programming*, Proc. 15th Int. Coll., Lecture Notes in Comp. Sci. 317, Springer Verlag, 1988, 362 – 378.

[13] Feder, J.: *Plex languages*, Information Sciences 3, 225 – 241, 1971

[14] Lin, W.C. / Fu, K.S.: *A syntactic approach to 3D object representation and recognition*, TR – EE 84 – 16, Purdue University, West Lafayette, Indiana, June 1984.

[15] Earley, J.: *An efficient context-free parsing algorithm*, Communications of the ACM, Vol. 13, No. 2, 94 – 102, Feb. 1970.

[16] Chang, S.K. et al., *Visual programming*, Plenum, 1986.

[17] Lin, W.C. / Fu, K.S.: *A syntactic approach to 3-D object representation*, IEEE Transaction on Pattern Analysis and Machine Intelligence, Vol. PAMI-6, No. 3, 351 – 364, May 1984.

[18] Habel, A. / Kreowski, H.–J.: *May we introduce to you: hyperedge replacement*, in [3], 15 – 26.

[19] Courcelle, B.: *Some applications of logic of universal algebra and of category theory to the theory of graph transformations*, Bulletin of the EATCS 36, 1988, 161 – 213.

[20] Haller, B.: *A parser for context-free plex grammars*, Diploma Thesis, Institute of Informatics and Applied Mathematics, University of Bern, Switzerland, 1989 (in German).

[21] Leung, J.Y.–T. / Witthof, J. / Vornberger, O.: *On some variations of the bandwith minimization problem*, SIAM J. Comp. 13, 1984, 650 – 667.

[22] Aho, A.V. / Peterson, T.G.: *A minimum distance error–correcting parser for context–free languages*, SIAM J. Comput., Vol. 1, No. 4, 305 – 312, Dec. 1972.

[23] Lu, S.Y. / Fu, K.S.: *Stochastic error–correcting syntax analysis for recognition of noisy patterns*, IEEE Transactions on Systems, Men, and Cybernetics, Vol. SMC-8, 380 – 401, 1978.

[24] Bunke, H. / Pasche, D.: *Parsing multivalued strings and its application to image and waveform recognition*, in Mohr, R. (ed.): *Proceedings of the Workshop on Syntactical and Structural Pattern Recognition*, Pont-à-Mousson, September 1988, World Scientific Publ. Co., Singapore, in print.

Introduction to PROGRESS,

an Attribute Graph Grammar Based

Specification Language

Andy Schürr[‡]

Lehrstuhl für Informatik III
Aachen University of Technology
Ahornstraße 55, D–5100 Aachen,
West Germany

Abstract: The language **PROGRESS** presented within this paper is the first **strongly typed** language which is based on the concepts of **PRO**grammed Graph **RE**writing Sy**S**tems. This language supports a **data flow oriented** style of programming (by means of attribute equations), an **object oriented** style of programming (by supporting multiple inheritance and dynamic bind of attribute designators to their value defining equations), a **rule based** style of programming (by using graph rewrite rules), and an **imperative** style of programming (by composing single graph rewrite rules to complex transformation programs). Both the language and its underlying formalism are based on an experience of about seven years with a **model oriented** approach to the specification of document classes and document processing tools (of the Integrated Programming Support EN**viroment IPSEN**). This approach, called **graph grammar engineering**, is characterized by using **attributed graphs** to model object structures. Programmed graph rewriting systems are used to specify operations in terms of their effect on these graph models. This paper informally introduces PROGRESS' underlying graph grammar formalism and demonstrates its systematic use by specifying parts of a desk calculator's functional behaviour.

1. Introduction

Modern software systems for application areas like office automation, and software engineering are usually highly interactive and deal with complex structured objects. The systematic development of these systems requires precise and readable descriptions of their desired behaviour. Therefore, many specification languages and methods have been introduced to produce formal descriptions of various aspects of a software system, such as the design of object structures, the effect of operations on objects, or the synchronization of concurrently executed tasks. Many of these languages use **special classes of graphs** as their underlying data models. Conceptual graphs /So 84/, (semantic) data base models /Me 82/, petri nets /GJ 82/, or attributed trees /Re 84/ are well-known examples of this kind.

[‡] Supported by Stiftung Volkswagenwerk

Within the research project **IPSEN** (an acronym for Integrated/Incremental Programming Support ENvironment) a **graph grammar based specification method** has been used to model the internal structure of software documents and to produce executable specifications of corresponding document processing tools, as e.g. syntax–directed editors, static analyzers, or incremental compilers /ES 85/. The development of such a specification, which is termed **'programmed graph rewriting system'** consists of two closely related subtasks. The first one is to design a graph model for the corresponding complex object structure. The second one is to program object (graph) analyzing and modifying operations by composing sequences of subgraph tests and graph rewrite rules.

Based on experiences of about seven years with this IPSEN specific approach to the **formal specification and systematic development** of software, we were able to adapt the original formalism (introduced in /Na 79/) to the requirements of this application area (cf. /En 86, Le 88a/). Furthermore, a method, called **graph grammar engineering**, has been developed for the construction of large rewriting systems in a systematic engineering–like manner (cf. /ES 85, EL 87/). Parallel to the continuous evolution of the graph grammar formalism and the graph grammar engineering method, the design of a **graph grammar specification language** is in progress. A first version of this language, termed **PROGRESS** (for **PRO**grammed Graph **RE**writing System Specification), was fixed a few months ago, and a **prototype of a programming environment** for this version of the language is under development.

The purpose of this paper is to survey the language PROGRESS. It is addressed to those readers who are familiar with the formalism of attribute (tree) grammars and tree rewriting systems (cf. /Re 84, MW 84/) but not with the formalism of (attribute) graph grammars and graph rewriting systems. Therefore, the next section is dedicated to an informal introduction of our **graph grammar formalism**. This section also introduces the running example which is used throughout the whole paper for demonstration purposes. Section 3 presents a typical cut out of the features of the **language PROGRESS** and demonstrates the supported way of 'programming'. Section 4 contains some remarks about **related work** and the last section discusses **future development plans** for this graph grammar based specification language.

2. The Graph Grammar Formalism

Writing about PROGRESS one has to explain and discuss topics at least at two different levels. The first one is that of the language's underlying **graph grammar formalism** which builds a fundament for the semantic definition of the language. The second one is that of **language design issues** and comprises tasks like defining the abstract syntax, the name binding rules, and the concrete representations for the new language. To avoid a confusion of these two levels within the paper, I decided to dedicate this section completely to the informal introduction of our graph grammar formalism and to defer the presentation of the language itself to the following section.

The first subsection introduces the formalism's underlying data model – **attributed, node and edge labeled, directed graphs** – and demonstrates the mapping of complex object structures onto this kind of graphs. The second subsection sketches the specification of the functional behaviour of object (graph) processing tools by means of **graph rewrite rules**. Within this section and for the remainder of the paper I use a subset of the well-known applicative programming language 'Exp' as a running example (variants of this example may be found in /RT 84, JF 84/).

2.1 Data Modeling with Attributed Graphs

This subsection introduces the class of **attributed, node and edge labeled, directed graphs** (in the sequel just called graphs). Based on the afore-mentioned 'Exp' language I will discuss how to represent the sentences of a certain programming language by the instances of a certain subclass of graphs.

Therefore, I start our explanations with the description of the language 'Exp'. This language characterizes all legal input sequences for a very primitive desk calculator. Landin's 'let' construct /La 66/ is used to name and reuse intermediate computation results. The usual scoping rules for block structured programming languages direct the binding of applied occurrences of names to their corresponding 'let'-definitions. Figure 1 defines the abstract syntax of the language in a manner of writing called the **operator / phylum** notation (cf. /No 87/). The somewhat artificial subphyla 'NIL__EXP', 'BINARY_EXP' etc. have been introduced to group operators with same properties and to emphasize the similarities between the abstract syntax description of a language and the PROGRESS declarations for a corresponding graph scheme (cf. section 3.).

> EXP ::= NIL_EXP | BINARY_EXP | CONST_EXP | DEF_EXP | NAME_EXP
>   (* 'NIL_EXP' etc. are subphyla of the phylum 'EXP'. *)
>
> NAME ::= NameDef (String)   (* Defining occurrence of an identifier. *)
>
> NAME_EXP ::= NameUse (String) (* Applied occurrence of an identifier. *)
>
> DEF_EXP ::= Let (NAME, EXP, EXP)(* Binds the first expression to applied occurrences
>   of the name within the second expression. *)
>
> BINARY_EXP ::= Prod (EXP, EXP) | Quot (EXP , EXP) | . . .
>
> CONST_EXP ::= DecConst (String)
>
> NIL_EXP ::= NilExp ()   (* Placeholder for unexpanded subexpressions. *)

Figure 1 : Abstract syntax of the language 'Exp'.

Figure 2 presents the **text and graph representation** of a typical sentence of the language 'Exp'. This sentence still contains one unexpanded subexpression with an undefined value represented by '(<EXP>)'. As a consequence, the values of the two (sub-) expressions starting with 'let x is ...' are undefined, too. The computation of the subexpression 'let y is 2 in x*y' yields the value '16' due to the fact that 'x' is bound to '8' (by the outermost 'let') and y to '2'.

Starting from the definition of the abstract syntax in figure 1, the **systematic development** of the graph representation has been directed by the following guide-lines (cf. also /En 86, EL 87/):

let x is 8   in ⌊ let x is ⌊let y is 2 in x * y⌋ in (<EXP>) / 4 ⌉

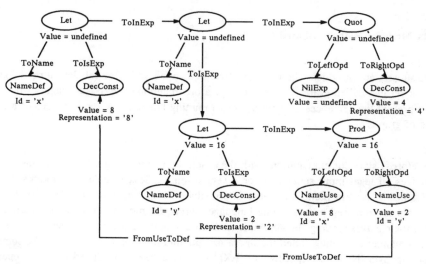

Figure 2 : Text and graph representation of an 'Exp' sentence.

- Map the contextfree syntactic structure of an 'Exp' sentence onto a tree-like graph structure which is equivalent to the sentence's abstract syntax tree. This structure builds the skeleton of the graph representation and it contains a separate **labeled node** for the root of any subexpression of the sentence. Distinguish different types of subexpressions by labeling their root nodes with the operators 'NilExp', 'Sum' etc. of the abstract syntax definition.

- Represent the abstract syntax tree's contextfree relationships, only implicitly defined within the abstract syntax, by **labeled edges** with arbitrarily chosen labels ('ToLeftOpd', ...) and introduce an additional type of edges for any kind of contextsensitive relationships. In our example we have to introduce edges labeled 'FromUseToDef' to bind applied occurrences of names to their value defining expressions.

- Use **(external) node attributes** to hold instances of phyla whose values are atomic from the current point of view and which encode properties inherent to a single node. This holds true for all phyla which do not appear on the left-hand side of any rule of the abstract syntax (phyla whose instances represent lexical units). Within our example we have to attach a 'String' attribute to all nodes labeled with one of the operators 'DecConst', 'NameDef', and 'NameUse'. To emphasize a distinction between strings representing identifiers and strings representing numbers, we introduce an attribute called 'Representation' for nodes labeled 'DecConst' and an attribute called 'Id' for nodes labeled either 'NameDef' or 'NameUse'.

- Use **(derived) node attributes** to encode node properties usually concerning aspects of dynamic semantics. Such an attribute is called 'derived' – instead of 'external' – if and only if its value is defined by a directed equation. Within this equation other attributes of the same node or of adjacent nodes may be referenced. Thus, we are able to establish functional attribute dependencies like: The 'Value' attribute of a 'Prod' node must be equal to the product of the 'Value' attributes of its two operand nodes which are the sinks of the two outgoing 'ToLeftOpd' and 'ToRightOpd' edges.

2.2 'Programming' with Graph Rewrite Rules

In the previous subsection I explained PROGRESS' underlying data model and discussed the matter of systematically deriving graph representations for the sentences of the language 'Exp'. This subsection deals with the subject of **specifying complex operations** mapping one sentence of the language 'Exp' onto another one of the same language. Thinking in terms of our data model these operations are nothing else but class preserving graph transformations.

So-called **subgraph tests** and **graph rewrite rules** are the basic building blocks for the definition of **complex graph transformations**. The former are boolean functions which test for the occurrence of a certain subgraph (pattern) within a host graph, and the latter are graph transformations which search for a certain subgraph within a host graph and replace this subgraph by another one. Usual control constructs known from imperative programming languages may be used to compose very complex graph transformations out of these tests and graph rewrite rules.

let x is 8   in ⌊ let x is ⌊let y is 2 in x * y⌉ in x / 4⌉

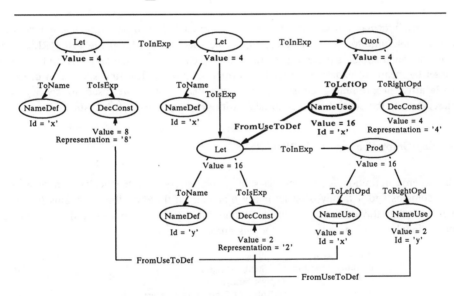

Figure 3 : Result of a rewrite rule's application to the graph of figure 2.

For a brief description of graph rewrite rules let us consider just one typical example. This is the rule 'replace an unexpanded subexpression by an applied occurrence of 'x' '. Figure 3 displays the effect of the application of such a rule onto the example displayed in figure 2. The execution of this rule (specified in section 3.2, figure 9) may be divided into the following five steps:

- **Subgraph test**: Select any subgraph within the host graph complying with the afore-mentioned requests, i.e. in this case any node labeled 'NullExp' within our tree-like graph structure. If there is none, return with failure.

- **Subgraph replacement**: Erase the selected subgraph including all incoming and outgoing edges, i.e. the node labeled 'NullExp' and one edge labeled 'ToLeftOpd', and insert the nodes and edges of the new subgraph, i.e. the node labeled 'NameUse'.

- **Embedding transformation**: Connect the new subgraph with the remainder of the host graph by incoming and outgoing edges, i.e. by two new edges labeled 'ToLeftOpd' and 'FromUseToDef'. So-called path expressions are used to determine the sometimes far–away located sources and targets of embedding edges within the host graph.

- **Attribute transfer**: Assign values to the external attributes of the nodes of the new subgraph, i.e. assign the value 'x' to the attribute 'Id' of the new node.

- **Attribute revaluation**: Compute the new values of all derived attributes within the new graph, i.e. the values of all 'Value' attributes, or at least of those derived attributes to which we have to assign new values.

3. The Specification Language PROGRESS

After this informal introduction to the basic principles of programmed graph rewriting systems, it's now time to focus our interest onto the **specification language PROGRESS** itself. For a first survey it should suffice to present only a typical subset of the language and to explain this subset by means of examples instead of formal definitions. The first subsection deals with the **declarative programming constructs** of PROGRESS which are used to specify graph schemes, whereas the second subsection deals with **graph queries and transformations**.

3.1 Defining Graph Schemes with PROGRESS

The scheme definition for a new class of graphs starts with the declaration of a set of attribute domains and a family of strict n–ary functions on these attribute domains (see figure 4). We shall skip the definition (implementation) of **attribute domains and functions** and assume a Modula–2 like host language for this purpose.

```
attribute_type Integer, String;
attribute_function undefined : Integer;
            empty : String;
            decToInt : ( String ) -> Integer;
            mult : ( Integer, Integer ) -> Integer;
            div : ( Integer, Integer ) -> Integer;

                . . .
```

Figure 4: Declaration of attribute domains and functions.

Based on these (incomplete) attribute declarations, type declarations of three different categories characterize our abstract data type's graph representation. **Node and edge type declarations** are used to introduce type labels for nodes and edges, whereas **node class declarations** play about the same role as phyla, their counterparts within the operator/phylum based description of the language 'Exp'.

```
class EXP
  derived Value : Integer;
end;

class NAME
  external Id : String;
end;

class DEF_EXP is_a EXP end;

edge_type ToName : DEF_EXP -> NAME;

edge_type ToIsExp : DEF_EXP -> EXP;

edge_type ToInExp : DEF_EXP -> EXP;

class NAME_EXP is_a NAME, EXP end;

edge_type FromUseToDef : NAME_EXP -> EXP;

class BINARY_EXP is_a EXP end;

edge_type ToLeftOpd, ToRightOpd : BINARY_EXP -> EXP;

class CONST_EXP is_a EXP
  external Representation : String;
end;

class NIL_EXP is_a EXP end;

class MARKER end;

edge_type ToMarkedExp : MARKER -> EXP;
```

Figure 5: Declaration of node classes, edge types, and attributes.

Primarily, **node classes** are used to denote coercions of node types with common properties (following the lines of IDL /Ne 86/). Thereby, they introduce the concepts of classification and specialization into our language, and they eliminate the needs for duplicating declarations by supporting the concept of **multiple inheritance** along the edges of a class hierarchy. Additionally, node classes play the role of second order types. Being considered as types of node types, they support the controlled use of **formal (node) type parameters** within generic subgraph tests and graph rewrite rules (cf. figure 8 and 9). The class 'EXP' e.g. is the type of all those node types, whose nodes possess the derived 'Integer' attribute 'Value' (see figure 5). This holds true for all node types with the exception of 'NameDef' and 'Cursor'. The class 'BINARY_EXP', a subclass of the class 'EXP', is the type of all those node types, whose nodes may be sources of edges typed 'ToLeftOpd' and 'ToRightOpd'.

In the presence of the smallest class, the empty set, and the largest class, comprising all node types, this family of sets (the **class hierarchy**) has to form a **lattice** with respect to the ordering of sets by inclusion (corresponding to the 'is_a' relationship). This request enforces a disciplined use of the concept of multiple inheritance. Furthermore, it was a precondition for the development of a system of type compatibility constraints for PROGRESS.

```
node_type NameDef: NAME end;

node_type NameUse : NAME_EXP
  Value := [ that -FromUseToDef->.Value | undefined ];
end;

node_type Let : DEF_EXP
  Value := that -ToInExp->.Value;
end;

node_type Prod : BINARY_EXP
  Value := mult( that -ToLeftOpd->.Value, that -ToRightOpd->.Value);
end;

node_type DecConst : CONST_EXP
  Value := decToInt( .Representation);
end;

node_type NilExp : NIL_EXP
  Value := undefined;
end;

node_type Cursor : MARKER end;
```

Figure 6: Declaration of node types, and attribute dependencies.

The main purpose of a **node type declaration** is to define the behaviour of the nodes of this type, i.e. the set of all directed attribute equations, locally used for the (re-) computation of the node's derived attribute values. Let's start our explanations of the definition of **functional attribute dependencies** with the declaration of the node type 'DecConst' (in figure 6). Its 'Value' attribute is equal to the result of the function 'decToInt' applied to its own 'Representation' attribute. To compute the 'Value' of a 'Prod' node, we have to multiply the corresponding attributes of its left and right operand, i.e. the sinks of the outgoing edges typed 'ToLeftOpd' and 'ToRightOpd', respectively. In this case we assume that any 'Prod' node is the source of exactly one edge of both types and we strictly prohibit the existence of more than one outgoing edge of both types (indicated by the key word 'that').

Due to the fact that we cannot always guarantee the existence of at least one edge, we had to introduce the possibility to define **sequences of alternative attribute expressions**. The general rule for the evaluation of such a sequence is as follows: try to evaluate the first alternative. If this fails due to the absence of an (optional) edge, then try to evaluate the next alternative. We request that at least the evaluation of the last alternative succeeds. Therefore, the 'Value' of a 'NameUse' node is either the 'Value' of its defining expression or, in the absence of a corresponding definition, 'undefined'.

3.2 Defining Graph Queries and Transformations with PROGRESS

Being familiar with PROGRESS' declarative style of programming, we are now prepared to deal with the operational constructs of the specification language. Some of these constructs form a partly textual, partly graphic **query sublanguage** for the definition of graph traversals which is very similar to data base query languages like /EW 83/. Elements of this sublanguage, in the sequel called path expressions, are mainly used to denote rather complex context conditions within subgraph tests and graph rewrite rules. Formal definitions of previous versions of this sublanguage have been published in /EL 87, Le 88a/.

```
path_op father : EXP -> EXP =
    <-ToLeftOpd- or <-ToRightOpd- or <-ToIsExp- or <-ToInExp-
end;


path_op nextValidDef : EXP -> DEF_EXP =
    (* Computes the root of the next valid surrounding definition. *)
        { not <-ToInExp- : father }
    &  <-ToInExp-
end;


path_op binding ( Id : String ) : EXP -> EXP =
    (* Computes the root of the next visible surrounding definition of the name 'Id' *)
        nextValidDef
    & { not definition(Id) : nextValidDef }
    &  -ToIsExp->
end;


path_op definition ( IdPar : String ) : DEF_EXP -> NAME  =  1 => 2 in
```

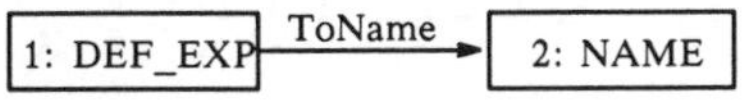

```
    condition 2.Id = IdPar ;
end;
```

(* Applied to a set of 'DEF_EXP' nodes it computes the set of all those appertaining 'NAME' nodes whose 'Id' attributes are equal to 'IdPar'. *)

Figure 7: Definition of graph traversals, using textual and graphical path expressions.

Path expressions may either be considered to be derived binary relationships or to be (node-) set-valued functions. The path expression '-ToLeftOpd->' e.g., used in figure 6 for the formulation of attribute dependencies, maps a set of nodes onto another node set; the elements of this second node set are sinks of 'ToLeftOpd' edges starting form nodes within the first node set. The expression '<-ToLeftOpd-' simply exchanges the roles of sinks and sources. Figure 7 contains the **textual definition** of three parameterized functional abstractions of path expressions, using conditional repetition ('{ <condition> : <path-expression> }'), composition ('&'), and union ('or'), and one example of a **graphic definition**. These path expressions specify the binding relationships between applied occurrences of identifiers ('NAME_EXP' nodes) and their corresponding value defining expressions ('EXP' nodes). The abstraction 'nextValidDef' e.g., applied to a set of nodes, computes the set of all those nodes that may be reached from elements of the first node set by: (1) following 'ToLeftOpd', 'ToRightOpd', and 'ToIsExp' edges form sinks to sources (cf. declaration of 'father'), as long as there is no incoming 'ToInExp' edge at any node on this path, (2) and finally following a 'ToInExp' edge from sink to source.

Skipping any further explanations concerning path expressions, we come to the specification (of a small subset) of the operations provided by the abstract data type 'ExpGraphs'. It is worthwhile to notice that node type parameters keep this part of the specification independent from the existence of any particular node type with the exception of the node type 'Cursor'. A unique node of this type is used to mark that subtree within a graph, called **current** subtree, that should be affected by the application of a graph rewrite rule. Thus, the declarations of figure 8 and 9 (together with the declarations of figure 5) may be used to specify a whole family of quite different desk calculators.

Let's start with the explanation of the operation 'Initialize' of figure 8. This operation creates the root of a tree–like subgraph if and only if the graph of interest contains no nodes of a type belonging to the class 'EXP'. We use the **test** 'ExpressionInGraph' to check this condition. The subsequent application of an instantiation of the graph rewrite rule (**production**) 'CreateExpressionRoot' never fails due to the fact that this rule has an empty matching condition. Being instantiated with any node type of the class 'NIL_EXP' its execution simply adds one node of this type, another node of the type 'Cursor', and an edge of the type 'ToMarkedExp' to a graph.

The production 'MoveCursorUp' is an example of a rewrite rule which **identifies (all) nodes** of its lefthand side with (all) nodes of its righthand side. An application of this rewrite rule neither deletes or modifies any node nor removes any not explicitly mentioned edge from the host graph. A successful application of this rule only redirects one edge of the type 'ToMarkedExp' so that its new sink is the father of its former sink, a condition that is expressed by the path operator 'father' within the rule's lefthand side.

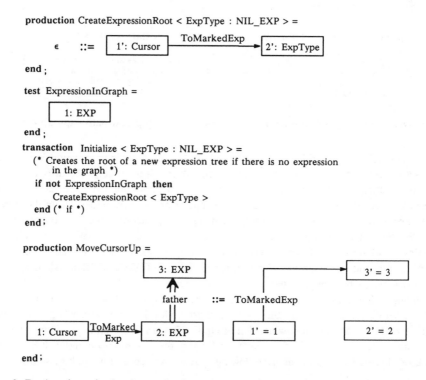

Figure 8: Declaration of simple productions, tests, and transactions.

(Complex) graph transformations (like 'Initialize' of figure 8 or 'CreateAndBindNameExp' of figure 9) are termed **transactions** in order to indicate their **atomic** character. Similar to single graph rewrite rules whole transactions either cause consistency preserving graph transformations or abort without any modifications of the graph they were applied to. A transaction has to abort if one of the transactions or graph rewrite rules it's composed of abort. **Abortion (failure)** of a single graph rewrite rule occurs if the rule's lefthand side doesn't match with any subgraph of the graph it's applied to.

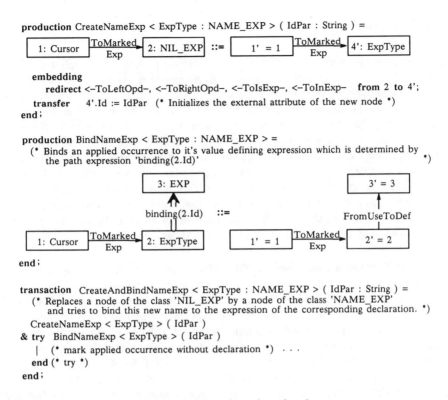

production CreateNameExp < ExpType : NAME_EXP > (IdPar : String) =

embedding
    redirect <–ToLeftOpd–, <–ToRightOpd–, <–ToIsExp–, <–ToInExp– **from 2 to** 4';
  transfer   4'.Id := IdPar (* Initializes the external attribute of the new node *)
end;

production BindNameExp < ExpType : NAME_EXP > =
  (* Binds an applied occurrence to it's value defining expression which is determined by
    the path expression 'binding(2.Id)'   *)

end;

transaction CreateAndBindNameExp < ExpType : NAME_EXP > (IdPar : String) =
  (* Replaces a node of the class 'NIL_EXP' by a node of the class 'NAME_EXP'
    and tries to bind this new name to the expression of the corresponding declaration. *)
  CreateNameExp < ExpType > (IdPar)
& try BindNameExp < ExpType > (IdPar)
    | (* mark applied occurrence without declaration *) · · ·
  end (* try *)
end;

Figure 9: Specification of the graph transformation of section 2.

Thus the application of the transaction 'CreateAndBindNameExp' (of figure 9) either fails or consists of two graph rewrite steps. The first one causes the replacement of the current 'NIL_EXP' node by a new node of the class 'NAME_EXP'. The second one either binds the new applied occurrence to its value defining expression ('**try** BindNameExp ...') or – in the case of failure of 'BindNameExp' – marks the new applied occurrence as to be erroneous. The execution of the second graph rewrite rule requires the evaluation of the non–trivial path operator 'binding(2.Id)' which determines the target of the 'FromUseToDef' edge, whereas the execution of the first graph rewrite rule contains an embedding transformation and an attribute transfer, the latter being the assignment of the new name's string representation to the corresponding node's 'Id' attribute. Both rewrite steps even may **trigger the revaluation of many derived attributes**. The application, explained in subsection 2.2 (transforming the graph of figure 2 to that of figure 3), for instance initiated the (re–) evaluation of four 'Value' attributes.

The above mentioned **embedding transformation** within the graph rewrite rule 'CreateNameExp' consists of one '**redirect**' clause. The purpose of this clause is to redirect any incoming edge belonging to the expression's abstract syntax tree skeleton from the former 'NIL_EXP' node to the new 'NAME_EXP' node. In addition to '**redirect**' clauses, '**remove**' and '**copy**' clauses may be used to remove edges from identically replaced nodes or to duplicate edges and attach them to the replacing subgraph's nodes.

4. Related Work

The work presented within this paper is the first attempt to design a **strongly typed** language which is based on the concepts of **programmed graph rewriting systems** (like that of /Gö 88/) and supports a **declarative style of programming** for the description of object structures (like IDL /Ne 86/, or the languages surveyed in /HK 87/). The development of the language's type concept has been influenced by polymorphic programming languages like HOPE /BM 81/ which combine the flexibility of typeless languages with the reliability of strongly typed languages. By relying on the concept of a stratified type system with an infinite hierarchy of type universes (cf. /CZ 84/) we were able to incorporate **typed type parameters** into our language and to avoid the theoretical pitfalls of reflexive type systems with the 'Type is the type of all types including itself' assumption (cf. /MR 86/). The concept of **specialization** enables us to build complex class hierarchies. The idea to restrict this concept to the construction of **class lattices** has already been used within rewrite systems (cf. /CD 87/) and logic programming languages (cf. /AK 84/).

In order to facilitate a comparison of PROGRESS with those languages based on the more popular formalism of attribute (tree) grammars, I have used the well-known expression tree example for demonstration purposes. Therefore, its necessary to emphasize that PROGRESS is **not restricted to** the specification of abstract data types with **tree-like representations** but even more adequate in the case of graph-like representations without any dominant tree-like sub-structure. Thus, it is almost impossible within the framework of attribute (tree) grammars to specify the restriction that a class hierarchy has to form a lattice, but it is a straight forward task to write an appropriate lattice test with PROGRESS (/He 89/). Comparing the PROGRESS approach to model tree-like structures with that kind of modeling inherent to other attribute (tree) grammar approaches, we discover at least three principle differences:

- Derivation trees corresponding to subsequent applications of rewrite rules (productions) are neither used to represent object structures themselves nor to represent additional informations about these structures (like /Ka 85, Sc 87/). As a consequence, directed attribute equations are used to describe functional **attribute dependencies between adjacent nodes of a given graph**, and not to describe attribute flows along the edges of a derivation tree. Therefore, we believe (in contrast to /KG 89/) that the specification of attribute dependencies and the specification of graph rewrite rules should be kept separate from each other.

- A common disadvantage of our approach and the proposal in /KG 89/ is that **we are not able to identify directions like 'up' and 'down'** within arbitrary graphs. Therefore, we cannot adopt the classification of what we call 'derived' attributes into 'inherited' and 'synthesized' ones (cf. /Re 84/). Thus, the development of non-naive (incremental) attribute evaluation algorithms is much more difficult than for the case of attribute (tree) grammars. For basic work on this problem the reader is referred to /AC 87, Hu 87/.

- A third difference is also a direct consequence of the fact that we are not restricted to the world of trees: Our data model refrains the specificator from the somewhat artificially different treatment of relationships representing either the **contextfree or the contextsensitive syntax** (usually called static semantics) of a sentence of the language 'Exp'. Using labeled directed edges for the representation of both kinds of relationships, we are not forced to put contextsensitive informations into complex structured attributes (being a severe handicap for any incremental attribute evaluation algorithm), like /RT 84/, or to escape to another formalism, like /HT 86/.

5. Summary

This paper contains a first presentation of the new specification language **PROGRESS** (for writing **PRO**grammed Graph **RE**writing System Specifications), its underlying formalism, and a model oriented approach to the specification of abstract data types. This approach, called **graph grammar engineering**, is a model oriented one due to the fact that we use

- **attributed, node and edge labeled, directed graphs** to model complex object structures,

- and **programmed graph rewriting systems** to specify operations in terms of their effect on these graph models.

Having fixed the presented version of PROGRESS' **programming–in–the–small part** a few months ago, we are now starting to evaluate first experiences with the language and to develop the language's **programming–in–the–large part**, supporting the decomposition of large specifications into separate, reusable, and encapsulated subspecifications. Therefore, the presented version of PROGRESS may be classified as the programming–in–the–small kernel of a (very high level programming) language offering **declarative and procedural** elements for

- a **data flow oriented** style of programming (by means of directed attribute equations),

- an **object oriented** style of programming (by supporting multiple inheritance and dynamic binding of attribute designators to their value defining equations),

- a **rule based** style of programming (by using graph rewrite rules),

- and an **imperative** style of programming (within transactions).

Last but not least, we have started to design and implement a **programming environment for PROGRESS**, based on our previous experiences with the construction of the Integrated Programming Support ENvironment **IPSEN**. This environment will be built on top of the non-standard data base management system **GRAS** (cf. /LS 88/) and will comprise at least a syntax-directed editor (see /He 89/) and an interpreter. Thus, in days to come we might be able to specify the functional behaviour of PROGRESS' environment with PROGRESS itself and to analyze and execute the environment's specification by means of its own implementation.

Acknowledgments

I am indebted to M. Nagl and C. Lewerentz for many stimulating discussions about PROGRESS, to B. Westfechtel for critical and constructive comments on this manuscript, to all students, willingly producing large specifications with PROGRESS, and especially to U. Cordts, P. Heimann, P. Hormanns, R. Herbrecht, M. Lischewski, R. Spielmann, C. Weigmann, and A. Zündorf for designing and implementing first parts of the PROGRESS programming environment.

References

/AC 87/   B.Alpern, A.Carle, B.Rosen, P.Sweeney, K.Zadeck: *Incremental Evaluation of Attributed Graphs*, T. Report CS–87–29; Providence, Rhode Island: Brown University

/AK 84/   H.Ait–Kaci: *A Lattice–Theoretic Approach to Computation Based on a Calculus of Partially–Ordered Type Structures*, Ph.D. Thesis; Philadelphia: University of Pennsylvania

/BM 81/   R.M.Burstall, D.B.MacQueen, D.T.Sannella: *HOPE – An Experimental Applicative Language*, Technical Report CSR–62–80; Edinburgh University

/CD 85/   R.J.Cunningham, A.J.J.Dick: Rewrite Systems on a Lattice of Types, in Acta Informatica 22, Berlin: Springer Verlag, pp. 149–169

/CZ 84/   R.Constable, D.Zlatin: *The Type Theory of PL/CV3*, in ACM TOPLAS, vol. 6, no. 1, pp. 94–117

/EL 87/   G.Engels, C.Lewerentz, W.Schäfer: *Graph Grammar Engineering – A Software Specification Method*, in Ehrig et al. (Eds.): Proc. 3rd Int. Workshop on *Graph Grammars and Their Application to Computer Science*, LNCS 153; Berlin: Springer Verlag, pp. 186–201

/En 86/   G.Engels: *Graphen als zentrale Datenstrukturen in einer Software–Entwicklungsumgebung*, PH.D. Thesis; Düsseldorf: VDI–Verlag;

/ES 85/   G.Engels, W.Schäfer: *Graph Grammar Engineering: A Method Used for the Development of an Integrated Programming Support Environment*, in Ehrig et al. (Eds.): Proc. TAPSOFT '85, LNCS 186; Berlin: Springer–Verlag, pp. 179–193

/EW 83/   R.Elmasri, G.Wiederhold: *GORDAS: A Formal High–Level Query Language for the Entity–Relationship Model*, in P.P.Chen (ed.): Entity-Relationship Approach to Information Modeling and Analysis, Amsterdam: Elsevier Science Publishers B.V. (North-Holland), pp. 49–72

/GJ 82/   H.J.Genrich, D.Janssens, G.Rozenberg, P.S.Thiagarajan: *Petri nets an their relation to graph grammars*, in Ehrig et al.: Proc. 2nd Int. Workshop on *Graph Grammars and Their Application to Computer Science*, LNCS 153; Berlin: Springer Verlag, pp. 115–142

/Gö 88/   H.Göttler: *Graphgrammatiken in der Softwaretechnik*, IFB 178; Berlin: Springer–Verlag;

/He 89/   R.Herbrecht: *Ein erweiterter Graphgrammatik–Editor*, Diploma Thesis; University of Technology Aachen;

/HT 86/   S.Horwitz, T.Teitelbaum: *Generating Editing Environments Based on Relations and Attributes*, in Proc. ACM TOPLAS, vol. 8, no. 4, pp. 577–608

/Hu 87/   S.E.Hudson: *Incremental Attribute Evaluation: An Algorithm for Lazy Evaluation in Graphs*, Technical Report TR 87-20; Tucson: University of Arizona

/HK 87/   R.Hull, R.King: *Semantic Database Modeling: Survey, Applications, and Research Issues*, in ACM Computing Surveys, vol. 19, No. 3, pp. 201–260

/JF 84/   G.F.Johnson, C.N.Fischer: *A Metalanguage and System for Nonlocal Incremental Attribute Evaluation in Language–Based Editors*, in Proc. ACM Symp. POPL '84

/Ka 85/   M.Kaul: *Präzedenz Graph–Grammatiken*, PH.D. Thesis; University of Passau

/KG 89/   S.M.Kaplan, St.K.Goering: *Priority Controlled Incremental Attribute Evaluation in Attributed Graph Grammars*, in Diaz, Orejas (Eds.): Proc. TAPSOFT '89, vol. 1, LNCS 351, Berlin: Springer Verlag, pp.306–320

/La 66/   P.J.Landin: *The next 700 programming languages*, Com. ACM 9, pp. 157–164

/Le 88/   C.Lewerentz: *Extended Programming in the Large in a Software Development Environment*, Proc 3rd ACM SIGPLAN/SIFSOFT Symp. on Practical Software Engineering Environments

/Le 88a/   C.Lewerentz: *Interaktives Entwerfen großer Programmsysteme*, PH.D. Thesis, IFB 194; Berlin: Springer–Verlag;

/LS 88/   C.Lewerentz, A.Schürr: *GRAS, a Management System for Graph–like Documents*, in C.Beeri et al. (Eds.): Proc. 3rd Int. Conf. on Data and Knowledge Bases; Los Altos, California: Morgan Kaufmann Publishers Inc., pp. 19–31

/Me 82/   A.Meier: *A Graph–Relational Approach to Geographic Databases*, in /ENR 82/, pp. 245–254

/MR 86/   A.R.Meyer, M.B.Reinhold: *'Type' is not a type*, Proc. 13th ACM Symp. POPL '86, pp. 287–295

/MW 84/   U.Möncke, B.Weisgerber, R.Wilhelm: *How to Implement a System for the Manipulation of Attributed Trees*, in U.Ammann (Ed.): Programmiersprachen und Programmentwicklung, IFB 77; Berlin: Springer Verlag

/Na 79/   M.Nagl: *Graph–Grammatiken: Theorie, Implementierung, Anwendungen*; Braunschweig: Vieweg–Verlag

/Na 85/   M.Nagl: *Graph Technology Applied to a Software Project*, in Rozenberg, Salomaa (Eds): The Book of L; Berlin: Springer–Verlag, pp. 303–322,

/Ne 86/   J.Newcomer: *IDL: Past Experience and New Ideas*, in Conradi et al. (Eds.): Advanced Programming Environments, LNCS 244; Berlin: Springer–Verlag, pp. 257–289

/No 87/   K.Normark: *Transformations and Abstract Presentations in Language Development Environment*, Technical Report DAIMI PB–222; Aarhus University

/Re 84/   T.Reps: *Generating Language–Based Environments*, PH.D. Thesis; Cambridge, Mass.: MIT Press

/RT 84/   T.Reps, T.Teilbaum: *The Synthesizer Generator*, in Proc. ACM SIGSOFT/SIGPLAN Symp. on Practical Software Development Environments, pp. 42–48

/Sc 87/   A.Schütte: *Spezifikation und Generierung von Übersetzern für Graph–Sprachen durch attributierte Graph–Grammatiken*, PH.D. Thesis; Berlin: EXpress–Edition

/So 84/:   J.F.Sowa: *Conceptual Structures: Information Processing in Minds and Machines*; Reading, Mass.: Addison–Wesley

/TS 86/   G.Tinhofer, G.Schmidt (Eds.): Proc. WG '86 Workshop on *Graph–Theoretic Concepts in Computer Science*, LNCS 246; Berlin: Springer–Verlag

On the Complexity of Optimal Drawings of Graphs

Franz J. Brandenburg

Lehrstuhl für Informatik, Universität Passau
Innstr. 33, D 8390 Passau, F.R. Germany

Abstract

We consider the problem of producing aesthetically nice drawings of graphs from the complexity point of view. The following questions are immediate:

    (1) How to formalize in algorithmic terms that a drawing is nice?

    (2) What are the computational costs for nice drawings?

    (3) Are there tools to beat the NP-completeness?

For (1) we propose grid embeddings of graphs and measure "nice" by algorithmic cost measures of the embeddings, e.g., area, expansion, edge length, etc. For (2) we prove that optimal embeddings with fixed costs are NP-complete, even for binary trees. This sharpens previous NP-completeness results of optimal embeddings from connected graphs to binary trees and extends them to other cost measures. For (3) we introduce placement graph grammars. These are special graph grammars enriched by a placement component. The placement component contains partial information on the relative positions of the vertices, which is a reduction of the placement information contained in any concrete grid drawing. Every derivation of a graph in the base graph grammar has an associated placement component. By an extension of the parsing process we can compute a placement of the vertices of each generated graph, which is consistent with the associated placement component, and is area minimal. For connected graphs of bounded degree this can be done in polynomial time.

Keywords

graph layout, embeddings, NP-completeness, graph grammars, placement graph grammars

Introduction

Graphs or nets are frequently used in computer science. They are a well-known tool both for formal descriptions and for visual illustrations. Examples are Petri nets, entity relationship diagrams, transition graphs, schedules, circuits, VLSI, and just an outline. This has good reasons. Graphs are very flexible and have a high expressive power, particularly when they are nicely drawn in the plane. A visual representation supports human cognitive capabilities, it activates our intuition, helps finding solutions and makes complex situations understandable. However, this works only if the drawn graph is readable, if its layout is nice, where nice comes from graph theoretic properties.

For this problem we must formalize what is means that a graph is aesthetically nice. Our approach are graph embeddings. The abstraction of nice aesthetics into a formal specification is done in to steps: First, there are constraints which must be satisfied. These restrict the space of solutions and methods. Secondly, there are formal cost parameters as aesthetics. These are optimized for best possible drawings. Examples for constraints are disjointness conditions such as injectivity and planarity, whereas area, expansion, width or edge length are typical aesthetics.

How hard is it to compute or produce an aesthetically nice drawing of a graph, or of a tree? That depends! Classical NP-complete graph problems such as Hamiltonian, clique, colorability, etc. easily reduce to nice drawings, where intuitively the drawing shows the solution. E.g. you see a Hamiltonian cycle on the spot, if it is drawn as an outer cycle of an n-gon. Hence, nice drawings may be NP-complete. On the other hand, one can efficiently construct layouts of graphs which are optimal up to some factor, i.e. $O(..)$ optimal. As an example consider the H-tree layout for complete binary trees or the standard level by level layout of ordered trees with all leaves on the boundary, see [27].

To the contrary, optimal embeddings are intractable. I.e., in general, the following decision question is NP-complete: Given a graph and a bound K. Is there a grid embedding with e.g. area $\leq$ K? Results of this type were known e.g. for connected graphs, forests or trees with special restrictions. See, e.g. [1, 5, 10-12, 14, 15, 17-19, 24]. We improve these results to binary trees. Hence for nice drawings of trees we are in the situation that there are efficient approximations and good practical algorithms (see [6, 20]), but sharp optimizations are intractable.

We wish to outwit the NP-completeness and search for general tools, which specify interesting classes of graphs and construct optimal layouts in polynomial time. As a first proposal we introduce placement graph grammars. These consist of two components, a base graph grammar and a placement component. The base graph grammar is such that it has a polynomial time syntax analysis. The placement component gives some information on the relative positions of the vertices in x- or in y-dimension. It is a subset of the "left of" and "below" relations that are given with any grid drawing. Intuitively, the placement component tells the following: if at an intermediate stage of a derivation process, a vertex of size (1×1) expands into a subgraph of size $(w \times h)$, then all vertices that are specified to be "to the right" of that vertex are shifted $w-1$ units to the right, and all vertices that are specified "below" are shifted $h-1$ units to the bottom. For every pair of vertices, one direction must be specified, and the other may be free. This leaves some flexibility for constructions.

1. Graph Drawing and Graph Embedding

We suppose familiarity with graph theory and complexity theory. See e.g. [9,11].

A *directed or undirected graph* $g = (V, E)$ consists of a finite set of vertices V and a finite set of edges E, each edge $e = (u,v)$ between vertices u and v. Labeled graphs shall be considered later. Special graphs are the *grids* and the *trees*. A grid graph is a rectangle of the discrete plane. Its vertices are identified with its coordinates, such that $v = (v.x, v.y)$, and the edges are horizontal or vertical lines of unit length.

A *path* p in a graph $g = (V, E)$ is an alternating sequence of vertices and edges, $p = v_0, e_1, v_1, e_2, v_2,..., e_n, v_n$ with $e_i = (v_{i-1}, v_i)$ for $1 \leq i \leq n$. In general, the vertices are distinct so that the paths are simple. A path is *directed*, if each edge $e_i = (v_{i-1}, v_i)$ is directed from v_{i-1} to v_i.

Two paths p and p' are *edge disjoint*, if they have no common edge, and are *vertex disjoint*, if they have no common vertex, except at their endpoints. Note that two edge disjoint paths may cross each other, whereas two vertex disjoint paths meet at most in their endpoints.

The *size* of a graph g is defined by the number of its vertices and is denoted by $|g|$; the *length* of a path p is the number of its edges and is denoted by $|p|$. If the vertices have weights $w_0, w_1,..., w_n$, then their sum defines the *weighted path length* of p.

Definition

Let $g = (V, E)$ and $g' = (V', E')$ be graphs, called the *guest graph* and the *host graph*, respectively. An *embedding* of g into g' is a mapping $f : g \rightarrow g'$, which maps the vertices of g one-to-one into the vertices of g' and maps the edges of g incidence preserving into the paths of g'.

An embedding f is *directed*, if for every edge $e \in E$, $f(e)$ is a directed path. f is *edge disjoint* resp. *vertex disjoint*, if the paths of distinct edges are edge disjoint and vertex disjoint, respectively.

An embedding f is a *grid embedding*, if the host graph g' is a grid.

Note that vertex disjoint embeddings preserve planarity and being a tree.

For embeddings and in particular for grid embeddings we consider the following cost measures:

Definition

Let $f : g \rightarrow g'$ be an (vertex disjoint) embedding of $g = (V, E)$.

    (1) area: size of the smallest rectangular grid graph that contains $f(g)$

    (2) expansion: ratio of the sizes of $f(g)$ and g, $|f(g)| / |g|$

    (3) maximal edge length: $\max \{ |f(e)| \mid e \in E\}$

    (4) total edge length: $\sum_{e \in E} |f(e)|$

    (5) density: size of the largest connected subgraph of $g' - f(g)$.

Our goal are formal descriptions of nice drawings of graphs. We claim that graph embeddings are a suitable approach to this problem. This holds in particular for grid embeddings. Our arguments are based on the flexibility of graph embeddings. These have three independent classes of parameters, which can be adapted to a particular problem and special needs.

First, there is the class of host graphs. Certainly, the particular structure and properties of a host graph g' imposes restrictions on the embedded graph f(g). For example, every vertex of a grid graph has degree four, and it has 4k vertices at distance k. Hence, there are (only) 2k(k-1) distinct vertices reachable within a radius of size k. On the other hand, there are many edge disjoint paths between two vertices of length at most k. A grid embedding of a graph means its comparison against such a structure. At another extreme consider trees. In a complete binary tree there are 2^{k+1} -1 vertices reachable from the root with distance at most k, and there exists exactly one edge disjoint path between every pair of vertices. The embedding of a graph g into a tree can be seen as imposing a hierarchical structure on g, and a cost measure for the embedding reflects by how much g fails being a tree.

Secondly, there are constraints of the embedding as a function, such as direction or edge or vertex disjointness. Direction is self-evident, and edge and vertex disjointness are restrictions on the capacity of the edges and vertices of the host graph. In particular, a vertex disjoint embedding preserves planarity and being a tree. Another reasonable constraint comes from a partition of the set of vertices of g and g', and embedding them componentwise. As an illustrative example distinguish inner and outer vertices and map the outer vertices to the outer face or to the boundary of a grid drawing. Or, demand that isomorphic subgraphs are embedded into isomorphic subgraphs. This rule can be used to preserve symmetries.

Finally, there is a rich collection of cost measures. These give access to a qualitative comparison of embeddings. Cost measures are usually defined by the size of certain objects, e.g. the size of distinguished subgraphs or the length of paths, or by structural properties, such as the number of crossings or the number of bends of paths. Their variety can be increased by considering combinations and weights on special substructures, and merged into a subtle balance.

A cost measure maps complex objects into the reals. It abstracts form many features and leads to a comparison of the objects by a single value. This implies a classification and gives a simplified view into the class of objects. However, one must be aware of an over-simplification. The chosen abstraction may give an incomplete picture of the real situation. In the field of visualization this may often be the case.

We are aware of the fact that our identification of nice drawings and graph embeddings is only an approximation of the real and complex drawing problem. It is an attempt to a formalization of aesthetics and will be insufficient and incomplete in many situations. Thus, our proposition should also be seen as a request for concrete examples (or counterexamples) of graph drawing problems, where a formal description as a graph embedding problem works properly (or fails).

PROPOSITION

Every completely specified graph drawing problem can be described as a graph embedding problem, and the aethetics of a drawn graph are formally reflected by cost measures for the embedding.

2. NP-complete Tree Layouts

The usefulness and applicability of a formal approach depends on its effectiveness. In the particular situation of graph drawings we ask for the computational costs of optimal graph drawings. Here, optimality means a sharp bound such that the costs satisfy an upper bound K. This contrasts the case of "good approximations", where the achieved solution is optimal up to some factor, i.e. O(..) optimal, and meets the upper and lower bound. E.g., for complete binary trees the H-tree layout is O(n) area optimal and for arbitrary binary trees there are O(n) area optimal grid embeddings, which are computable in O(n) time, see [27]. This situation changes completely for sharp bounds.

Definition

INSTANCE: Given a graph g and some K ≥ 0.

PROBLEM: Is there a (directed) grid embedding $f : g \rightarrow g'$ that satisfies the following property ?

RECTANGLE:	g' is a fixed rectangle
AREA:	f(g) has area ≤ K
EXPANSION:	f(g) has expansion ≤ K ?
MAXIMAL EDGE LENGTH:	f(g) has maximal edge length ≤ K ?
TOTAL EDGE LENGTH:	f(g) has total edge length ≤ K ?
DENSITY:	f(g) has density ≤ K ?

Obviously, there are trade-offs among these problems. E.g., the bounds for RECTANGLE, AREA, TOTAL EDGE LENGTH and DENSITY imply upper bounds for EXPANSION, and EXPANSION and TOTAL EDGE LENGHT are equivalent for vertex disjoint embeddings. For the least possible bounds, i.e. K = 1 for EXPANSION and MAXIMAL EDGE LENGTH and K = |g| for TOTAL EDGE LENGTH, these problems are equivalent and require sharp layouts without any extra vertex. Similarly, AREA = |g| and DENSITY = 0 are equivalent and require that f(g) completely fills a rectangle and has no holes. Clearly, RECTANGLE is more restrictive than AREA, since it also bounds the width and the height of the smallest rectangle enclosing the embedding f(g).

These problems (and many others) have been studied previously in the literature and have been proved NP-complete for various classes of graphs. E.g. for connected graphs the NP-completeness is shown in [17] for RECTANGLE, in [14,15, 17, 24] for AREA and in [17, 24] for EXPANSION. The NP-completeness of MAXIMAL or TOTAL EDGE LENGTH has been shown for connected graphs in [19, 24], for ternary trees in [1] and for binary trees in [12].

Combining techniques from [1] and [12] we obtain the NP-completeness for binary trees and each of the above problems:

THEOREM 1

For binary trees and (directed, vertex disjoint) embeddings, the following problems are NP-complete:
RECTANGLE, AREA, EXPANSION, MAXIMAL and TOTAL EDGE LENGTH, DENSITY .

Proof.

The containment in NP is obvious.

For the NP-completeness reduce the Bhatt and Cosmadakis problem of Gregori [12] to our problems. Gregori's result of the NP-completeness of unit length grid embeddings of binary trees proves the NP-completeness of MAXIMAL EDGE LENGTH and of TOTAL EDGE LENGTH. It follows easily that the embedding may or may not be directed or edge disjoint or vertex disjoint. Simultaneously, we obtain the NP-completeness for EXTENSION.

For RECTANGLE, AREA and DENSITY modify the trees from Bhatt and Cosmadakis [1] as follows. See Fig. 2 in [1] and Fig. (1). Omit the singular outer vertices. Insert new completely filled rows between the rows of the vertical spines of the Bhatt and Cosmadakis trees. This guarantees that at most a single vertice remains as a hole, which is a connected subgraph of the minimal size grid that is not covered by the tree.

If the horizontal spine now has length w and each vertical spine has height h, then there is a vertex disjoint embedding of the tree T into a rectangle R of width w and height h, or into a rectangle of area $w*h$, or with $(h-3)/4$ holes each of size one if and only if T has a unit length embedding, and is embedded as illustrated in Fig. 2. (This is one possible solution from Fig. 2 in [1].)

The tree T is a ternary tree. To come down to binary trees, we expand each vertex of T into the tree T49 with 49 vertices. Two T49 trees are connected by unit length edges at the leaves illustrated by "o". T49 is an extension of the complete binary tree with 31 vertices. Fig. 3 shows its unique embedding into a 7×7 grid. The uniqueness follows from Gregori's arguments on complete binary trees with 31 vertices. ◻

From this result we can conclude that optimal grid embedding are NP-hard for all "nontrivial" classes of graphs. In the spirit of our Proposition we conclude that optimal drawings of graphs are NP-hard, in general. This insight can also be obtained from the fact that many classical NP-complete problems can be seen as graph drawing problems. The intractability of grid drawings of binary trees with only one cost parameter under optimization indicates that the problem of computing and constructing optimal graph drawings under more complex constraints including symmetries or subgraph isomorphism is likely to be a hard problem. From this fact we also learn that good heuristics for graph drawing algorithms are nontrivial, and it may be even harder to explain where and why a known graph drawing algorithm fails producing nice drawings.

To complete the picture of NP-complete drawings of trees recall that minimal width eumorphous tree layouts are NP-complete, too [25]. These are level by level grid drawings of ordered binary trees, which preserve subtree isomorphism and center the father over its sons.

3. Placement Graph Grammars

Next we introduce placement graph grammars. These are a general tool for polynomial time constructions of optimal drawings of graphs. A placement graph grammar is a graph grammar from the NLC graph grammar family (see e.g. [7]). It replaces vertices by graphs from a finite set of productions and establishes connections to the neighbors of the replaced vertex using edge labels and vertex labels.

A placement graph grammar is composed of two parts, a base graph grammar and a placement component. This parallels syntax directed translation schemes. Conversely, the partition is induced by distinct edge labels. Base edges are considered as undirected and unlabeled. Placement edges are directed and labeled by {x, y}, where x means "from left to right" and y means "from top to bottom". Note that placement edges are given implicitly by a grid drawing of a graph. See Fig. 1-3.

We wish to achieve polynomiality. Hence, we must choose a class of base graph grammars with an effective polynomial time syntax analysis. A characterization of the "maximal" class has been obtained in [2]; for convenience we restrict ourselves here to a model close to boundary graph grammars introduced by Rozenberg and Welzl [21].

Definition

Let Σ and Δ be alphabets, which shall be used as vertex labels and as edge labels, respectively.
A *labeled graph* $g = (V, E, m)$ consists of *vertices* V, a *vertex labeling function* $m : V \to \Sigma$ and *directed, labeled edges* $E = \{(u, a, v) \mid u, v \in V \text{ with } u \neq v \text{ and } a \in \Delta\}$. An edge $e = (u, a, v)$ with label a is called an *a-edge* from vertex u to vertex v. Vertices u and v are *neighbors*, if there is an edge (u, a, v) or (v, a, u). Notice that there may be multiple edges with distinct labels and direction between two vertices. For convenience, loops (v, a, v) are excluded.
For a graph g let $V(g)$, $E(g)$ and $m(g)$ denote its set of vertices, its set of edges and its vertex labeling function, so that $g = (V(g), E(g), m(g))$.

Next we need some graph theoretic terms which are relevant for graph grammars.

Definition

Let $g = (V, E, m)$ be a labeled graph and let $V' \subseteq V$ be a set of vertices of g.
The *set of neighbors* of V' is $neigh(V', g) = \{u \in V \mid u \notin V' \text{ and } u \text{ is a neighbor of some } v \in V'\}$.
The *induced subgraph* of V' is $ind(V', g) = (V', E', m')$ with $E' = \{(u, a, v) \in E \mid u, v \in V'\}$ and $m'(v) = m(v)$. Similarly, a subset of the edges $E' \subseteq E$ induces the subgraph (V, E', m). In particular, if A is a subset of the set of edge labels Δ, then $E_A = \{(u, a, v) \in E \mid a \in A\}$ is the set of a-edges of g, and $g_A = (V, E_A, m)$ is the subgraph induced by A. g_A is the restriction of g to A. A set of edge labels A induces a partition of the graph g into edge disjoint components with the same sets of vertices, such that $g = g_A \cup g_{\Delta-A}$.
For a set of edge labels A, an *A-path* in g is a directed path consisting of A-edges, only. Denote this by A-path(u,v). Thus, A-path(u,v) if and only if there is a directed path from u to v in g_A.

Next we turn to graph grammars. We define a general type with terminal and nonterminal vertex labels, terminal edge labels and directed edges. See e.g. [2, 7, 8, 16, 22].

Definition

A *graph grammar* is a system GG = (N, T ∪ Δ, P, S), where N, T and Δ are the alphabets of nonterminal vertex labels, of terminal vertex labels and of terminal edge labels, respectively. S ∈ N is the axiom, and P is a finite set of productions. A *production* is of the form p = (A, R, C), where A ∈ N is the left-hand side, R is the right-hand side and is a nonempty graph with vertex labels from N ∪ T and edge labels from Δ, and C is the *connection relation* of p. C consists of tuples (a, d, B, a', d', u), where a, a' ∈ Δ are edge labels, d, d' ∈ {in, out} specify the direction of the edges, B ∈ N ∪ T is a vertex label, and u ∈ V(R) is a target vertex in R.

Definition

A direct derivation step g ⇒$_{(v, p)}$ g' is informally defined as follows. Select a vertex v of g with label A and a production p = (A, R, C) and replace v by (an isomorphic copy of) R. Then establish connections between the neighbors of v and vertices of R as specified by the connection relation C. Thus V(g') = V(g) - {v} ∪ V(R) with V(g) ∩ V(R) = Ø and e = (t, a, u) ∈ E(g') if and only if e ∈ E(g) and u ≠ v ≠ t or e ∈ E(R) or t ∈ neigh(v, g), u ∈ V(R), (b, d', B, a, d, u) ∈ C, m(g)(t) = B, there is an b-edge with direction d' from v to t and (t, a, u) has direction d. See [2] for a formal definition.

The *language* generated by GG consists of all graphs with terminal vertex labels and derivable from the axiom, L(GG) = {g | S ⇒* g, g = (V, E, m) and m(v) ∈ T for every v ∈ V}.

For placement graph grammars we distinguish base edes and placement edges. As a simplification, base edges are considered as undireced and unlabeled. For notational convenience we use the label "*". Placement edges are labeled by {x,y}. Their purpose is expressed by the notions of a partially ordered graph and its legal grid embedding. In a natural way, the edge labels {*} and {x,y} induce a partition of a graph g = g$_*$ ∪ g$_{xy}$ into a base component and a placement component. We shall omit the subscript "*" and write the subscript xy for {x, y} and x for {x}.

Definition

A graph g$_{xy}$ with edge label alphabet {x,y} is a *partially ordered graph,* if
      (i) g$_x$ and g$_y$ are acyclic
      (ii) for every pair of vertices (u, v) with u ≠ v,
            x-path(u, v) or x-path(v, u) or y-path(u, v) or y-path(v, u) holds.
A graph g = g$_*$ ∪ g$_{xy}$ is a *placement graph,* if g$_{xy}$ is partially ordered.
g$_{xy}$ is the *placement component associated* with g$_*$.

Consider some basic properties of partially ordered graphs g$_{xy}$.
The acyclicity implies that x-path(u, v) holds iff x-path(v, u) does not hold, and y-path(u, v) holds iff y-path(v, u) does not holds. Clearly, there may be a cycle of the form x-path(u, v), y-path(v, u).

Moreover, g_{xy} is connected when the direction of the x-edges and y-edges is ignored. From the acyclicity, g_x has vertices without incoming x-edges and vertices without outgoing x-edges. These are called *x-sources* and *x-sinks*, respectively. Accordingly, g_y has *y-sources* and *y-sinks*. Notice that a vertex may belong to more than one of these classes. Hence, g_x and g_y can be ordered level by level. These orders induce a lexicographic order on the vertices of g_{xy}, by ordering vertices first according to their y-level, and for the same y-level according to the x-level.

If x is interpreted as "from left to right" and y is interpreted as "from top to bottom", then (i) means that the placement component g_{xy} is consistent with this interpretation. (ii) is a completeness condition. It says that for every pair of vertices there is a directed x-path from u to v or from v to u or a directed y-path from u to v or from v to u. (ii) means that for every pair of vertices {u,v} either "u lies left of v" or "u lies above v", or vice versa. If u is positioned at the origin of the plane or the grid, then x-path(u, v) implies that v lies in the right halfspace, y-path(u, v) implies that v lies in the lower halfspace and if both x-path(u, v) and y-path(u,v) holds, then v lies in the lower right quadrant.

Hence, for every pair of vertices the placement component determines one dimension for their relationship. The other dimension may be left unspecified. For a concrete grid drawing, both dimensions must be determined and moreover, there must be fixed coordinates for the vertices. Thus g_{xy} gives some partial information on the relative positions of the vertices. This information is universal in the sense that it given for any pair of distinct vertices. It is incomplete, since we know only a "right of" *or* "below", i.e. only the x-relation or the y-relation. Notice that there is no specification of the routing of the edges.

The interpretation of {x,y} is made precise by legal grid embeddings.

Definition

Let $g \cup g_{xy}$ be a placement graph. Let g' be a grid graph of an appropriate size, whose vertices are specified by their coordinates such that v = (v.x, v.y).

A grid embedding f : g → g' of the base component g is a *legal placement* w.r.t g_{xy} if for every pair of vertices (u, v)

x-path(u, v) implies f(u).x < f(v).x and y-path(u, v) implies f(u).y < f(v).y.

Legal embeddings are strict in the sense that an x-path from vertex u to vertex v with vertices u = $v_0, v_1, ..., v_{k-1}, v_k = v$ implies that each v_i has its own x-coordinate. Hence, u and v are at least k units apart in x-dimension. Conversely, if every vertex has its own x- and y-coordinate, then a legal placement is possible, even if there is an x-path and a y-path between any pair of vertices, in which case we place the vertices on the second diagonal.

Lemma 1

Let g be a graph and let g_{xy} be a placement component associated with g.
Then there is a legal placement of g w.r.t g_{xy} on an (n × n) grid with n = |g|.

Example 1

Consider a grid. For each unit square, let the horizontal lines be x-edges and the vertical lines be y-edges, and let each diagonal by both and x-edge and a y-edge.

Every graph g drawn in the discrete plane has an associated placement component g_{xy} such that the drawing is a legal placement. Let $g_{xy} = (V(g), E_{xy}, m(g))$, where $(u, x, v) \in E_{xy}$, if u is drawn to the left of v, and $(u, y, v) \in E_{xy}$, if u is drawn on top of v.

In general, this is an "over-specification". G_{xy} has more x-edges and y-edges than required by the completeness property (ii) of partially ordered graphs.

Example 2

Consider partially ordered graphs $g_{xy} = g_x \cup g_y$, where g_y is a binary tree and x-edges are added to satisfy property (ii). E.g., draw an x-edge from vertex u to vertex v, if u and v belong to neighbored subtrees and u is on the rigthmost path on the left subtree and v on the leftmost path of the right subtree. See Fig. 3. (x-edges are shown as dotted lines).

We are now ready to define placement graph grammars. According to the edges $\Delta = \{*, x, y\}$ a placement grammar splits into a base- or "*"-component and a placement or "xy"-component. Both have special properties.

Definition

A *placement graph grammar* is a graph grammar PGG = $(N, T \cup \Delta, P, S)$ with $\Delta = \{*, x, y\}$.

$\{*\}$ and $\{x, y\}$ split PGG into a *base graph grammar* GG = $(N, T \cup \{*\}, P', S)$ and a *placement component* $G_{xy} = (N, T \cup \{x, y\}, P_{xy}, S)$ where P' and P_{xy} consist of the base and placement parts of the productions.

For every production $p = (A, R, C)$ let $p' = (A, R', C')$ and $p_{xy} = (A, R_{xy}, C_{xy})$ be such that $R = R' \cup R_{xy}$ is a placement graph and $C = C' \cup C_{xy}$ is split according to the edge labels.

C' contains all pairs (B, u) (or (a, d, B, a', d', u) with $a = a' = *$) and C_{xy} contains all tuples (a, d, B, a', d', u) with $a, a' \in \{x, y\}$.

Moreover, C' is supposed to be neighborhood preserving (see [21]). E.g. it contains a pair (B, u) for every vertex label B.

The xy-connections are special. They preserve the label and the direction, and an incoming x-edge is transferred to each x-source of the right hand side R_{xy} and an outgoing x-edge is transferred to each x-sink, and similarly for the y-edges. This type of embedding has been used by Slisenko [23].

Formally, (a, d, B, a', d', u) in C_{xy} implies $a = a'$ and $d = d'$. If $(a, d) = (x, in)$, then u is an x-source of R_{xy} and (x, in, B, x, in, u') is in C_{xy} for every x-source u' of R_{xy}. If $(a, d) = (x, out)$, then u is a x-sink and (x, out, B, x, out, u') is in C_{xy} for every x-sink u' of R_{xy}. The according relations hold for (y, in) and (y, out).

Thus C_{xy} is completely specified by the graph R_{xy}.

Example 3

Consider some simple placement graph grammars for binary trees. For the producions we use an elegant graphic description as illustrated in Fig. 4. The base productions (a) and (b) are as follows: The left hand side is drawn as an outer box with label A. The right hand side graph is shown inside the box. Each connection is indicated by a line from a vertex label outside the box and is connected to its target vertex inside the box. In general, these lines may be labeled and directed.

Some choices for a placement component for (a) are shown in (c) and (d) with full lines for y-edges and dotted lines for x-edges. The a-vertex of production (b) is simultaneously a source and sink for x and y. Using (c) as placement component, the tree of Fig. 3 will be drawn as by an inorder traversal with every vertex having its own x-coordinate. The constructed area minimal placement has maximal width. Using (d), the grammar draws binary trees level by level such that each vertex of a right subtree appears properly to the right of each vertex of a left subtree. The root is arbitrarily on top of each subtree. Fig. 5 illustrates some legal and area minimal grig drawings of the graoh of Fig. 3.

The purpose of placement graph grammars are polynomial time drawings of graphs. This shall be achieved by a polynomial time syntax analysis of a graph g in the base graph grammar and the translation into the grid under the control of an associated placement component g_{xy}. In general, graph grammars have an intractable membership problem, see [2]. Hence, we need further restrictions on placement graph grammars.

As a global property, we require that GG is confluent (or has the finite Chruch Rosser property). A graph grammar is *confluent,* if rewriting steps on distinct vertices can be done in any order. Formally, $g \Rightarrow_{(u,p)} h \Rightarrow_{(v,q)} k$ and $g \Rightarrow_{(v,q)} h' \Rightarrow_{(u,p)} k'$ with distinct vertices $u \neq v$ implies $k = k'$. See [2, 4, 8] for further discussions.

The connection relation of the placement component inherits an incoming x-edge to each x-source of the right hand side graph R_{xy} and inherits each outgoing x-edge to each x-sink, and similarly for the y-edges. It is easily seen that this implies confluence and guarantees that no connections go lost.

Lemma 2

If PGG = GG $\cup$ GG$_{xy}$ is a placement graph grammar, then PGG, GG, and GG$_{xy}$ are confluent and neighborhood preserving.

The placement-component grammar GG$_{xy}$ generates partially ordered graphs. This follows from the fact that each right hand side graph R_{xy} is a partially ordered graph and the connection relation C_{xy} preserves the acyclicity of the induced x- and y- subgraphs and the completeness condition (ii).

Lemma 3

If a graph g' is generated by a placement graph grammar PGG,
then $g' = g \cup g_{xy}$ is a placement graph. The placement component g_{xy} is connected.

Lemma 4

Let PGG = GG $\cup$ GG$_{xy}$ be a placement graph grammar. Let S $\Rightarrow$* g with g $\in$ L(GG) be a derivation of some graph g in the base graph grammar GG. Then there is a unique placement component g$_{xy}$ associated with the derivation such that g $\cup$ g$_{xy}$ $\in$ L(PGG).

For an analysis of languages of graphs generated by placement graph grammars we can employ the rich theory of graph grammars. See e.g. [7]. In parallel with contextfree string grammars, our graph grammars are recursive. The recursion stems from the repeated replacement of some A-labeled vertex and gives raise to a "pumping lemma", see [13]. Schuster [22] has shown a very strong and powerful structural property of graphs with bounded degree that are generated by a confluent graph grammar. These graphs have a bounded separator, i.e., they can recursively be partitioned into two parts of almost the same size and this partition cuts only a bounded number of edges. See [27].

THEOREM 2

Let PGG = GG $\cup$ GG$_{xy}$ be a placement graph grammar.

Let g be a base graph, which is connected and of bounded degree and g $\in$ L(GG).

There exists a polynomial time algorithm which for every such base graph g constructs a grid embedding f : g $\rightarrow$ g', such that f is legal for some xy-placement component g$_{xy}$ associated with g and f(g) has minimal area.

Proof.

We consider an extension of the bottom up parsing algorithm of Rozenberg and Welzl [21]. (For an extension to arbitrary confluent base graph grammars see [2, 22]).

Let PPG = (N, T $\cup$ {*, x, y}, P, S) and g = (V(g), E(g), m(g)) .

A placement parsing element for g is a tuple PPE = (A, V, p, w, h, π_1,..., π_r), where A $\in$ N $\cup$ T is a vertex label, V $\subseteq$ V(g) is a subset of the vertices of g, p = (A, R, C) with p = p' $\cup$ p$_{xy}$ is a production, w and h are integers, and π_1,..., π_r are (pointers to) PPEs with r = |R|.

The meaning of a placement parsing element is as follows: there is a derivation A $\Rightarrow$* d in GG with d = ind(V, g) and there is a placement component d$_{xy}$ associated with this derivation which induces a legal placement of ind(g, V) in a rectangle of width w and height h. w and h are minimal for all derivations A $\Rightarrow$* d in PGG. The production p is applied to A and π_1,..., π_r are PPEs addressed by this step, and recursively trace the history of the derivation process.

The set of PPEs is recursively constructed as follows:

(1) For every vertex v $\in$ V(g) (m(v), {v}, -, 1, 1, -) is a PPE; ("-" is a dummy element or nil).

(2) Let p = (A, R, C) be a production with p = p' $\cup$ p$_{xy}$. Let V(R) = {v$_1$,...,v$_r$}.

Let q$_1$,...,q$_r$ be PPEs with q$_i$ = (A$_i$, V$_i$, p$_i$, w$_i$, h$_i$, π_{i1},..., π_{iri}) and V$_i$ $\cap$ V$_j$, = $\emptyset$ for i $\neq$ j.

Associate each PPE q$_i$ with some vertex v$_i$ of R by some bijection, 1 $\leq$ i $\leq$ r .

Let V = $\cup_i$ V$_i$. V is the new goal.

For (A, V) and {q$_1$,...,q$_r$} define graphs α and β with

$V(\alpha) = \{(A,V)\} \cup neigh(V, g)$, $E(\alpha) = \{((A,V), u) \mid u \in neigh(V, g)\}$ and

$m(\alpha)((A, V)) = A$, and $m(\alpha)(u) = m(u)$ for $u \in neigh(V, g)$.

$V(\beta) = \{q_1,...,q_r\} \cup (\cup_i\ neigh(V_i, g) - \cup_i V_i)$,

$E(\beta) = \{(q_i, q_j) \mid$ for $i \neq j$ there is $u \in V_i$, $u' \in V_j$ and $(u, u') \in E(g)\}\ \cup$

$\qquad \{(q_i, u) \mid u \in neigh(V_i, g) - \cup_i V_i\}$, and

$m(\beta)(q_i) = A_i$, $1 \leq i \leq r$, $m(\beta)(u) = m(u)$, otherwise.

Let $\alpha \Rightarrow_{((A, V),\ p')} \beta$

be a neighborhood preserving derivation step in GG such that q_i is the copy of v_i, $1 \leq i \leq r$.

Let $w = \max\{$ weighted x-path length in R_{xy} with $weight(v_i) = w_i\}$

and $h = \max\{$ weighted y-path length in R_{xy} with $weight(v_i) = h_i\}$.

If there is no PPE $(A', V', p', w', h', \pi_1',..., \pi_r')$ with $A' = A$, $V' = V$, $w' \leq w$ and $h' \leq h$

then $(A, V, p, w, h, \pi_1,..., \pi_r)$ is a (new) PPE

(and delete all (old) PPEs with $A' = A$, $V' = V$, and $(w, h) < (w', h')$).

(3) There are no other PPEs.

The correctness and complexity of this procedure can be proved along the lines of the proof of Thereom 6.3 in [21]. See also [2, 22].

(i) $g \in L(GG)$ if and only if there is a PPE $(A, V, p, w, h, \pi_1,..., \pi_r)$ with $A = S$ and $V = V(g)$.

(ii) If $(A, V, p, w, h, \pi_1,..., \pi_r)$ is a PPE,

then there is a derivation $A \Rightarrow^* d$ with $d = ind(V, g)$, and there is a placement component d_{xy} associated with the derivation such that d has a grid embedding with area $(w*h)$, which is legal for d_{xy}, and (w, h) are minimal.

(iii) If g is connected and of bounded degree, then there are only polynomially many PPEs.

This follows from the fact that there are only polynomially many components (A, v) for PPEs, and for every PPE, $(A, V, p, w, h, \pi_1,..., \pi_r)$, $1 \leq w, h \leq |g|$ and $r = O(1)$.

(iv) The set of all PPEs can be computed in time polynomial in the number of PPEs.

Thus the running time of the procedure is polynomial.

(v) For $g \in L(GG)$, given a PPE $(S, V(g), p, w, h, \pi_1,..., \pi_r)$ (which is unique in the components $(S, V(g), w, h)$), a minimal area grid embedding $f: g \to g'$ can effectively be constructed in linear time by tracing back the derivation via p and $\pi_1,..., \pi_r$.

If PGG is ambiguous there can be several area minimal legal placements. $\qquad$ □

This theorem should be seen as a first step towards a theory of syntax directed constructions of grid drawings.

Acknowlegement:

I thank the anonymous referee for his careful reading of the submitted manuscript and his useful comments.

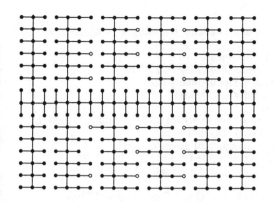

Fig. 1

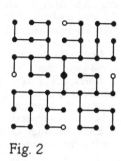

Fig. 2

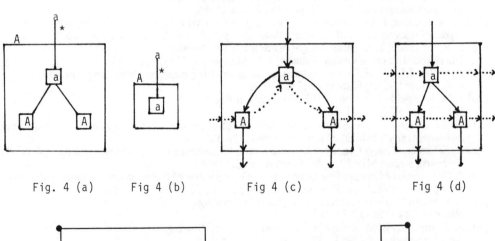

Fig. 3

Fig. 4 (a) Fig 4 (b) Fig 4 (c) Fig 4 (d)

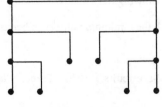

Fig. 5 (a)

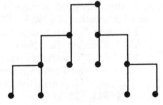

Fig. 5 (b)

References:

[1] S.N. Bhatt, S. S. Cosmadakis, "The complexity of minimizing wire lenghts in VLSI layouts", Inform. Proc. Letters 25 (1987), 263-267

[2] F.J. Brandenburg, "On polynomial time graph grammars" Proc. 5 STACS, Lecture Notes in Computer Science 294 (1988), 227-236

[3] F.J. Brandenburg, "Nice drawings of graphs are computationally hard", Proc. 7th Interdisciplinary Workshop in Informatics and Psychology, Lecture Notes in Computer Science (to appear)

[4] B. Courcelle, "An axiomatic definition of context-free rewriting and its application to NLC graph grammars", Theoret. Comput. Sci. 55 (1987), 141-181

[5] D. Dolev, F.T. Leighton, H. Trickey, "Planar embedding of planar graphs", Advances in Computing Research, vol .2, F.P. Preparata (ed.), JAI Press Inc. 1984

[6] P. Eades, R. Tamassia, "Algorithms for Drawing Graphs: An Annotated Bibliography" Networks (to appear)

[7] H. Ehrig, M. Nagl, G. Rozenberg, A. Rosenfeld (Eds.), "Graph Grammars and Their Application to Computer Science", Lecture Notes in Computer Science 291 (1987)

[8] J. Engelfriet, G. Leih, E. Welzl, "Boundary graph grammars with dynamic edge relabeling", J. Comput. System Sci. (to appear)

[9] S. Even, "Graph Algorithms", Computer Science Press, Maryland (1979)

[10] M.J. Fischer, M.S. Paterson, "Optimal tree layout", Proc. 12 ACM STOC (1980), 177-189

[11] M.R. Garey, D.S. Johnson, "Computers and Intractability: a Guide to the Theory of NP-Completeness", Freeman and Company, San Francisco (1979)

[12] A. Gregori, "Unit-Length embedding of binary trees on a square grid" Inform. Proc. Letters 31 (1989), 167-173

[13] D. Janssens, G. Rozenberg, "On the structure of node label controlled graph languages", Inform. Sci. 20 (1980), 191-216

[14] D. S. Johnson, "The NP-completeness column: An ongoing guide" J. Algorithms 3 (1982), 89-99

[15] D. S. Johnson, "The NP-completeness column: An ongoing guide" J. Algorithms 5 (1984), 147-160

[16] M. Kaul, "Syntaxanalyse von Graphen bei Präzedenz-Graph-Grammatiken" Technical Report MIP-9610, Universität Passau (1986)

[17] M. R. Kramer, J. van Leeuwen, "The complexity of wire-routing and finding minimum area layouts for arbitrary VLSI circuits", Adv. Comput. Research 2 (1984), 129-146

[18] F. S. Makedon, I. H. Sudborough, "Graph Layout Problems" Überblicke Informationsverarbeitung, B.I., Mannheim (1984), 145-183

[19] Z. Miller, J.B. Orlin, "NP-completeness for minimizing maximum edge length in grid embeddings", J. Algorithms 6 (1985), 10-16

[20] E.M. Reingold, J. Tilford, "Tidier drawing of trees" IEEE Trans SE 7(1981), 223-228

[21] G. Rozenberg, E. Welzl, "Boundary NLC graph grammars -basic definitions, nornal forms and complexity", Inform. Contr. 69 (1986), 136-137

[22] R. Schuster, "Graph Grammatiken und Graph Einbettungen: Algorithmen und Komplexität", Technical Report MIP 8711, Universität Passau, (1987)

[23] A.O. Slisenko, "Context-free grammars as a tool for describing polynomial-time subclasses of hard problems", Inform. Process. Letters 14 (1982), 52-56

[24] J. A. Storer "On minimal-node-cost planar embeddings" Networks 14 (1984), 181-212

[25] K. J. Supowit, E. M. Reingold, "The complexity of drawing trees nicely" Acta Informatica 18 (1983), 377-392

[26] R. Tamassia, "On embedding a graph in the grid with the minimum number of bends" SIAM J. Comput. 16 (1987), 421-444

[27] J. D. Ullman, "Computational Aspects of VLSI" Comput. Science Press, Rockville, Md. (1984)

[28] L. Valiant, "Universality considerations in VLSI circuits", IEEE Trans. Computers, C-30 (1981), 135-140.

BOUNDS TO THE PAGE NUMBER OF PARTIALLY ORDERED SETS[&]

Maciej M. Sysło[@]
Technische Universitaet Berlin, FB 3 – Mathematik
Str. des 17. Juni 135, 1000 Berlin 12

ABSTRACT

In this paper we initiate study of a new poset invariant, *the page number*, well-known for graphs. Lower and upper bounds are derived and then they are used to evaluate and to bound the exact value of the page number for some families of posets. Several problems are posed.

1. Introduction

A book consists of a *spine* (which is a line) and a number of *pages*. Each page is a half-plane that has the spine as its boundary. Thus, any half-plane is a one-page book and a plane with a distinguished (vertical) line is a 2-page book. The embedding of a graph $G=(V,E)$ in a book consists of two steps:

1. place the vertices V on the spine in some order, and
2. assign each edge to one of the pages in such a way, that the edges assigned to one page do not cross.

Instead of taking a linear order of vertices, we may place them on a circle. Then the edges of the graph become chords of the circle and Step 2 is to color the chords with few colors so that no two intersecting chords get the same color – this is equivalent to color the corresponding circle graph, see [2]. It is well-known, that the

---

[&]   This research was partially supported by the grant RP.I.09 from the Institute of Informatics, University of Warsaw.

[@] A Fellow of the Alexander von Humboldt-Stiftung (Bonn). On leave from Institute of Computer Science, University of Wrocław, Przesmyckiego 20, 51151 Wrocław, Poland.

circle graphs are exactly the overlap graphs [4]. Hence, let $\Pi=(\pi_1,$ $\pi_2,\ldots,\pi_n)$ be an order of the vertices of G along the spine. Now, each edge $\{u,v\}$ is interpreted as an open interval I_{uv} whose left and right ends are respectively the smaller and the greater of two integers $\pi^{-1}(u)$ and $\pi^{-1}(v)$. We construct the overlap graph $O(G,\Pi)$ of the family $\mathfrak{F}=\{I_{uv}: \{u,v\}\in E(G)\}$ and Step 2 above is now to color the graph $O(G,\Pi)$ optimally.

The *page number* $pn(G)$ of G is the smallest number k such that G has a book embedding on k pages.

The problem of embedding graphs in books was first investigated by Bernhart and Kainen [1]. The recent interest in such graph embeddings is motivated by their applications in VLSI design of fault-tolerant processor arrays, see Chung et al. [2].

In some applications mentioned in [2], for instance in sorting using stacks and queues in parallel, the set of feasible orderings of vertices on the spine is restricted to a certain subfamily of permutations. If this family coincides with the set of all linear extensions of a partially ordered set then the book embedding problem for general graphs reduces to that for (Hasse) diagrams – covering graphs of ordered sets, see [9].

Let $P=(P,\leq)$ be a partially ordered set (simply called a *poset* and denoted by P) and let $\mathcal{L}(P)$ denote the set of all linear extensions of P (a *linear extension* of P is a total order of P which preserves the relation $\leq$). The *Hasse diagram* $H(P)$ of P is the graph whose vertex set corresponds to P and two vertices $u,v\in P$ are adjacent in $H(P)$ if and only if either u covers v or v covers u in P. A poset P is *connected* if $H(P)$ is a connected graph. In what follows we assume that all posets are connected.

A *book embedding of a poset* P *with respect to* $L\in\mathcal{L}(P)$ is the embedding of $H(P)$ with its vertices (i.e., elements of P) placed on the spine in accordance with L. Let $O(P,L)$ denote the corresponding overlap graph. The *page number* $pn(P,L)$ *of* P *with respect to* L is the smallest number k of pages such that $H(P)$ has a book embedding on k pages, or equivalently

$$pn(P,L) = \chi(O(P,L)), \qquad (1)$$

where χ denotes the chromatic number of a graph.

The *page number* $pn(P)$ of P is defined as follows:

$$pn(P) = \min\{pn(P,L): L \in \mathcal{L}(P)\}.$$

There exist complete characterizations of graphs with page number 1 and 2 [1]. It is also known that $pn(G) \leq 4$ for every planar graph G and this bound is tight (see Yannakakis [12]). From the algorithmic point of view, recognizing if a (planar) graph G has page number 2 is as difficult as finding if a maximal planar graph has a Hamiltonian cycle, which is an NP-complete problem (see Wigderson [11]). Moreover, optimal book embedding of G when the vertex ordering is a part of the problem instance is also NP-complete, since the coloring problem for circle graphs is NP-complete, see [3].

The purpose of this paper is to initiate study of book embeddings of posets. We provide for posets counterparts of some of the results obtained for graphs and present some new ideas and results which hold only for posets. The most important ones are obtained by using lower and upper bounds to the page number.

When this paper was completed, the author received a copy of [10] which also contains some bounds to the page number of posets, however in terms of other invariants.

2. Lower and upper bounds

Let $P=(P, \leq)$ be a poset with n elements and m edges in $H(P)$. Each linear extension $L \in \mathcal{L}(P)$ can be expressed as a direct sum of chains of P, i.e.

$$L = C_0 + C_1 + \ldots + C_k, \tag{2}$$

where C_i ($0 \leq i \leq k$) is a chain of P. The number of breaks (i.e. non-comparabilities) between consecutive elements in L is denoted by $s(P,L)$ and called the *jump number of P with respect to L*. The *jump number $s(P)$* of P is equal to minimum of $s(P,L)$ and the *bump number $b(P)$* of P is equal to maximum of $s(P,L)$, in both cases taken over all L in $\mathcal{L}(P)$. The problem of finding $s(P)$ is NP-complete even for bipartite posets, whereas the bump number can be calculated in polynomial time (see [6]).

Let us consider a book embedding of P with the elements placed on the spine according to L. An edge of $H(P)$ which belongs to a chain of P in L is called a *spine edge*. A spine edge can be placed on any page of a book embedding of P. It is clear that for every two chains C_i and C_j ($i<j$) in (2), no page contains more than one edge between elements in C_i and in C_j. Otherwise, let $e_1=(u,v)$ and $e_2=(x,y)$ be two edges in $H(P)$, where $u,x \in C_i$, $v,y \in C_j$ and $u \leq x$. Then, if $y \leq v$ then (u,v) is a transitive edge and if $u<x$ and $v<y$ then e_1 and e_2 cross. In general, we cannot have a cycle on one page formed by some spine edges and the non-spine edges placed on that page. Therefore, if L consists of $k+1$ chains then a page in any book embedding of P according to L may contain at most k non-spine edges of $H(P)$. The number of spine edges is equal to $n-1-k$, hence

$$\lceil (m - (n - 1 - k))/k \rceil \leq pn(P,L). \tag{3}$$

Thus, we obtain

LEMMA 1. The page number of a poset P satisfies

$$\lceil (m - n + 1)/b(P) \rceil + 1 \leq pn(P). \tag{4}$$

Another lower bound to $pn(P)$ can be derived from (1). Note that an l-clique in $O(P,L)$ corresponds to a special matching $M(P,L)=\{e_1,e_2, ...,e_l\}$ of $H(P)$, where $e_i=(x_i,y_i)$, called a *sequential matching with respect to L*, for which we have $x_1<x_2<...<x_l<y_1<y_2<...<y_l$ in L. It is clear that in any book embedding of P according to L, no two edges of M can be placed on the same page, see also [2]. Hence,

LEMMA 2. The page number of a poset P with respect to $L \in \mathcal{L}(P)$ satisfies

$$\omega(O(P,L)) \leq pn(P,L) \tag{5}$$

where ω denotes the clique number of a graph.

Although the best lower bound of type (3) is obtained for $k=b(P)$ as in Lemma 1 (since we intend to minimize pn), a poset may attain the page number for a linear extension with less than $b(P)$ jumps. For instance, for a poset P in Fig. 1(a), we have $pn(P)=2$ as shown in Fig. 1(b) and $b(P)=5$, but one can easily check that for every linear extension L of P with 5 jumps (in which d must immediately follow a) we have $\omega(O(P,L))=3$. In spite of this example, we have the following conjecture:

CONJECTURE 1. A linear extension with the largest number of jumps can be used to produce a book embedding on a near-optimal number of pages.

We return to discuss relations between $s(P)$ and $pn(P)$ in the next Section.

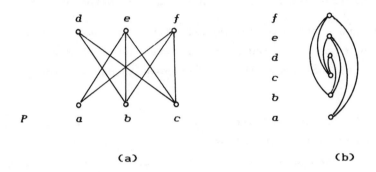

Fig. 1. A poset P and its 2-page embedding

Regarding an upper bound to pn, similarly as for graphs [1], we have only a trivial one which follows from an observation that, given a vertex v, one can always place on one page all edges of $H(P)$ incident with v.

LEMMA 3. The page number of a poset P satisfies

$$pn(P) \leq c(P),$$

where $c(P)$ denotes the size of a smallest vertex cover in $H(P)$.

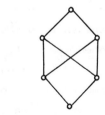

Fig. 2. A 3-page poset with no sequential 3-matchings

Figure 2 shows a poset Q for which the lower bound of Lemma 1 is equal to 3 (in fact, all linear extensions of Q have the same number of jumps) and the upper bound of Lemma 3 is also equal to 3. Hence, $pn(Q)=3$. This is also an example, that the bound in Lemma 2 in not tight, since Q contains no sequential 3-matching. i.e. $\omega(O(P,L))<3$ for every $L\in\mathfrak{L}(P)$.

It turns out that the lower bound of Lemma 2 is quite helpful in evaluating and calculating the page number. The clique number of an overlap (or circle) graph can be found by a polynomial-time algorithm due to Gavril(1973), see [4] and we pose the following problem

Problem 1. Design a polynomial-time algorithm for the following problem: given a poset P, find a linear extension L of P with the smallest $\omega(O(P,L))$, or prove that the corresponding decision problem is NP-complete.

Many characterization and algorithmic problems for graphs can be simplified by restricting the class of objects to 2-connected graphs. In particular, it was shown in [1] that the page number of a graph is equal to the maximum of the page number of its 2-connected components. The proof of this fact easily follows from an obvious observation that for every fixed vertex v of a graph G, there is an optimal book embedding of G in which v is the highest element on the spine (and another embedding in which v is the lowest one). For a poset P, we may only expect that its maximal elements can be placed at the top or minimal elements - at the bottom. Unfortunately, even this is not true in general. Let us consider a poset P shown in Fig. 3(a) which has page number 2 (see Fig. 3(b)). We now prove that $pn(P,L) \geq 3$ for every linear extension L of P which terminates with y or z. Without loss of generality we may assume that the elements $\{a,b,c,y,z\}$ appear in L in the order $(abcyz)$. The edges $\{(a,y),(b,z)\}$, $\{(a,y),(c,z)\}$ and $\{(b,y),(c,z)\}$ form sequential 2-matchings with respect to L. Therefore, if we want to have a 2-page embedding of P, we have to place (a,y) and (b,y) on one page and (b,z) and (c,z) on the other. If such a partial configuration is to be augmented to a complete 2-page embedding of P, then x must be placed above z. Hence, $pn(P,L)$ is equal

to at least 3 for every L

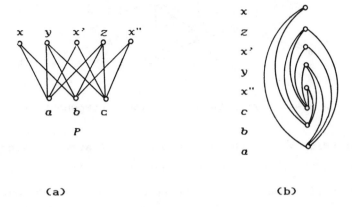

(a)   (b)

Fig. 3. Another poset P and its 2-page embedding

which terminates with z. We argue similarly in the case of y as the last element.

We now derive a decomposition lemma for the page number of posets. Let P be a poset. A subposet Q of P is a *component* of P if the subgraph of $H(P)$ corresponding to Q is a block in $H(P)$ (i.e., contains no cut-vertices) and Q contains exactly one cut-vertex of P. A component Q of P is *simple* if this cut-vertex is either the only maximal element or the only minimal element of Q. The poset in Fig. 4 contains three components, none of which however is simple. As a counterpart of the decomposition theorem for the page number of graphs we have the following equality.

LEMMA 4. For every poset P we have

$$pn(P) = \max\{\max\{pn(Q):Q\in\mathcal{P}\}, pn(P-U_{Q\in\mathcal{P}}Q)\}, \qquad (6)$$

where $\mathcal{P}$ is a set of simple components of P.

PROOF: Let P be a poset, Q be a simple component of P, and z be a cut-vertex of P which belongs to Q. Without loss of generality we may assume that z is a minimal element in Q. In every book embedding of Q, z is its lowest element on the spine. Moreover, in any book embedding of $P-Q$, one can make enough room on the spine to place all the elements of Q

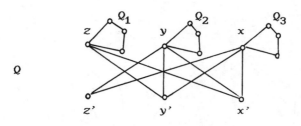

Fig. 4. A poset Q with three non-simple components

between z (including z) and the next element of P above z. Hence, by the inductive argument, we reach the equality.

We now show that Lemma 4 may not be true for a family $\mathscr{F}$ containing non-simple components. Let us consider the poset in Fig. 4, whose diamond components Q_1, Q_2, Q_3 (see Fig. 7) and $Q_0 = Q - (Q_1 \cup Q_2 \cup Q_3)$ are of page number 2, and we shall show that $pn(Q) = 3$. The proof consists of two steps. First, we demonstrate that for every linear extension L of Q such that Q_0 has a 2-page embedding with respect to L: (1) there is an element among x, y, z which has its edges from Q_0 embedded on both pages, and (2) this element is followed in L by at least one of the two other elements. To show (1), let us assume that each of x, y, z has all its edges in Q_0 embedded on one page. Hence, in all possible cases, there is a page which contains a cycle - a contradiction (see Theorem 1).

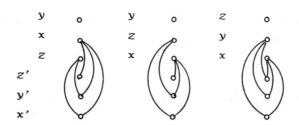

Fig. 5. Attempts to embed P of Fig. 4 on two pages

Condition (2) can be checked by inspection. It is sufficient to consider three positions for z and assume that each of the two first elements of x, y, z in L have all its edges on one of the pages. No such configuration however can be augmented to a 2-page embedding of the whole Q_0, see Fig. 5.

To complete the proof, it is sufficient to show that for an element of Q_0 which satisfies (1) and (2), there is not enough room on the two pages to add the diamond attached to it in Q. To this end, one has to consider all possible 2-page embeddings of a diamond (see Fig. 6) and notice that for any of them, there will be two elements of Q_0 adjacent in $H(Q)$, one inside and one outside of the embedded diamond.

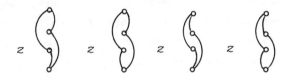

Fig. 6. Two-page embeddings of a diamond

Using the above Lemmas, in the next Section we evaluate the page number for some classes of posets.

3. Page number of some posets

It is clear that a graph embedded on one or two pages must be planar. Bernhart and Kainen [1] completely characterized one- and 2-page graphs as those which are outerplanar or which are subgraphs of Hamiltonian planar graphs, respectively. The latter fact, together with the NP-copleteness of the Hamiltonian cycle problem for maximal planar graphs [11] provides a straightforward NP-completeness proof of the recognition problem of 2-page embeddable graphs.

In this section, we completely characterize 1-page posets and provide some families of 2-page embaddable ones. A complete characterization of 2-page posets and complexity of the recognition problem for k-page posets remain open.

It follows from Lemma 1, that if $pn(P)=1$ than $m-n+1=0$. Note that the Hasse diagram $H(P)$ of a poset P satisfies $m-n+1=0$ if and only if it (as a graph) contains no cycles, or equivalently, if it is a tree.

On the other hand, if P is a 2-element chain then its page number is 1 and applying Lemma 4 inductively one can easily see that every tree can be embedded on one page. Hence,

THEOREM 2. A poset P is of page number 1 if and only if its Hasse diagram $H(P)$ is a tree.

A graph G is *unicyclic* if it contains exactly one cycle.

THEOREM 1. A poset P whose Hasse diagram is unicyclic has page number two.

PROOF: If $H(P)$ is unicyclic then $m-n+1=1$. Therefore, by Lemma 1, $pn(P) \geq 2$. On the other hand, if e is an edge of the cycle in $H(P)$, then the graph $G=H(P)-e$ is a tree. We can embedd G on one page and then add e on a new one. Hence, $pn(P)=2$.

Fig. 7. The diamond D and poset Y

It follows from Theorem 2 that every crown can be embedded on two pages.

Notice that a graph can be the Hasse diagram of many non-isomorphic posets. For instance, graph $K_{1,3}$ is the diagram of the Y poset (see Fig. 7), of the complete bipartite poset consisting of one minimal element and three maximal elements, and of the complete bipartite poset consisting of three minimal elements and one maximal element. A number i assigned to posets is called a *diagram invariant* if $i(P)=i(Q)$ for every two posets P and Q with isomorphic Hasse diagrams. Theorems 1 and 2 imply the following partial answer to another question posed by Nowakowski [9], whether the page number is a diagram invariant:

COROLLARY 1. The page number is a diagram invariant in the class of posets with at most one cycle in their diagrams.

Unfortunately, this result cannot be extended even to the family of all posets with page number 2. For instance, as we showed in the previous section, the poset Q in Fig. 4 is of page number 3 and the poset with the isomorphic diagram in which the diamonds share with Q_0 their minimal or maximal elements, by Lemma 4, has page number 2.

A number i assigned to posets is called a *comparability invariant* if $i(P)=i(Q)$ for every two posets P and Q with isomorphic comparability graphs. A number of important poset invariants, e.g., the jump and the bump numbers, are comparability invariants, see [5] for details. The page number however is not a comparability invariant as show the posets in Fig. 7: their comparability graphs are isomorphic to K_4-e, but $pn(Y)=1$ and $pn(D)=2$.

The next posets of interest are complete bi- and tri-partite posets. Let $P_{k,l}$ denote the complete bipartite poset with k minimal elements and l maximal elements.

THEOREM 3. We have $pn(P_{k,l}) = \min\{k,l\}$.

PROOF : Let us assume that $\min\{k,l\}=k$. The inequality $pn(P_{k,l}) \leq k$ follows from Lemma 3. On the other hand, one can easily see that every linear extension of $P_{k,l}$ contains a sequential k-matching, hence, by Lemma 2, $k \leq pn(P_{k,l})$.

Let $P_{j,k,l}$ denote a complete tripartite poset with j minimal elements, l maximal elements and k elements, each of which is greater than every minimal and smaller than every maximal one.

THEOREM 4. We have $pn(P_{j,k,l}) = \min\{j+l,k\}$.

PROOF : Let us denote $r=\min\{j+l,k\}$. The covering number $c(H(P_{j,k,l}))$ is not greater than r, since we can cover all the edges either by all minimal and maximal elements or by the elements which lie in between. On the other hand, the graph $H(P_{j,k,l})$ contains a sequential r-matching with respect to every linear extension of $P_{j,k,l}$, which has r end-vertices on the middle level and the other vertices on the other two levels.

We now turn again our attention to relations between the jump and the page numbers of a poset.

THEOREM 5 (Le Tu Quoc Hung). If P is a poset of jump number 1 then $pn(P) = 2$.

PROOF : Figure 8 shows a structure of a connected poset of jump number 1. The elements of P can be partitioned into two chains C_1 and C_2. Moreover, all the other covering relations are of the form $x<y$, where $x \in C_1$ and $y \in C_2$, and for two such relations $x_1<y_1$ and $x_2<y_2$, where $x_1, x_2 \in C_1$ and $y_1, y_2 \in C_2$, we have $x_1<x_2$ and $y_1<y_2$. Assume that a poset P contains k covering relations which are not in $C_0 \cup C_1$. Then P has a sequential k-matching with respect to the linear extension $C_0 + C_1$. To avoid such a matching, we build a linear extension of P which uses the relations of that matching.

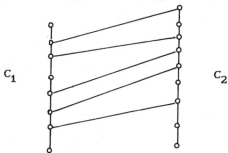

C_1 $\qquad\qquad\qquad\qquad\qquad\qquad$ C_2

Fig. 8. General structure of a poset of jump number 1

Formally, let us define

$$L = D_1 + D_2 + \ldots + D_k + D_{k+1} + D_{k+2},$$

where $D_1 = \{p \in C_2 : p < y_1\}$, $D_i = \{p \in P_i : p < y_i\}$ for $i = 2, 3, \ldots, k$, $D_{k+1} = \{p \in P_{k+1} : p < \max C_2\}$, $D_{k+2} = \{p \in P_{k+2} : p < \max C_1\}$ and $P_1 = P$, $P_i = P - U_{j<i} D_j$ ($i = 2, 3, \ldots, k+2$). The chains D_1, D_{k+1} and D_{k+2} may be empty. The only edges which are not included in L correspond to the covering relations of the form $x_i < \min D_{i+2}$ and $\max D_i < y_i$ for $i = 1, 2, \ldots, k$. We embed P on two pages in accordance with L by placing the non-spine edges of the former type on one page and the latter ones — on the other.

It is quite surprising that there is to bound on the page number for posets with the jump number greater than 1. We provide a family of such posets with jump number 2.

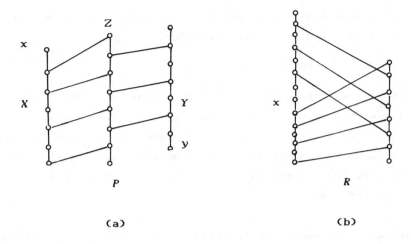

(a)   (b)

Fig. 9. A poset with jump number 2
and with unbounded page number

First, let us consider the poset P shown in Fig. 9(a). It is easy to see that its page number is equal to 2 and is attained for the linear extension constructed similarly as in the proof of Theorem 5. Let us now consider the poset R obtained from P by identifying x with y. We denote the resulting vertex also by x. The poset R has the jump number equal to 2. In every linear extension of R, all elements of X have to precede x and all elements of Y have to follow x. The elements of Z can be placed before or after x. Let Z_1 denote a segment of consecutive elements of Z which appear before x, and $Z_2 = Z - Z_1$. It is clear that the elements in Z_1 form a sequential

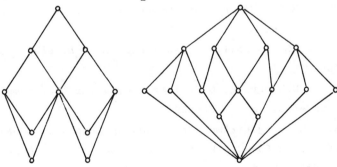

Fig. 10. Planar posets of page number 3 (D. Kelly and
R. Nowakowski, Ottawa 1987).

matching with a subset of Y and the elements in Z_2 - with a subset of X. Hence, $pn(R) \geq \max\{|Z_1|, |Z_2|\}$, and therefore

$$pn(R) \geq \lceil k/2 \rceil.$$

Similarly as for graphs, it is natural to consider the book embeddings of planar posets [8]. It is obvious that every 2-page poset is planar [8]. Some planar posets of page number 3 are shown in Figs. 4 and 10. It is quite easy to construct a planar posets which needs 4 pages (Le Tu Quoc Hung, 1989) but we are not aware of any planar poset which requires 5 pages.

Acknowledgements. The author is indebted to his colleagues Le Tu Quoc Hung, Wiktor Piotrowski and Grzegorz Stachowiak for helpful discussions and thanks the referees for helpful suggestions.

REFERENCES

[1] Bernhart, F., P.C.Kainen, The book thickness of a graph, *J. Combin. Theory* B **27**(1979), 320-321.

[2] Chung, F.R.K., F.T.Leighton, A.L.Rosenberg, Embedding graphs in books: a layout problem with application to VLSI design, *SIAM J. Algebraic and Discrete Methods* **8**(1987), 33-58.

[3] Garey, M.R., D.S.Johnson, G.L.Miller, C.H.Papadimitriou, The complexity of coloring circular arcs and chords, *SIAM J. Algebraic and Discrete Methods* **1**(1980), 216-227.

[4] Golumbic, M.C., *Algorithmic Graph Theory and Perfect Graphs*, Academic Press, New York 1980.

[5] Habib, M., Comparability invariants, *Ann. Discrete Math.* **23**(1984) 371-386.

[6] Habib, M.,R.Moehring, G.Steiner, Computing the bump number is easy, *Order* **5**(1988), 107-129.

[7] Heath, L.S., *Algorithms for embedding graphs in books*, Ph.D. Dissertation, TR 85-028, Dept. Comput. Sci., Univ. North Carolina at Chapel Hill, 1985.

[8] Kelly, D., Fundamentals of planar ordered sets, *Discrete Math.* **63**(1987), 197-216.

[9] Nowakowski, R., Problems 3.7 and 3.8 in: Rival, I.(ed.), *Algorithms and Order*, Kluver, Dordrecht 1989, p. 478.

[10] Nowakowski, R., A.Parker, Ordered sets, pagenumbers and planarity, manuscript, 1989.

[11] Wigderson, A., The complexity of the Hamiltonian circuit problem for maximal planar graphs, Report 298, EECS Dept. Princeton University, 1982.

[12] Yannakakis, M., Four pages are necessary and sufficient for planar graphs, in: Proceedings of the 18th ACM Symposium on Theory of Computing, 1986, pp. 104-108.

Beyond Steiner's problem:
A VLSI oriented generalization

Gabriele Reich[1]   Peter Widmayer[1]

Abstract

We consider a generalized version of Steiner's problem in graphs, motivated by the wire routing phase in physical VLSI design: given a connected, undirected distance graph with groups of required vertices and Steiner vertices, find a shortest connected subgraph containing at least one required vertex of each group. We propose two efficient approximation algorithms computing different approximate solutions in time $O(|E| + |V| \log |V|)$ and in time $O(g \cdot (|E| + |V| \log |V|))$, respectively, where $|E|$ is the number of edges in the given graph, $|V|$ is the number of vertices, and g is the number of groups. The latter algorithm propagates a set of wavefronts with different distances simultaneously through the graph; it is interesting in its own right.

1 Introduction

A number of optimization problems in different application areas can be modelled by Steiner's problem in graphs: given a connected, undirected distance graph with two types of vertices, the *required* vertices and the *Steiner* vertices, find a shortest connected subgraph that contains all required vertices. This subgraph is called a *Steiner minimal tree*; it need not contain all Steiner vertices, but may contain the desired ones. Since Steiner's problem is NP-complete for arbitrary distance graphs [Karp], efficient approximation algorithms have been proposed (see e.g. [Kou81, Kou87, Mehl88, Taka, Widm86, Wu86].

One of the problems that lend themselves to a formulation as Steiner's problem is the wire routing phase in physical VLSI design. Here, after the placement of components on a chip, sets of pins on the component boundaries are to be connected (wired) within the remaining free chip space. For each set of pins sharing the same electrical signal (a *net*), a Steiner minimal tree (or an approximation thereof) is sought, with the pins as required vertices, and the vertices of a grid-like graph, defined by the positions of pins and component boundaries, as Steiner vertices (see e.g. [Wu87]).

In this paper, we consider a more general problem than Steiner's, motivated by the flexibility of pin positions on component boundaries after the placement phase, and we propose two different approximate solutions. Even for components with predetermined interior layout (e.g. from a library), each pin can take any one out of several given positions, representing the fact that entire components placed on a chip can be flipped or rotated by the routing algorithm [Schi, Widm88]. Figure 1 shows an example: (a) depicts the space reserved for a certain component by

---

[1]Authors' address: Institut für Informatik, Universität Freiburg, Rheinstraße 12, 7800 Freiburg, West Germany

the placement algorithm, (b) shows the template for the component with one pin on its boundary, and (c) shows the pin in all possible four positions, depending on how the template is oriented within the reserved chip space. Note that for square components, each pin may have up to eight possible positions.

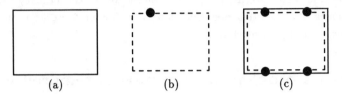

(a)   (b)   (c)

Figure 1: Orienting a component.

For each net, the objective is to find a shortest connecting network, where each component may be oriented as needed. That is, if we regard the set of all possible pin positions of a component as a group of pins, a Steiner minimal tree (approximation) with one pin of each group is sought.

In more general terms, this leaves us with *Steiner's problem with groups of required vertices*: given a connected, undirected distance graph with *groups of required vertices* and Steiner vertices, find a shortest connected subgraph that contains at least one required vertex of each group.

Figure 2 shows an example graph for four groups of four, eight, two, and four pins, respectively, and a shortest connecting subgraph for them. When no distances are shown explicitly in our examples, the distance between vertices is the Manhattan distance with obstacles.

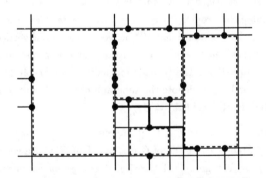

Figure 2: Four groups of pins, and a shortest connecting network.

In spite of its practical significance, to our knowledge Steiner's problem with groups of required vertices has not been investigated as yet. Since Steiner's problem in graphs is a special case of ours with only one vertex per group, the NP-hardness of the former implies the NP-hardness of the latter. Clearly, a Steiner minimal tree approximation algorithm should not be directly applied to the problem with groups of required vertices. In Section 3 of this paper, we propose two different efficient approximation algorithms for this problem. The first algorithm computes a Steiner minimal tree approximation and then applies local changes; it is asymptotically as fast as the underlying Steiner minimal tree approximation of [Mehl88], i.e., runs in time $O(|E| + |V| \log |V|)$, where E is the set of edges and V the set of vertices of the given graph. In the second algorithm, we try to achieve a better approximation. Here, a sequence of shortest paths is computed by simultaneously propagating a set of different wave fronts, a technique that is interesting in its own right. For a graph with g groups of required vertices, the proposed approximate solution can be computed in time $O(g \cdot (|E| + |V| \log |V|))$ in the worst case; for many VLSI wire routing problem instances, the algorithm runs much faster. In Section 2 we define the problem more rigorously. Section 4 offers a conclusion.

2 Steiner's problem with groups of required vertices

Let $G = (V, E, l)$ be an *undirected, connected distance graph*, where V is the set of *vertices* in G, $E \subseteq V \times V$ is the set of *edges* in G, and l is a distance function which maps each edge $e \in E$ to a non-negative number $l(e)$, the *length* of edge e. For some positive integer g, let V be partitioned into sets $R_1, \ldots, R_g$ of *required vertices* and a set S of *Steiner vertices*, with $R_i \neq \emptyset$, $R_i \cap S = \emptyset$, $R_i \cap R_j = \emptyset$, where $1 \leq i, j \leq g$, $i \neq j$, and $\bigcup_{i=1}^{g} R_i \cup S = V$. For $1 \leq i \leq g$, R_i is called *group i of required vertices*; graph G is called a *group Steiner graph*, *gSt graph* for short.

A *simple path* $p(v_1, v_k)$ in G is a sequence of vertices $v_1, v_2, \ldots, v_k$ of V, with $v_i \neq v_j$ for $i \neq j$, such that for all i, $1 \leq i < k$, $(v_i, v_{i+1}) \in E$ is an edge of the graph G. $V(p(v_i, v_j))$ is the set of *vertices on the path* $p(v_i, v_j)$, and $E(p(v_i, v_j))$ is the set of *edges on this path*. An *open path* $op(v_1, v_k)$ is the same as a path $p(v_1, v_k)$, except that v_k does not belong to $V(op(v_1, v_k))$; especially, the edge set of the open path includes the edge to v_k. The *length of a path* $l(p(v_i, v_j))$ is the sum of the lengths of the edges on $p(v_i, v_j)$. A *shortest path* $sp(v_i, v_j)$ is a path with minimum length among all the different paths from v_i to v_j. The *total length* $l(G)$ of a graph G is the sum of the lengths of all edges. The *distance* $d(V, V')$ between two sets of vertices V and V' is defined as $d(V, V') = \min\{l(sp(v, v')) | v \in V, v' \in V'\}$. Note that d is a distance function, i.e., is non-negative, symmetric, and satisfies the triangle inequality. For graph G with vertex set V and edge set E, let $G - V'$ for $V' \subseteq V$ denote the subgraph of G with vertex set $V \setminus V'$ and edge set $E \setminus E'$, where $E' = \{(v_i, v_j) | \{v_i, v_j\} \cap V' \neq \emptyset\}$.

A connected subgraph of G is a *tree* T of G, if the removal of any edge in T causes T to become disconnected. A *group Steiner tree* $gStT$ of G is a tree T of G whose vertex set contains at least one required vertex r_i for each group i, $1 \leq i \leq g$. T is a *group Steiner minimal tree* $gStMT$ if it is a $gStT$ of minimal length. In case each group contains exactly one vertex, Steiner's problem with groups of required vertices degenerates to the classical Steiner problem in graphs, and a group Steiner minimal tree is a Steiner minimal tree.

In the next section, we show how an approximate $gStMT$ for a gSt graph can be computed efficiently.

3 Approximation algorithms for the $gStMT$ problem

We present two algorithms for computing a group Steiner tree, serving as an approximate group Steiner minimal tree, with different runtime characteristics and different approximations. In the first algorithm, we tentatively ignore the fact that required vertices occur in groups, and compute a Steiner tree for the given graph with required vertex set $R = \bigcup_{i=1}^{g} R_i$. In a cleanup phase on the resulting tree, we discard several of the multiple occurrences of required vertices of the same group. This approach is conceptually simple, makes use of known algorithms for approximating Steiner trees in graphs, and is very worst-case efficient; however, it does not always yield a good $gStT$ (see Figure 3(a)). In the second algorithm, we propose to incrementally compute a $gStT$ by repeatedly including into the partial $gStT$ obtained so far a nearest required vertex in another group. This algorithm is asymptotically less efficient than the first one, but tends to deliver better solutions (see Figure 3(b)).

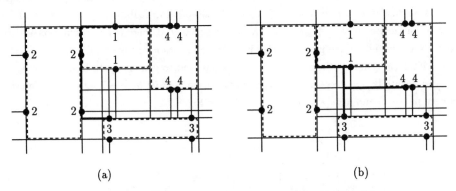

(a)   (b)

Figure 3: Two different approximate group Steiner minimal trees.

3.1 A Steiner minimal tree approximation with cleanup

The structure of the algorithm we propose here is the following:

Algorithm Steiner minimal tree approximation with cleanup

Input: A gSt graph $G = (V, E, l)$, a positive integer g, required vertex sets R_i, $1 \leq i \leq g$, Steiner vertex set S.

Output: A $gStT$ T for G.

Method:

1. Let $G' = (V, E, l)$ with required vertex set $R = \bigcup_{i=1}^{g} R_i$ and Steiner vertex set S be the Steiner graph corresponding to G.

2. Compute a Steiner tree $T = (V_T, E_T, l_T)$ for G', according to some known approximation algorithm for Steiner minimal trees in graphs.

3. Cleanup tree T.

After Step 2 of this algorithm, the computed tree T contains the whole set of required vertices. Since we are interested in a shortest possible tree with at least one vertex per group, we might want to compute a $gStMT$ for tree T. However, in sharp contrast to the complexity of Steiner's problem, the problem of computing a $gStMT$ for a tree is NP-complete, even if there are no Steiner vertices (see [Reich]). Therefore, we aim at an approximate rather than an exact solution.

The idea for the cleanup heuristic is to delete a required vertex that is not the only vertex of its group in the tree — a *redundant* vertex — whenever this can be done without disconnecting the tree. Hence, only redundant leaves may be deleted. Whenever a redundant leaf is deleted, there may be a path of Steiner vertices that becomes unnecessary, since it only connects the leaf with the remainder of the tree. Hence, an entire dangling path can be deleted. The step of deleting a redundant leaf, together with its dangling path, can be carried out repeatedly, until no more deletion is possible.

More precisely, call any vertex that is either a Steiner vertex of degree more than two or a required vertex an *articulation vertex*. Let $A_T \subseteq V_T$ be the set of all articulation vertices of tree T. Two articulation vertices $v, v' \in A_T$, $v \neq v'$, are called *adjacent*, if there is no articulation vertex v'', $v \neq v'' \neq v'$, on the unique path $p(v, v')$. For any articulation vertex v and any edge e incident with v, let $av(v, e)$ denote the articulation vertex adjacent to v on the path from v through e, let $pav(v, e)$ denote the predecessor of $av(v, e)$ on this path, and let $ae(v, e)$ be the *articulation edge* between $pav(v, e)$ and $av(v, e)$. For an example, see Fig. 4.

Figure 4: Two different kinds of articulation vertices, their predecessors, and the articulation edges for leaf r. Required vertices are shown as □, Steiner vertices as •.

The vertices of A_T partition T into non-branching subtrees — the *subpaths* — that will be considered for deletion, starting at the leaves. For leaf r, the open path $op(r, av(r))$ is called the *dangling path* containing r, denoted by $dp(r)$. To allow for an efficient cleanup of T, we do not only keep track of A_T, but also of the set $AE_T \subseteq E_T$ of all articulation edges of T. In the cleanup of T, we repeatedly remove from T all vertices and edges of a longest dangling path $dp(r)$ for some leaf r that is not the only vertex of its group in T. Recall that since $dp(r)$ is an open path, $ae(r)$ will be removed, but $av(r)$ will not.

We use a priority queue Q to store and retrieve dangling paths in order of decreasing length. A *deletemax* operation on Q returns a longest dangling path stored in Q and deletes it from Q. A dangling path is represented uniquely by its leaf, with which we associate the necessary path information. Note that during the computation, required vertices may become leaves, as dangling paths are deleted.

To be more specific, assume that a gSt graph is given in the form of an adjacency list ADL (see e.g. [Mehl84]). This is an efficient representation for many fast Steiner minimal tree approximation algorithms, especially for the algorithm proposed by Mehlhorn (see [Mehl88]). We decide to use Mehlhorn's algorithm for Step 2 of our algorithm, since no generally faster algorithm is known; recall that it runs in time $O(|E| + |V| \log |V|)$. Let us now show how to implement the cleanup step of our algorithm, such that it can be applied to the approximate Steiner minimal tree resulting from Step 2 within time $O(|E| + |V| \log |V|)$, i.e., without increasing the (order of the) computational complexity. Let the Steiner tree computed by the algorithm of Mehlhorn be given in ADL form. For the cleanup, we use a constant number of extra record fields for each edge and each vertex. For vertex v, we will use $v.degree$ for the number of edges currently incident with v, $v.group$ for a reference to a *group description* (empty if v is a Steiner vertex), and $v.pos$ for a reference to the occurrence of v in Q.

For edge $e = (v, v'')$ associated with vertex v in the ADL, $e.l$ is the length of edge e. If $v \in A_T$, $e.av$ stores a reference to $av(v, e)$, $e.ae$ stores a reference to $ae(v, e)$ in the edge list of $e.av$, and $e.lp$ stores $l(p(v, e.av))$. Note that each edge is stored twice in the ADL, once with each of its two incident vertices. This is essential whenever an edge directly connects two articulation vertices: the two entries for the edge point to both vertices. To make the deletion of an edge from the ADL fast, we maintain for each edge $e = (v, v'')$ an additional link from the occurrence of e in the edge list of v to the occurrence of e in the edge list of v'', and vice versa. For leaf r and its only incident edge e in T, $e.lp$ is the key for storing $dp(r)$ in Q.

A *group description* for group g_i contains a field $g_i.degree$ for the sum of $v.degree$ over all vertices v in the group.

Then, the cleanup of T is achieved as follows:

> *cleanup tree $T = (V_T, E_T, l_T)$:*
>   A: Initialize extra fields for T.
>   B: Priority queue $Q :=$ empty.
>   C: **foreach** $v \in A_T$ **do**
>   *store leaf* v in Q.
>   D: **while** Q not empty **do**
> $r := $ *deletemax* from Q;
>   **if** r is a redundant leaf **then**
>   *reduce* T at r.

Let us specify two steps of this algorithm in more detail:

> *store leaf v in Q:*
>   **if** v is a leaf in T **then**
>   (* let e be the edge incident with v *)
>   *store*$(v, e.av, e.lp)$ in Q.

reduce T at r:

    (* r is a redundant leaf in T; let e be the edge incident with r *)

    $r' := r$;

    $r := e.av$;

    remove dangling path from r' to r;

    if two paths have been combined between v' and v'' **then**

        store leaf v' in Q;

        store leaf v'' in Q;

    else

        store leaf r in Q.

Theorem 3.1: Algorithm *Steiner minimal tree approximation with cleanup* computes a *gStT* for a given *gSt* graph $G = (V, E, l)$ in time $O(|E| + |V| \log |V|)$ in the worst case.

Proof: It should be clear that the algorithm computes a *gStT*. To see that the bound on the runtime holds, consider the three main steps of the algorithm in turn. Step 1 is merely a reinterpretation of the input, and Step 2 can be performed by any standard Steiner tree approximation algorithm, the fastest one being Mehlhorn's algorithm with a runtime of $O(|E| + |V| \log |V|)$ in the worst case [Mehl88]. Let us look at Step 3 in more detail. In the cleanup of tree T, the computation of A_T and AE_T, the subpaths between adjacent articulation vertices and their lengths, and the vertex and group degrees can be accomplished in time $O(|V_T|)$ in the worst case, by a depth-first search starting at an arbitrary leaf of T. To substantiate this claim, let us show in more detail how A_T and AE_T can be computed. Let r be the leaf where the depth-first search starts, and let e be the edge incident with r. Let the depth-first search traverse tree T until a Steiner vertex with degree > 2 or a required vertex, say v', is reached via some edge, say e'. Thereby, the lengths of the traversed edges are summed up to *sum*. We keep track of this information as follows:

    $e_r :=$ the entry for edge e in the edge list of vertex r;

    $e_r.av := v'$;

    $e_r.ae := e'$;

    $e_r.lp := sum$;

    $e'_{v'} :=$ the entry for edge e' in the edge list of vertex v';

    $e'_{v'}.av := r$;

    $e'_{v'}.ae := e$;

    $e'_{v'}.lp := sum$;

The priority queue Q can be implemented as a Fibonacci heap on the lengths of the stored paths, with the operations *deletemax*, *insert*, *increasekey*, and *makeheap* obtained from [Fred] by changing the heap organization from a min-heap to a max-heap. Recall that a *deletemax* operation finds and deletes an entry with maximum key from the heap. An *increasekey* operation increases the key of an entry to the specified value, if the new value is greater than the old one. All bounds given in the following for Fibonacci heap operations are in the amortized sense (see [Fred]). Step B of the cleanup is a *makeheap* operation, offered by Fibonacci heaps in time $O(1)$. The *store* operation can then be implemented as follows, where we assume that *store* orders its first and second arguments according to a global order on the vertices, so as to avoid distinguishing between (v, v') and (v', v):

> *store*(v, v', l) in Q:
> > **if** v is already in Q **then**
> > > *increasekey*(v, v', l) in Q
> >
> > **else** *insert*(v, v', l) into Q.

The case that v is already in Q can only occur when two subpaths are combined; then, because the length of the combined paths is larger than that of its components, an *increasekey* operation will suffice.

In an *ADL*, all $|V_T|$ vertices can be visited in $O(|V_T|)$ time. Vertex v is a leaf iff $v.degree = 1$. The parameters for a *store* operation with v can be determined in $O(1)$ time, as well as the fact whether v is already in Q, from the corresponding fields of vertices and edges in the *ADL*.

Since an *insert* as well as an *increasekey* operation costs only $O(1)$ time, Step C altogether consumes $O(|V_T|)$ time. Since vertices and edges are only deleted from T (and never inserted), a vertex that has been deleted from Q will not be inserted into Q again. Hence, the total of all entries ever in Q is $O(|V_T|)$, and therefore the total cost of all *deletemax* operations in Step D is $O(|V_T| \log |V_T|)$. Vertex r is a redundant leaf, iff $r.group.degree \neq r.degree$, a condition that can be tested in constant time. Since this test is only performed for vertices deleted from Q, we get $O(|V_T|)$ time altogether for the test. The operation *remove dangling path* can be implemented in the following way:

> *remove dangling path* from r' to r:
> > *delete* $dp(r')$ from T;
> > decrement $r.degree$ by 1;
> > decrement $r.group.degree$ by 1;
> > decrement $r'.group.degree$ by 1;
> > **if** $r.degree = 2$ **and** r is a Steiner vertex **then**
> > > *combine two paths* at r.

The deletion of a vertex or an edge from T takes only constant time in the *ADL*, since the position in the *ADL* is known from the path traced in T. Because any edge or vertex can be deleted from T only once, this takes at most $O(|V_T|)$ time altogether. Decrementing counters is possible in the same time bound.

Let us now have a closer look at the operation *combine two paths*. Consider the example shown in Figure 5.

For vertex r under consideration, let e and e' be the two incident edges. Then, the *combine* operation can be implemented as follows:

combine two paths at r:

$$e.ae.lp := e.ae.lp + e'.lp;$$
$$e.ae.av := e'.av;$$
$$e.ae.ae := e'.ae;$$
$$e'.ae.lp := e'.ae.lp + e.lp;$$
$$e'.ae.av := e.av;$$
$$e'.ae.ae := e.ae.$$

Here, $e.ae$ and $e'.ae$ is taken with respect to $e.av$ and $e'.av$, respectively. Using this implementation, combining two paths takes only constant time, and therefore the entire tree cleanup can be accomplished in $O(|V_T| \log |V_T|)$ time. $\qquad\square$

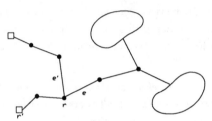

Figure 5: Combining two paths at r.

3.2 A group Steiner minimal tree approximation

Even though the cleanup algorithm operating on Steiner trees ignoring the partition of required vertices into groups is very fast, it need not necessarily be the algorithm of choice, because the quality of the approximate solution may suffer from the straightforward approach. Let us therefore consider an approximation algorithm that is aware of groups in all stages of the tree construction.

Our algorithm for constructing a group Steiner tree repeatedly adds to the tentative tree a shortest additional path to a required vertex whose group does not yet occur in the tree; initially, the tentative tree is just a single required vertex. For Steiner minimal trees in graphs (without groups), an approximation algorithm based on adding shortest paths has been presented in [Taka]. The approximate solution of the algorithm we propose degenerates to the one in [Taka] for graphs without groups; since efficiency is one of our major concerns, our algorithm is faster than that of [Taka], even for graphs without groups.

In general, whenever a shortest additional path from the current tree T to a required vertex r has been found, the path from r to T is added to T, if T does not yet contain a required vertex in the same group as r. Otherwise, adding r is unnecessary, and r is essentially treated as a Steiner vertex. After connecting at least one required vertex per group with T, we check for each leaf r' in T, whether T contains a required vertex r'' of the same group as r'. This can be the case if a shortest path containing r'' has been added to T later than r'. In that case, we delete the dangling path at r' from T, repeatedly if necessary. As a consequence, at most one required vertex per group is a leaf in T, and if T contains a non-leaf required vertex of group g_i, then no vertex of this group is a leaf in T.

Let us now describe intuitively the basic idea of how to implement our algorithm efficiently. Initially, the tentative Steiner tree T consists of an arbitrary required vertex r_1. Around r_1 we grow a wave that hits and passes vertices in increasing order of their distance to r_1. If a wave hits a required vertex r_2, where the group of r_2 is different from that of r_1, the wave stops growing, and a wave offspring begins to grow around the path from r_1 to r_2, i.e., around the new part of T. For wave W_i, we define the distance $d(W_i)$ as the distance between the last vertex hit by W_i and the part of T around which W_i grows. Coexisting waves with different distances have to grow so as to ensure that shortest additional paths will be found. That is, as long as the order of the wave distances does not change, a wave with smallest distance grows. If the distance of a growing wave becomes large enough to dominate the distance of some other wave, the two waves are merged into one.

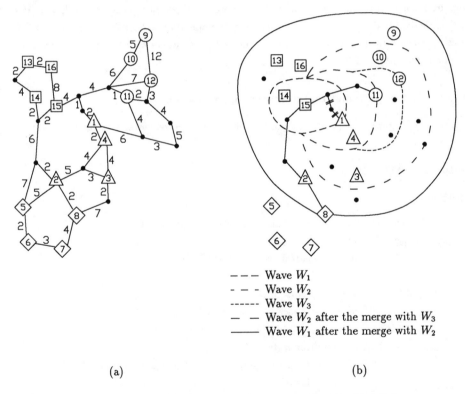

--- Wave W_1
- - - Wave W_2
------ Wave W_3
— — Wave W_2 after the merge with W_3
—— Wave W_1 after the merge with W_2

(a)   (b)

Figure 6: Construction of a Steiner minimal tree approximation by growing waves.

Figure 6 depicts an example graph (a) with waves (b), where each group consists of four required vertices; Steiner vertices are shown as •, the four groups are indicated by □, ○, ◇, and △. Let the starting vertex be the vertex with number 15. The first wave W_1 grows around vertex 15, passing through Steiner vertices, until it reaches required vertex 1; $d(W_1)$ is 7. A new part T_2 of the tree is created, consisting of two Steiner vertices and vertex 1. Around these three vertices W_2 grows until it reaches vertex 11; $d(W_2)$ then is 5. The vertex 11 together with one Steiner vertex builds a new partial tree T_3, around which W_3 starts growing. In W_3 two Steiner vertices and vertex 12 are inserted, and $d(W_3)$ is 5. The next vertex reached during the growth is vertex 10, but the distance of vertex 10 to T_3 is 6, which is more than $d(W_2)$. Therefore, we merge W_2 and W_3 into W_2, and W_2 continues growing. Then, two Steiner vertices and the vertices 3 and 10 are inserted into W_2, each with distance 6, and in the following an additional Steiner vertex with distance 7 is inserted into W_2. The next vertex to be inserted into W_2 is a Steiner vertex with distance 8. That is more than $d(W_1)$, so W_1 and W_2 are merged, and the only remaining wave W_1 around T grows further until vertex 8 is reached. Now a cleanup is performed and during this cleanup vertex 1 and one Steiner vertex are removed from T, because vertex 1 is a redundant leaf.

At any stage of the tree construction, a subset of the vertices has been selected to form the tentative tree T. The vertices of $\bigcup_i W_i$ have been determined to be candidates for shortest additional paths extending T. The subset of the vertices that can be reached through a single edge from some vertex in a wave or in T are stored in a priority queue Q; the key of an entry in Q is the vertex distance to T. The priority queue operation *deletemin* ensures that we always grow the wave with smallest distance, or we merge the two waves with the two smallest distances. Then our algorithm has the following structure:

Algorithm group Steiner minimal tree approximation

Input: A gSt graph $G = (V, E, l)$, a positive integer g, required vertex sets R_i, $1 \leq i \leq g$, Steiner vertex set S.

Output: A $gStT$ T for G.

Method:

    1. *initialize* (* all necessary variables and data structures *);
    2. **do** $g - 1$ **times**
        (* find a path to a nearest required vertex v_j outside T, where
            the group of v_j is not yet selected, and include it into T *)
        2.1 *add subpath neighbors to Q*;
            (* of the most recently selected subpath *)
        2.2 **loop**
            2.2.1 $v_j := $ *deletemin* from Q;
            2.2.2 **if** $v_j \in R$, where the group of v_j has not been selected **then**
                exit **loop**;
            2.2.3 *add vertex neighbors of v_j to Q*;
        2.3 *update tree T by including v_j and its shortest path to T*;
    3. *cleanup tree T*. (* see Section 3.1 *)

Again, we implement the priority queue as a Fibonacci heap (F-heap) [Fred]. In contrast to Section 3.1 we use the F-heap as a min-heap. Hence, a *deletemin* operation deletes an entry with minimum key, and a *decreasekey* operation decreases the key of an entry to the new value. Note, that for the operations *decreasekey* and *delete* (not *deletemin*) the position of the entry in the heap must be known. As mentioned in Section 3.1, we use an *ADL* for storing G. In addition to the extra fields for each vertex, described in Section 3.1, we use the fields $v.key$ for the length of a tentative shortest path from the tentative tree to v, and $v.pred$ for the predecessor of v on a tentative path containing v. For the edges we do not need any extra fields at all. An extra field $g_i.selected$ for each group indicates, whether a vertex of group g_i has already been selected. To implement the partial trees T_i and the tentative tree T, we use unsorted linked lists with pointers to the first and last list elements.

Let us now describe the operations of algorithm *group Steiner minimal tree approximation* in more detail.

initialize:
    $T := \{r_1\};$
    foreach $v \in V$ **do**
        $v.key := \infty;$
        $v.pos := nil;$
    $r_1.key := 0;$
    $i := 1;$ (* i is the number of partial trees *)
    $T_i := T;$
    create a new empty F-heap Q.

add subpath neighbors to Q:
    foreach $v_i \in T_i$ **do**
        add vertex neighbors of v_i to Q.

add vertex neighbors of v_i to Q:
    foreach $(v_i, v_j) \in E$ **do**
        if $(v_j \notin T)$ **and** $(v_j.key > v_i.key + l(v_i, v_j))$ **then**
            $v_j.key := v_i.key + l(v_i, v_j);$
            $v_j.pred := v_i;$
            $put(v_j, v_i, v_j.key).$

$put(v_j, v_i, k)$:
    if v_j not stored in Q **then**
        insert(v_j, v_i, k) into Q;
        set $v_j.pos$ to the position of this entry in Q
    else
        decreasekey(v_j, v_i, k) in Q.

update tree T by including v_j and its shortest path to T:

    (* create a new partial tree T_{i+1} for the shortest path from v_j to T *)

    $T_{i+1} := \emptyset$;

    $v := v_j$;

    repeat

        $T_{i+1} := T_{i+1} \cup \{v\}$;

        $v.key := 0$;

        $v := v.pred$

    until $v \in T$;

    $T := T \cup T_{i+1}$;

    $i := i + 1$.

Theorem 3.2: The algorithm *group Steiner minimal tree approximation* computes a $gStT$ for a given gSt graph $G = (V, E, l)$ with g groups in time $O(g \cdot (|E| + |V| \log |V|))$ in the worst case.

Proof: We will analyze the steps of the algorithm in turn. The operation *initialize* costs $O(|V|)$ time. Steps 2.1 through 2.3 are performed $g - 1$ times. Each time, in Step 2.1 the at most $O(|V|)$ vertices of T_i can be found by scanning the linear list for T_i in time $O(|V|)$; at most $O(|E|)$ neighboring vertices are put into Q. Since each *put* operation, each test, and the necessary bookkeeping can be done in time $O(1)$, Step 2.1 costs at most $O(|E|)$ time. In Step 2.2, the interior of the loop is performed at most $|V|$ times, because each vertex may be stored in Q at most once, in the interior of the loop a vertex is deleted from Q, and no vertex is inserted into Q after it has been deleted, unless tree T is updated, which is not the case within Step 2.2. Since the size of F-heap Q is $O(|V|)$, each *deletemin* can be carried out in $O(\log |V|)$ time. Step 2.2.2 can be carried out in $O(1)$ time by testing if $v_j.group \neq empty$ and if $v_j.group.selected$. Apart from Step 2.2.3, Step 2.2 hence costs at most $O(|V| \log |V|)$ time. By similar arguments, Step 2.2.3 costs no more than $O(|E|)$ time, summed up over the entire Step 2.2. Hence, Step 2.2 within Step 2 costs $O(|E| + |V| \log |V|)$ time. Since Step 2.3 affects $O(|V|)$ vertices each that can be handled in $O(1)$ time per vertex, and inserting an element into a list, representing a partial tree, or merging of two lists can be done in time $O(1)$, the interior of Step 2 costs no more than $O(|E| + |V| \log |V|)$ time. This clearly dominates the time needed in Step 3 for the cleanup of T, which is at most $O(|V| \log |V|)$ (see Theorem 3.1), although we must transform the linked list representing T to the ADL, needed for the cleanup. So, together with the number $g - 1$ of iterations, we get a total runtime of $O(g \cdot (|E| + |V| \log |V|))$. $\qquad\qquad\square$

4 Conclusion

We have considered a generalized version of Steiner's problem in graphs with groups of required vertices, where a shortest connected subgraph containing at least one required vertex of each group is sought. The two approximation algorithms that we have proposed are just a very first step towards investigating the problem. Especially, we are as yet unable to give tight bounds on the qualities of the approximate solutions. From a VLSI designer's point of view, in some settings the freedom of choosing pin positions during the routing phase is even larger than what we have considered: a pin may be located anywhere within specified intervals on a component boundary, or a component may be stretched to some extent to bring pins closer together [Schi]. On the other hand, some graphs representing VLSI designs seem to be structurally simple enough to allow for good approximate solutions.

Acknowledgement

We wish to thank Wolfgang Rosenstiel for bringing the VLSI problem to our attention and for many valuable discussions, Stefan Schiller for sharing with us his practical experience in implementing a different approximation algorithm, and Andreas Hutflesz for his constructive criticism.

References

[Fred]   M.L. Fredman, R.E. Tarjan:
           Fibonacci Heaps and Their Uses in Improved Network Optimization Algorithms, Journal of the ACM, Vol. 34, 1987, 596 – 615.

[Karp]   R.M. Karp:
           Reducibility among Combinatorial Problems, Complexity of Computer Computations, New York, 1972, 85 – 103.

[Kou81]   L. Kou, G. Markowsky, L. Berman:
           A Fast Algorithm for Steiner Trees, Acta Informatica, Vol. 15, 1981, 141 – 145.

[Kou87]   L. Kou, K. Makki:
           An Even Faster Approximation Algorithm for the Steiner Tree Problem in Graphs, 18th Southeastern Conference on Combinatorics, Graph Theory, and Computing, 1987, 147 – 154.

[Mehl84]   K. Mehlhorn:
           Data Structures and Algorithms, Vol. 2: Graph Algorithms and NP-Completeness, Springer-Verlag, 1984.

[Mehl88]   K. Mehlhorn:
           A Faster Approximation Algorithm for the Steiner Problem in Graphs, Information Processing Letters, Vol. 27, 1988, 125 – 128.

[Reich]   G. Reich, P. Widmayer:
           On Minimum Group Spanning Trees, Technical Report, Institut für Informatik, Universität Freiburg, 1989.

[Schi]   S. Schiller:
           Yet Another Floorplanner, Diplomarbeit, Institut für Angewandte Informatik und Formale Beschreibungsverfahren, Universität Karlsruhe, 1989.

[Taka]   H. Takahashi, A. Matsuyama:
           An Approximate Solution for the Steiner Problem in Graphs, Math. Japonica, Vol. 24, 1980, 573 - 577.

[Widm86] P. Widmayer:
           On Approximation Algorithms for Steiner's Problem in Graphs, Graph-Theoretic Concepts in Computer Science, WG 86, Lecture Notes in Computer Science, Vol. 246, 1986, 17 - 28.

[Widm88] P. Widmayer, L.S. Woo, C.K. Wong:
           Maximizing Pin Alignment in Semi-Custom Chip Circuit Layout, Integration, the VLSI journal, Vol. 6, 1988, 3 - 33.

[Wu86]   Y.F. Wu, P. Widmayer, C.K. Wong:
           A Faster Approximation Algorithm for the Steiner Problem in Graphs, Acta Informatica, Vol. 23, 1986, 223 - 229.

[Wu87]   Y.F. Wu, P. Widmayer, M.D.F. Schlag, C.K. Wong:
           Rectilinear Shortest Paths and Minimum Spanning Trees in the Presence of Rectilinear Obstacles, IEEE Transactions on Computers, Vol. C-36, 1987, 321 - 331.

A FAST SEQUENTIAL AND PARALLEL ALGORITHM FOR THE COMPUTATION OF THE k-CLOSURE OF A GRAPH

Ingo Schiermeyer

Lehrstuhl C für Mathematik
Technische Hochschule Aachen
Templergraben 55, D-5100 Aachen
West-Germany

Key Words: sequential and parallel graph algorithms, graph
properties, hamiltonian graphs, computational complexity,
polylog parallel algorithm, random graphs

Abstract:

In 1976 Bondy and Chvátal introduced the k-closure $C_k(G)$ of a
graph and described an algorithm which constructs it in $O(n^4)$ time.
We present an algorithm which requires $O(n^3)$ time. However, in the
average case, $C_k(G)$ is computed by this algorithm for almost all
integers k (in the asymptotic sense) with $0 \le k \le 2n - 2$ in $O(n^2)$
time. We next present a parallel algorithm which requires
$O(n^2 \log n)$ time. In the average case, this algorithm computes
$C_k(G)$ for almost all integers k in $O(\log n)$ parallel time.

1. Introduction

Let G be a simple and undirected graph with vertex set
$V(G)$, $|V(G)| = n \ge 3$, and edge set $E(G)$. The degree of a vertex
$v \in V(G)$ is denoted by $d_G(v)$, the complement of G by $\bar{G}$ with
$V(\bar{G}) = V(G)$ and $e \in E(\bar{G})$ iff $e \notin E(G)$.

In 1976 Bondy and Chvátal [BoC 76] introduced the k-closure
$C_k(G)$ of a graph G. $C_k(G)$ is obtained from G by recursively joining
pairs of non-adjacent vertices whose degree-sum is at least k until
no such pair remains. Furthermore, they showed its applications for
14 different graph properties. Another application was studied by
Gurgel and Wakabayashi [GuW 86]. The most well-known of these
provides a sufficient condition for hamiltonian graphs, stating
that if $C_n(G)$ is complete, then G contains a Hamilton cycle. This
condition is very powerful and at the same time provides an
algorithm for the detection of a Hamilton cycle in G in polynomial

time. On the one hand this condition detects considerably more hamiltonian graphs than the six previous conditions based on degrees [Di 52, Or 60, Pó 62, Bo 69, Ch 72, La 72], on the other hand almost all graphs have a complete n-closure [Sc 88]. This was confirmed emphatically by empirical examinations for random graphs [Fa 89], stating that for $n \geq 18$ more than 99 % of all hamiltonian graphs were detected by this condition.

The problem "Hamilton cycle" is NP-complete, however, the k-closure $(0 \leq k \leq 2n - 2)$ of a graph can be computed in $O(n^4)$ time [BoC 76]. This time complexity has been improved to $O(n^3)$ by Szwarcfiter [Sz 87]. The basic idea used in his algorithm is the following: If two non-adjacent vertices u and v are joined which have degree-sum k-i, $1 \leq i \leq k$, in the (original) graph G, then i pairs of non-adjacent vertices containing either u or v must have been joined before. Having joined these i pairs of vertices, u and v then can be found in $O(n)$ time. This idea will be described more detailed in section 2.

In this paper we present a sequential algorithm which computes the k-closure in $O(n^3)$ time in the worst case as well, however, in the average case it requires $O(n^2)$ time for almost all k (in the asymptotic sense). We also present a parallel algorithm which computes the k-closure in the average case for almost all k in $O(\log n)$ time. In the worst case it requires $O(n^2 \log n)$ time.

Let k be an integer, $0 \leq k \leq 2n - 2$. $C_k(G)$ can recursively be defined as follows. If G is complete $(G = K_n)$ or if $d_G(u) + d_G(v) < k$ for any non-edge $uv \in E(\bar{G})$, then $C_k(G) := G$. Otherwise, $C_k(G) := C_k(G + uv)$, for some $uv \in E(\bar{G})$ such that $d_G(u) + d_G(v) \geq k$. $C_k(G)$ is unique for all graphs and all integers n,k such that $n \geq 3$ and $0 \leq k \leq 2n - 2$ [BoM 76].

2. The sequential algorithm

In the prepairing step, let G be a graph and k an integer, $0 \leq k \leq 2n - 2$. Compute the degree $d_G(v)$ of each $v \in V(G)$, then sort all vertices by decreasing degrees. This can be done in $O(n \log n)$ time, e.g. by Heapsort. Now let $D = (v_1 v_2 ... v_n)$ be a permutation of the vertices such that $d_G(v_1) \geq d_G(v_2) \geq ... \geq d_G(v_n)$. For each vertex v_i, $1 \leq i \leq n$, now determine (corresponding to the order of D) the set $N(v_i)$ of all non-adjacent vertices $x \in V(G) \setminus \{v_i\}$ within a total amount of $O(n^2)$ time. In every set $N(v_i)$, $1 \leq i \leq n$, the corresponding degrees then form a decreasing sequence.

In the initial step, let u_i be the first element of $N(v_i)$ (if not empty), $1 \leq i \leq n$. Successively compute $d_G(v_i) + d_G(u_i)$, $1 \leq i \leq n$. If $d_G(v_i) + d_G(u_i) \geq k$ for some i, $1 \leq i \leq n$, then increase $d_G(v_i)$ and $d_G(u_i)$ by one. Eliminate v_i and u_i from the set $N(u_i)$ and $N(v_i)$, respectively, and add them to the set S ($S = S \cup \{v_i, u_i\}$, $S = \emptyset$ in the beginning). Finally, add $v_i u_i \in E(\bar{G})$ to $E_k(G)$. Thus the initial step requires a total amount of $O(n)$ time.

In the general step, if $S = \emptyset$, the algorithm terminates ($C_k(G) := G$). Otherwise, choose an element u from S and eliminate it in S, compute $d_G(u) + d_G(v)$ by successively choosing all vertices $v \in N(u)$. If $d_G(u) + d_G(v) \geq k$ for some $v \in N(u)$, then increase $d_G(u)$ and $d_G(v)$ by one. Add u and v to S as well as $uv \in E(\bar{G})$ to $E_k(G)$. Finally, eliminate u and v from $N(v)$ and $N(u)$, respectively, and repeat the general step. Every general step requires $O(n)$ time. Altogether, the algorithm adds $|E(C_k(G))| - |E(G)|$ edges ($O(n^2)$), thus it requires a total amount of $O(n^3)$ time.

In the initial step, the algorithm exactly joins all pairs of non-adjacent vertices with degree-sum at least k. All vertices incident with at least one of these edges are added to S. In the general step, if the algorithm adds an edge $uv \in E(\bar{G})$, this implies that the degree of u or v has been increased by at least one in the previous steps. Hence u or v belongs to S and the correctness of the algorithm then follows by induction.

3. The parallel algorithm

In the initial step, compute the degree $d_G(v_i)$ of each $v_i \in V(G)$, $1 \leq i \leq n$, according to $d_G(v_i) = \sum_{j=1}^{n} A(i,j)$ where $A(i,j)$, $1 \leq i, j \leq n$, denotes the adjacency matrix of the graph G. The sum of n integers (in this case $A(i,1)$, $A(i,2)$,...,$A(i,n)$) can be computed in $O(\log n)$ parallel time using $\lceil n/\log n \rceil$ processors [DeS 83, KiL 86]. Hence all degrees can be computed in $O(\log n)$ parallel time using $n \lceil n/\log n \rceil$ processors.

In the general step, $\binom{n}{2}$ processors are used, one processor corresponding to each pair of vertices u and v. For each processor, the input contains the degrees $d_G(u)$ and $d_G(v)$. Two integers, $a(u)$ and $a(v)$, are contained in the output with $a(u) = a(v) = 1$ iff $d_G(u) + d_G(v) \geq k$ and $a(u) = a(v) = 0$ iff $d_G(u) + d_G(v) < k$ or $uv \in E(G)$. For each vertex $w \in V(G)$ the degree $d_G(w)$ is now increased by the number of all added edges which are incident to w, i.e., by the sum of all $n - 1$ integers of the type $a(w)$. As already mentioned above these sums can be computed for all vertices in $O(\log n)$ parallel time using $n\lceil n/\log n\rceil$ processors. Hence the general step requires $O(\log n)$ parallel time using $\binom{n}{2} + n\lceil n/\log n\rceil$ processors. Altogether the algorithm this time requires $O(n^2 \log n)$ parallel time using $\binom{n}{2} + 2n\lceil n/\log n\rceil$ processors. The algorithm terminates as soon as in a general step no edges are added. The correctedness of the algorithm then follows immediately.

4. Average time complexity of the algorithms

In order to determine the time complexity of both algorithms in the average case we make use of some results concerning random graphs. Assuming that all (labeled) graphs with n vertices occur with the same probability we then may restrict on random graphs with edge probability $p = \frac{1}{2}$. Let δ denote the minimum degree and Δ the maximum degree of all vertices of a graph G, respectively. Then

(1) $P((1 - \varepsilon) \frac{n}{2} < \delta \leq \Delta < (1 + \varepsilon) \frac{n}{2}) \to 1$ for $n \to \infty$,

i.e., for almost all graphs

(2) $(1 - \varepsilon) \frac{n}{2} < d_G(v) < (1 + \varepsilon) \frac{n}{2}$

for all vertices $v \in V(G)$ [Pa 85].

Let u be an arbitrary vertex, $0 \leq d_G(u) \leq n - 1$, and $X(u)$ the number of all vertices $x \in V(G) \setminus \{u\}$ which are non-adjacent to u $(ux \in E(\bar{G}))$ such that $d_G(x) \geq k - d_G(u)$. Writing $d_G(u)$ in the form $d_G(u) = \frac{n-1}{2} + x\frac{\sqrt{n-1}}{2}$, $x \in \mathbb{R}$, we then obtain for the expectation of $X(u)$

(3) $E(X(u)) = (\frac{n-1}{2} - x\frac{\sqrt{n-1}}{2}) \sum_{i=k-d_G(u)}^{n-2} \binom{n-2}{i}(\frac{1}{2})^{n-2}$,

stating that in the case k = n [Sc 88]

(4) $E(x(u) \sim \frac{n}{2} \cdot c(x)$ with $0 < c(x) < 1$.

Even for all integers k and arbitrary $c > 0$ with $0 \leq k \leq n + c \cdot \sqrt{n}$
(4) for remains true.

In the case of the parallel algorithm the expectation of the
degree $d_G(u)$ after adding all non-edges $uv \in E(\bar{G})$ with
$d_G(u) + d_G(v) \geq k$ is

(5) $d_G(u) \sim \frac{n}{2} (1 + c(x))$,

implying that

(6) $d_G(u) > \frac{n + c\sqrt{n}}{2}$ $\forall n \geq n_o(x)$.

Hence for all vertices $v \in V(G)$ the expectation of $d_G(v)$ is
greater than $\frac{n + c\sqrt{n}}{2}$ implying that in the second general step all
missing edges are added to $E_k(G)$, i.e., $C_k(G) = K_n$. Thus, for all
integers k and arbitrary $c > 0$ with $0 \leq k \leq n + c\sqrt{n}$, we obtain
the following theorems.

Theorem 1A   Almost all graphs have a complete k-closure
$(C_k(G) = K_n)$.

Theorem 2A   The parallel algorithm computes $C_k(G)$ in $O(\log n)$
parallel time for almost all graphs.

In the case of the .sequential algorithm, the condition
$d_G(u) + d_G(v) \geq k$ for the first vertex v in the (actual) set N(u)
is satisfied $\sim \frac{n}{2} \cdot c(x)$-times. Hence every such general step
requires $O(1)$ time, remembering that in the beginning the degrees
of the vertices of N(u) form a decreasing sequence.
Now let $d_G(u)$ of an arbitrary $v \in V(G)$ be increased according
to (4) such that

(7) $d_G(u) \sim \frac{n}{2} (1 + c(x))$.

(2) implies that

(8) $d_G(v) > \frac{n}{2} (1 - \frac{c(x)}{2})$

and therefore

(9) $d_G(u) + d_G(v) > n(1 + \frac{c(x)}{4}) > n + c \cdot \sqrt{n}$ $\forall n \geq n_o(x)$.

Thus the condition $d_G(u) + d_G(v) \geq k$ is now satisfied for all vertices $v \in N(u)$. Similarly as above we obtain that every such general step requires $O(1)$ time as well. Considering the time complexities of the prepairing step ($O(n^2)$)) and the initial step ($O(n)$) we obtain the following theorem for all integers k and arbitrary $c > 0$ with $0 \leq k \leq n + c \cdot \sqrt{n}$.

Theorem 3A The sequential algorithm computes $C_k(G)$ in $O(n^2)$ time for almost all graphs.

For all integers k and all ε with $0 < \varepsilon < 1$ and $n(1 + \varepsilon) \leq k \leq 2n - 2$ we now deduce similar results as mentioned in these three theorems. Corresponding to (2) for almost all graphs $d_G(v)$ is smaller than $(1 + \varepsilon) \frac{n}{2}$ for all $v \in V(G)$ implying that

$$(10) \quad d_G(u) + d_G(v) < n(1 + \varepsilon)$$

for all $uv \in E(\bar{G})$, i.e., the parallel algorithm terminates after the first general step and the sequential algorithm after the initial step, respectively. Thus, for all integers k and all ε with $0 < \varepsilon < 1$ and $n(1 + \varepsilon) \leq k \leq 2n - 2$ we obtain the following theorems.

Theorem 1B $C_k(G) = G$ for almost all graphs.

Theorem 2B The parallel algorithm computes $C_k(G)$ in $O(\log n)$ parallel time for almost all graphs.

Theorem 3B The sequential algorithm computes $C_k(G)$ in $O(n^2)$ time for almost all graphs.

Joining the intervals of the integers - on the one hand $0 \leq k \leq n + c \cdot \sqrt{n}$, on the other hand $n(1 + \varepsilon) \leq k \leq 2n - 2$ - it turns out that the average time complexities of both the parallel and the sequential algorithm hold for almost all k (in the asymptotic sense).

We conclude with a problem which has been posed by Dahlhaus, Hajnal and Karpinski [DaHK 87]. For $0 \leq k \leq 2n - 2$, $n \geq 3$, and all graphs G let "k-closure" denote the problem of determining $C_k(G)$.

Problem Does "n-closure" belong to NC,

i.e., can $C_n(G)$ be computed in polylog parallel time on a polynomial number of processors?

Only recently Khuller [Kh 88] showed that "n-closure" is P-complete with respect to log-space transformations (hence not likely to belong to NC).

References

[Bo 69] J. A. Bondy, Properties of Graphs with Constraints on Degrees, Studia Sci. Math. Hungar.4 (1969) 473-475.

[BoC 76] J. A. Bondy and V. Chvátal, A Method in Graph Theory, Discrete Math. 15 (1976) 111-135.

[BoM 76] J. A. Bondy and U.S.R. Murty, Graph Theory with Applications, Macmillan, London and Elsevier, New York, 1976.

[Ch 72] V. Chvátal, On Hamiltonian's Ideals, J. Comb. Theory 12 B (1972) 163-168.

[DaHK 87] E. Dahlhaus, P. Hajnal and M. Karpinski, Bonn Workshop on Foundations of Computing 1987, private communication.

[DeS 83] E. Dekel and S. Sahni, Binary trees and parallel scheduling algorithms, IEEE Trans. Comput. C-32, 307-315.

[Di 52] G. A. Dirac, Some Theorems on abstract Graphs, Proc. London Math. Soc. 2 (1952) 69-81.

[Fa 89] B. Faber, Empirische Testmethoden an Zufallsgraphen für Eigenschaften im Problemkreis "Hamiltongraphen" und ihre Implementierung, Diplomarbeit an der RWTH Aachen, 1989.

[GuW 86] M. A. Gurgel and Y. Wakabayashi, On k-leaf-connected Graphs, J. Comb. Theory 41B (1986) 1-16.

[Kh 88] S. Khuller, On Computing Graph Closures, TR 88-921, Cornell University.

[KiL 86] G. A. P. Kindervater and J. K. Lenstra, Parallel Computing in Combinatorial Optimization, Preprint, 1986.

[La 72] M. Las Vergnas, Thesis, University of Paris VI (1972).

[Pa 85] E.M. Palmer, Graphical Evolution: An Introduction to the Theory of Random Graphs, Wiley-Interscience Series in Discrete Mathematics, New York, Toronto, Singapore, 1985.

[Pó 62] L. Pósa, A Theorem concerning Hamilton Lines, Magyar Tud. Akad. Mat. Kukato Int. Közl. 8 (1963) 335-361.

[Or 60] O. Ore, Note on Hamilton Circuits, Amer. Math. Monthly 67 (1960) 55.

[Sc 88] I. Schiermeyer, Algorithmen zum Auffinden längster Kreise in Graphen, Dissertation an der RWTH Aachen, 1988.

[Sz 87] J. L. Szwarcfiter, A Note on the Computation of the k-closure of a Graph, Information Processing Letters 24 (1987) 279-280.

ON FEEDBACK PROBLEMS IN DIGRAPHS

Ewald Speckenmeyer

Abteilung Informatik, Universität Dortmund

Postfach 500 500, D-4600 Dortmund 50

Abstract: A subset $F \subseteq V$ of vertices of a digraph $G=(V,A)$ with n vertices and m arcs, is called a feedback vertex set (fvs), if G-F is an acyclic digraph (dag). The main results in this paper are:
(1) An $O(n^2m)$ simplification procedure is developed and it is shown that the class of digraphs, for which a minimum fvs is determined by this procedure alone, properly contains two classes for which minimum fvs's are known to be computable in polynomial time, see [13,15].
(2) A new $O(n^3 \cdot |F|)$ approximation algorithm MFVS for the fvs-problem is proposed, which iteratively deletes in each step a vertex with smallest mean return time during a random walk in the digraph. It is shown that MFVS applied to symmetric digraphs determines a solution of worst case ratio bounded by $O(\log n)$. The quality of fvs's produced by MFVS is compared to those produced by Rosen's fvs-algorithm, see [11], for series of randomly generated digraphs.

INTRODUCTION

Let $G:=(V,A)$ be a digraph with vertex set $V=\{v_1,\ldots, v_n\}$ and arc set $A \subseteq V \times V$. By n we always denote the number of vertices and by m the number of arcs of G. By definition G may contain selfloops $(v,v) \epsilon A$, but no parallel arcs. Two arcs (u,v), (v,u) ϵA, $u \neq v$, are called antiparallel. For a subset $V' \subseteq V$ we define G-V' to be the digraph, which is induced by the vertex set V-V' in G. A subset $F \subseteq V$ is called a **feedback vertex set (fvs)** of G, if G-F is an acyclic digraph (dag). A minimum fvs F of G is a fvs of G of minimal cardinality and we define $f(G):=|F|$. The problem of determining a fvs of cardinality f(G) of G is called the **fvs-problem**.

Related to the fvs-problem is the **feedback arc set-problem (fas-problem)** where we look for a minimum arc set $B \subseteq A$ such that the digraph G-B, resulting from G by deleting all the arcs of B from G, is a dag. It is well known that the fas-problem can be reduced to the fvs-problem simply by solving the fvs-problem for the corresponding arc-digraph of G. The fvs-problem can be reduced to the fas-problem as well, simply by splitting each vertex v of G into vertices iv and ov, where an arc runs from iv to ov and all incoming arcs into v are going into iv and all arcs leaving v are leaving ov. Then a minimum fas in the resulting digraph G' can be transformed into a minimum fvs of G. Because of this analogy we concentrate on the fvs-problem in this paper.

Solving the fvs-/ fas-problem often occurs as an algorithmic problem in computer science. We mention here only three applications. In the area of operating systems the problem of breaking deadlocks is formulated as fvs-problem, see [11], e.g.. Proving partial correctness of programs by means of Floyd's method requires to determine first a fvs, as small as possible, of the corresponding flowchart digraph, see [3]. Checking whether a VLSI-circuit C satisfies certain design rules, relies on first deleting as few links as possible from C s.t. the resulting circuit C' is a combinational circuit, i.e. C' is a dag. Here we meet the fas-problem, see [1].

Unfortunately the fvs- and the fas-problem, both, belong to the class of NP-hard problems, see [4]. So one is often content finding sufficiently small, but not necessarily optimal, fvs's/ fas's for digraphs in "reasonable" time, or one is interested in finding large "interesting" subclasses of digraphs, for which the fvs-/ fas-problem can be solved in polynomial time. Concerning these aims there are only few general results, see [5,6,11,13,15]. For this reason Wang, et al. denote the fvs-problem to be "the least understood of the classic (NP-complete, E.Sp.) problems", see [15].

In this paper we show the following results.

In chapter 2 we develop a $O(n^2(n+m))$ simplification procedure, which first determines all vertices, which can "easily" be seen to belong to a minimum fvs of G, and then all parts from the remaining digraph are deleted, which are "irrelevant" for the fvs-problem. We denote the class of digraphs, for which a minimum fvs is found by applying this simplification procedure only, the class of **fvs-simple** digraphs. Then we show that the class of fvs-simple digraphs properly contains the class of **reducible flow graphs**, which has been considered by Shamir in [13], and the class of **cyclically reducible graphs**, introduced and studied by Wang, et al. in [15]. The algorithm by Wang, et al. for solving the fvs-problem for cyclically reducible graphs has the same running time as our simplification procedure, where Shamir's fvs-algorithm for reducible flow graphs based upon depth first search (dfs) runs in time $O(n+m)$, only.

In chapter 3 a Markovian approximation algorithm for the fvs-problem in digraphs, MFVS, is proposed, and it is shown that MFVS always determines a minimum fvs of G, if $f(G)=1$. Then we show that MFVS applied to symmetric digraphs G determines a fvs F of worst case ratio $|F|/f(G)=O(\log n)$, and MFVS may produce fvs's F for certain classes of symmetric digraphs G satisfying $|F|/f(G)=c \cdot \log n$, for some $c>0$. We conjecture that MFVS applied to arbitrary digraphs always determines fvs's of worst case ratio $O(\log n)$.

The idea behind MFVS is as follows. In order to determine a (minimum) fvs F of a digraph G it is sufficient to decompose G into its strongly connected components, which can be treated independently. Each strongly connected component H with at least two vertices is processed as follows. For each vertex $v \in V(H)$ the mean return time $r(v)$ to vertex v during a random walk in H is determined. Then a vertex v with smallest mean return time $r(v)$ is added to F and deleted from H. Then $H-\{v\}$ recursively is processed like G.

The most time consuming part of MFVS is to determine the mean return times $r(v)$, for all $v \in V$. By a result from the theory of Markov chains $r(v)$ is obtained from the solution of a system of $|V(H)|$ linear equations with $|V(H)|$ variables, which can be solved in $O(|V(H)|^3)$ steps. Therefore the running time of MFVS is bounded by $O(|F| \cdot n^3)$.

We will mention here that to our knowledge there is no polynomial time approxima-tion algorithm for the fvs-problem, which has been proven, always to determine a fvs F for an input-digraph G of worst case ratio $|F|/f(G) \leq g(n)$, for some bounding function $g(n)$ satisfying $g(n)=o(n)$.

MFVS has been implemented in the programming language C on a Sun-workstation and the cardinalities of the fvs's produced by MFVS for samples of random digraphs have been compared to those found by a modified version of Rosen's approximation algorithm for the fvs-problem, see [11]. We call Rosen's algorithm ROSEN and its modified version n-ROSEN. ROSEN, with input-digraph G=(V,A), first applies dfs to G, in order to partition A into the set of backarcs B and the set of dag-arcs D:=A-B. Obviously the subset $U \subseteq V$ of vertices into which backarcs run is a fvs of G. Then ROSEN determines a minimal fvs $F \subseteq U$. The running time of ROSEN is bounded by $O(n+m)$. The size of the fvs F of G determined by ROSEN essentially depends on the startvertex for dfs and on the set of backarcs found by dfs. Therefore we apply ROSEN n times to G with each vertex as dfs-startvertex, and output a fvs F of smal-lest size among the n fvs's. This algorithm is called n-ROSEN and its running time is bounded by $O(n(n+m))$. The test results shown in the table in chapter 3 exhibit that MFVS determines fvs's for our test series, which are significantly better than those computed by n-ROSEN.

In chapter 4 we consider the fvs-problem for tournaments T=(V,A), i.e. for digraphs, where for every two vertices u, $v \in T$ either $(u, v) \in A$ or $(v, u) \in A$. In the introduc-tory chapter of their book, [2], Erdös and Spencer argue that the probability of a random tournament to contain an acyclic subtournament with more than $O(\log n)$ vertices asymptotically is going to zero. This implies immediately that the probability of a random tournament to have a fvs of less than $n-O(\log n)$ vertices asymptotically is going to zero.

We will show here however, that the fvs-problem, when restricted to tournaments, remains NP-hard. (Note that the Hamiltonian cycle problem for tournaments T is solvable in linear time, because T is Hamiltonian, iff T is strongly connected, see [10].) Next we show that the fvs-problem for tournaments can be approximated by a worst case ratio of 3.

We end with some concluding remarks.

Results on the fvs-problem for undirected graphs can be found, e.g., in [9,14]. In [9] an approximation algorithm of worst case ratio $O(\log^{1/2} n)$ has been developed, and in [14] the fvs-problem for cubic graphs is studied. In [6] a linear time algorithm for the fas-problem in directed series parallel graphs is described. For notions and results about graph algorithms, which are implicitly used in this paper, the reader is referred to [8].

2. A SIMPLIFICATION PROCEDURE

Let G=(V,A) be a digraph, for which a fvs F of smallest possible cardinality is to be determined. The NP-hardness of the fvs-problem, which lends strong evidence for the hypothesis that the fvs-problem becomes computationally intractable even for moderate input sizes, indicates that it is desirable to "simplify" the input digraph G in a first step as far as possible. We formulate in the following algorithmic scheme the kinds of action of the procedure SIMPLIFY.

procedure SIMPLIFY;

<u>Input</u>: digraph G=(V,A);

0. F:=∅;

1. <u>while</u> there is some vertex v of G, which can "easily" be recognized to belong
      to a minimum fvs of G
      <u>do</u> F:=F∪{v}; G:=G-{v} <u>od</u>;

2. decompose G into strongly connected components and let $G_1,...,G_k$ be the s.c.c.'s
      with at least 2 vertices;

3. <u>while</u> there is some G_i and arc $(u,v) \in A(G_i)$ s.t. every cycle containing u walks
        along (u,v) or every cycle containing v comes along (u,v)
      <u>do</u> contract u and v into a single node <u>od</u>;

<u>Output</u>: digraphs $G_1,...,G_k$ and a subset F⊆V of vertices;

We will discuss the statements 1. to 3. of SIMPLIFY. Determining a minimum fvs of a digraph we certainly can restrict to it's s.c.c.'s, thus justifying statement 2. The while condition of statement 3. expresses the fact that after having applied statement 2. to G either vertex u has outdegree 1 or vertex v has indegree 1. Contracting u and v into a single new vertex w means that with the exception of (u,v) all arcs entering u or v now are entering w and all arcs leaving u or v now are leaving w. Note that if moreover G contains the arc (v,u), then contraction produces the selfloop (w,w). It should be mentioned however that selfloops will not be produced because of our realization of statement 1., which will be described next.

Let v be a vertex and C be a cycle of G. Then we say $P_G(v,C)$ holds, iff v lies on C and all vertices u on C don't belong to a cycle of G-{v}. Finally we say $P_G(v)$ holds iff there is some cycle C of G s.t. $P_G(v,C)$ holds.

Obviously, if $P_G(v)$ holds, then there is a minimum fvs F of G containing v, because

there is some cycle C containing v and at least one vertex u of C has to be taken
into F, but on the other hand all cycles, which are destroyed by u, are destoyed
by v, too, for all u of C. This leads to the following realization of action 1.
of SIMPLIFY:

1. <u>while</u> there is some vertex $v \epsilon V(G)$ s.t. $P_G(v)$ holds
    <u>do</u> $F:=F \cup \{v\}$; $G:=G-\{v\}$ <u>od</u>;

<u>Lemma 1</u>: Simplify computes in $O(|F| \cdot n(n+m))$ steps for an input-digraph G the output
$G_1, \ldots, G_k$, and F, satisfying

$$f(G) = |F| + \sum_{i=1}^{k} f(G_i).$$

<u>Proof</u>: The equality obviously follows from the discussion of the statements 1. to 3.
The most time consuming statement of SIMPLIFY is statement 1. Note that $P_G(v)$
holds, for some vertex $v \epsilon V(G)$, iff $(v,v) \epsilon A(G)$ or if there is some vertex $u \neq v$ s.t.
u lies on a cycle of G, but u does not lie on a cycle of $G-\{v\}$. This can be checked
by applying Tarjan's strongly connected components algorithm twice, to G and to
$G-\{v\}$, in time $O(n+m)$, see [8] , e.g.. With this the time bound easily follows. □

We use the procedure SIMPLIFY for defining the **class of fvs-simple digraphs**. A digraph G is called to be **fvs-simple**, if there is some computation of SIMPLIFY with
input G yielding as output a minimum fvs F of G.
It can easily be seen that for a fvs-simple digraph the equality $f(G)=c(G)$ holds,
where $c(G)$ denotes the maximal number of vertex-disjoint cycles in G.
Next we will show that SIMPLIFY, when applied to an arbitrary fvs-simple digraph
$G=(V,A)$ always determines a minimum fvs F of G. For this reason we define for each
vertex $v \epsilon V$ the set of vertices $X(v,G)$, consisting of all the vertices u of G,which
lie on some cycle C of G satisfying $P_G(v,C)$.
Without proof we mention, that in case of $X=X(v,G) \neq \emptyset$, there is always a partition
of V into three subsets X, Y, Z, s.t. the vertices of $X':=X-\{v\}$ induce a dag of G,
there are no arcs from Y to X', from X' to Z, and from Y to Z, see figure 2.1.

G:

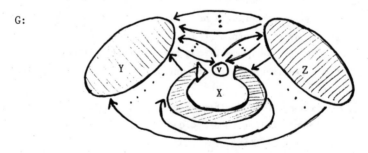

Figure 2.1

Note that $X(v,G) \neq \emptyset$ iff $P_G(v)$ holds.

If $X(v,G) \neq \emptyset$ and v is deleted from G, then all vertices from $X(v,G)$ can be deleted from G, because none of them lies on a cycle of $G-\{v\}$.

Theorem 1: Let $G=(V,A)$ be a fvs-simple digraph, then SIMPLIFY outputs a minimum fvs F of G.

Proof: It is sufficient to show the following. Let $v \in V$ be chosen arbitrarily and let $H:=G-X(v,G)$, then H is fvs-simple, if G is fvs-simple.
Suppose w.l.o.G. $X(v,G) \neq \emptyset$.
By definition G is fvs-simple, if there is some sequence $u_1, \ldots, u_k$ of vertices of G, s.t. the sequence $G_0, G_1, \ldots, G_k$ of digraphs defined by $G_0:=G$, and $G_i:=G_{i-1}-X(u_i, G_{i-1})$, $1 \leq i \leq k$, satisfies G_k is a dag.
By induction on k we show that in case of $X(v,G) \neq \emptyset$ either H is a dag or there is some $u \in V(H)$ satisfying $X(u,H) = \emptyset$.
If $X(v,G) \neq \emptyset$ then $X(v,G)=X(u_1,G)$ or $X(v,G) \cap X(u_1,G) = \emptyset$.
In the first case $H=G_1$ and in the second case $X(u_1,H) \neq \emptyset$ holds.
This proves our theorem. $\blacksquare$

In [16] the notion of fvs-simple digraphs is extended and a Church Rosser property like Theorem 1 remains valid.

We now compare the class of fvs-simple digraphs with the class of cyclically reducible graphs, defined by Wang, et al., see [15].
Let $G=(V,A)$ be a digraph and $v \in V$. Define $Y(v,G)$ to be the subset of vertices $u \in X(v,G)$, s.t. there is no path in $G-\{v\}$ from u to some vertex w lying on a cycle of $G-\{v\}$.
Note that $Y(v,G) \neq \emptyset$ but $Y(v,G)$ does not induce a cyclic subdigraph of G is possible.
Then G is called a **cyclically reducible graph**, if there is some sequence $v_1, \ldots, v_k$ of vertices of G s.t. the sequence of digraphs $G_0, G_1, \ldots, G_k$, defined by $G_0:=G$, $G_i:=G_{i-1}-Y(v_i, G_{i-1})$, $1 \leq i \leq k$, satisfies the following: $Y(v_i, G_{i-1})$ induces a cyclically subdigraph of G, for all $1 \leq i \leq k$, and G_k is a dag.
Actually, Wang, et al. define cyclically reducible graphs a bit different. Our definition however is equivalent to theirs.
Then the following holds

Theorem 2: The class of cyclically reducible graphs is properly contained in the class of fvs-simple digraphs.

Proof: By definition every cyclically reducible graph is a fvs-simple digraph. Now consider the digraph G in figure 2.2, taken from [15].

G is fvs-simple by first deleting v and then w. But G is not cyclically reducible, because there is no vertex y s.t. Y(y,G) induces a cyclic subdigraph of G. □

G:

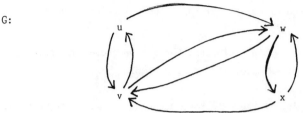

Figure 2.2

We next compare the class of fvs-simple digraphs with the class of reducible flow graphs, defined in [13].

A digraph G=(V,A) is a **reducible flow graph**, if there is some node v∈V, from which G can be searched completely and every possible dfs of G started in v produces the same set of backarcs (for the definitions of dfs and backarcs we refer to [8]).

Consider the digraph H in figure 2.3, taken from [13].

H:

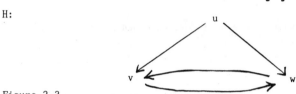

Figure 2.3

H is not a reducible flow graph, but H is both, a fvs-simple digraph and a cyclically reducible graph, as can easily be seen. The digraph G of figure 2.2 is a reducible flow graph (start dfs at v). This implies that the classes of reducible flow graphs and of cyclically reducible graphs are not comparable, as mentioned in [15] already. We can show however the following.

Theorem 3: The class of reducible flow graphs is properly contained in the class of fvs-simple digraphs.

Proof: Because of the discussion of the digraph H of figure 2.3 it remains to show that every flow graph is fvs-simple.

Shamir's fvs-algorithm for reducible flow graphs G=(V,A) works as follows, see [13]. In a first step the algorithm visits G by dfs starting from some vertex v, from which G can be searched completely and there is a unique set B⊆A of backarcs determined by every possible dfs from v. The vertices of G are numbered by nr from 1 to n in the order in which they are first visited. Obviously each cycle in G contains some backarc. Therefore the set of vertices, into which backarcs run, forms a fvs

of G. It is shown that reducible flow graphs share the following property: every
simple cycle of G contains exactly one backarc. A vertex v of the actual digraph
G is called **active**, if there is some backarc b=(u,v) s.t. there is a path from v
to u in G-B. Then Shamir's algorithm determines fvs F of G as follows.

while there is some active vertex v of G
do determine an active vertex v of G with maximal nr(v);
   $F:=F \cup \{v\}$; $G:=G-\{v\}$
od;

In the proof that Shamir's algorithm determines minimum fvs's for the flow graphs,
the proof of theorem 3 in [13], it is shown that the current maximal active vertex
v of G determined during the current while-loop has the following property. There is
some simple cycle C containing v in the actual digraph G s.t. every cycle C' containing some vertex u of C also contains v. This exactly means, however, that $P_G(v)$
holds. By inductive application of this argument the theorem follows. □

3. A MARKOVIAN APPROXIMATION ALGORITHM

In [11], Rosen gives an example of a digraph G satisfying f(G)=1, but for which Rosen's fvs approximation algorithm, an extension of Shamir's fvs-algorithm for reducible flow graphs described in the proof of theorem 3, may possibly determine a
fvs with n-1 vertices. In order to obtain an approximation algorithm for the fvs-problem, yielding fvs's F , whose size does not deviate as far from f(G) as Rosen's
algorithm does, one might try the following approach:

algorithm VERSUCH;

repeat
1. apply SIMPLIFY to G;
2. delete a vertex v from G, lying on a shortest cycle and add v to F
until a fvs F of G has been found;

In order to demonstrate that VERSUCH may behave bad, consider the digraph $Z_{r,s}$ with
$n=r \cdot s$ vertices, $r,s \geq 3$, where $V(Z_{r,s})$ is partitioned into s layers $L_0, L_1, \ldots, L_{s-1}$,
and each L_i has exactly r vertices. The arc set of $Z_{r,s}$ is defined as follows:
for u in L_i and v in $L_{(i+1) \bmod s}$, $Z_{r,s}$ contains the arc (u,v).
Note that $Z_{r,s}$ is strongly connected, all simple cycles of $Z_{r,s}$ have length exactly s,
and $f(Z_{r,s})=r$. Then it is not hard to see that SIMPLIFY effects the current digraph
Z during a repeat-loop only, if some layer of Z contains one vertex, only.
So there may be first determined (r-2)s+1 many vertices   by statement
2. of VERSUCH for the fvs F of $Z_{r,s}$, before one application of SIMPLIFY stops the
computation, yielding a fvs of (r-2)s+2 many vertices, wheras $f(Z_{r,s})=r$.

So, e.g., in case of r=3, VERSUCH may determine a fvs F of $Z_{r,s}$ satisfying

$$|F|/f(Z_{r,s})=(s+2)/3 \geq n/9.$$

We will develop now a more promising approximation algorithm for the fvs-problem, which we call **Markovian fvs-algorithm** (MFVS). Let G=(V,A) be a digraph, w.l.o.g. G is strongly connected and V={1,...,n}. Denote by od(i) the outdegree of vertex i. We associate with G a finite homogeneous **Markov chain** M with **state set** V and 1-step **transition probabilities** $p_{i,j}$, where $p_{i,j}=od^{-1}(i)$, if $(i,j) \in A$, $p_{i,j}=0$ otherwise, for all $(i,j) \in V \times V$, where $P=(p_{i,j})_{i,j \in V}$ denotes the 1-step **transition matrix** of M. (Because the state set of M is equal to the vertex set V of G, we say vertices instead of states.) Obviously P reflects the 1-step transition probabilities during a random walk in G. For if we are at vertex i at time t, then we will be at vertex j at time t+1 with probability $p_{i,j}$.

From the 1-step transition probabilities $p_{i,j}$ the r-step transition probabilities $p_{i,j}^{(r)}$, $r \geq 0$, can be determined, where $p_{i,j}^{(r)}$ denotes the probability of the event that after r steps of a random walk in G started at vertex i we arrive at vertex j. Using the Chapman-Kolmogorov equations, we obtain $p_{i,j}^{(0)}=1$, if i=j, $p_{i,j}^{(0)}=0$, otherwise, and $p_{i,j}^{(r+1)} = \sum_{k \in V} p_{i,k}^{(r)} \cdot p_{k,j}$, for $r \geq 0$ and i,j$\in$V.

By $P^{(r)}$ we denote the r-step transition matrix. Obviously $P^{(r)}=P^r$.

Denote by $h_i^{(r)}$ the probability of the event that after exactly r steps of a random walk started at vertex i of G we return to i for the first time. Finally the **mean return time to vertex i** in random walks in G is defined by $r_i = \sum_{r=0}^{\infty} r \cdot h_i^{(r)}$.

Because G is a finite strongly connected digraph, the corresponding transition matrix P is irreducible and all vertices of V are positive recurrent, see [12], e.g.. Then the following result, known as the main theorem of the theory of Markov chains, holds, see [12].

Theorem 4: Let G=(V,A) be a strongly connected digraph. Then there is a unique non-negative vector $p=(p_1,...,p_n) \in (0,1)^n$ satisfying

a) $p_j = \sum_{i \in V} p_i \cdot p_{i,j}$, for all j$\in$V, and $1 = \sum_{i \in V} p_i$, and

b) $r_j = p_j^{-1}$, for all j$\in$V.

The solution vector p is called the **stationary distribution** of the corresponding Markov chain. □

Because of b) the mean return time r_i to node i in G is the inverse of the ith component, p_i, of the stationary distribution vector p, and by a) the stationary distribution vector is obtained solving a system of n linear equations. (Because P is a stochastic matrix, one of the first n equations from b) is superflous.)

Then MFVS is formulated as follows.

algorithm MFVS;

<u>Input</u>: digraph $G=(V,A)$;

1. $F:=\emptyset$;
2. decompose G into its strongly connected components and push those components consisting of more than 1 vertex on the stack;
3. <u>while</u> stack is not empty
4. <u>do</u> pop top digraph $G=(V,A)$ from the stack;
5.   determine a vertex $i\epsilon V$ of smallest mean return time r_i in G;
6. $F:=F\cup\{i\}$; $G:=G-\{i\}$;
7.   decompose G into its strongly connected components and push those components consisting of more than 1 node on the stack
   <u>od</u>;

<u>Output</u>: vertex set F;

<u>Lemma 2</u>: MFVS outputs in $O((n+m)+|F|n^3)$ steps a fvs F of an input digraph G.

<u>Proof</u>: F is obviously a fvs of G.
Statement 2 is Tarjan's strong connectivity algorithm with time bound $O(n+m)$. The while-loop is executed $|F|$ times, where during every loop a system of n linear equations has to be solved, which can be done by the Gauss -algorithm in time $O(n^3)$. $\square$

We will mention, that MFVS determines a minimum fvs for the digraphs $Z_{r,s}$, defined above.

A digraph $G=(V,A)$ containing for every arc $(i,j)\epsilon A$ the antiparallel arc $(j,i)\epsilon A$, is called a **symmetric digraph**.

<u>Theorem 5</u>: Let $G=(V,A)$ be an input digraph for MFVS.
a) If $f(G)=1$, then MFVS outputs a minimum fvs F of G.
b) If G is a symmetric digraph, then MFVS outputs a fvs F of G satisfying $|F|/f(G) = O(\log n)$, and this bound is best possible.

<u>Proof</u>: Suppose, w.l.o.g., G to be strongly connected.
<u>ad a</u>): Every vertex i of G satisfying the property that $G-\{i\}$ is a dag, must have the same mean return time $r_i=\rho$ and for all other vertices j holds $r_j>\rho$. From this a) follows.
<u>ad b</u>): We will first show that the stationary distribution vector $p=(p_1,\dots,p_n)$ of the Markov chain $M=(V,P)$ corresponding to G satisfies $p_i=od(i)/m$, for all $i\epsilon V$. Because of theorem 4 a) we have to show: (i) $\sum_{i\epsilon V} p_i=1$, and (ii) $p_j=\sum_{i\epsilon V} p_i\cdot P_{i,j}$, for all $j\epsilon V$.

(i) follows imediately from the definition of p_i. In order to show (ii) remind, that $p_{i,j} = od^{-1}(i)$, if $(i,j) \epsilon A$, otherwise $p_{i,j} = 0$. Then

$$\sum_{i \epsilon V} p_i \cdot p_{i,j} = m^{-1} \cdot \sum_{i \epsilon V} od(i) \cdot p_{i,j} = m^{-1} \cdot \sum_{i:(i,j)\epsilon A} od(i) \cdot od^{-1}(i) = m^{-1} \cdot od(j) = p_j,$$

where the last inequality holds, because G is symmetric and so the number of incoming arcs into vertex j is equal to od(j).

During every while–loop of MFVS a vertex i with minimal mean return time r_i in the actual digraph G is chosen to be added to F, and because of theorem 4 b) $r_i = p_i^{-1} = m/od(i)$, the vertex i chosen, will be of highest outdegree in G.
Because G is symmetric, MFVS stops computation until for every arc $(u,v) \epsilon A$ at least one of its endpoints belongs to F, for otherwise there would remain a cycle of length 2.

This means, that MFVS applied to symmetric digraphs G determines a subset F of vertices, covering all arcs of G by vertices, following the strategy of choosing always a vertex i of highest actual degree (we can neglect the directions of the arcs) into F.

This greedy heuristic for this problem has been analysed by Johnson, see [7], where it is shown that the ratio between the number of vertices determined by this strategy and the number of vertices of a minimum solution for an input problem is bounded by O(log n), and this bound is best possible. This proves our theorem. □

We conjecture that theorem 5 b) holds for arbitrary digraphs, and if it is true, MFVS can be used as well to determine solutions for the fas–problem of worst case ratio O(log n).

P. Heusch and F. Höfting have implemented MFVS and n–ROSEN (described in the introduction) at the computer science department of the University of Paderborn.
The fvs's produced by these two heuristics have been compared in tests, where the mean over 50 randomly chosen digraphs from $\mathcal{G}(n,p)$ was taken, each time. $\mathcal{G}(n,p)$ denotes the class of digraphs G with n vertices, and for every two vertices i and j, $i \neq j$, $(i,j) \epsilon A(G)$ holds, independently with probability p. The results are summarized in table 4.1.

$\mathcal{G}(n,p)$	MFVS	n–ROSEN	Improvement of MFVS over n–ROSEN in %
n=70			
p=0.05	17.25	21.14	17.1
0.10	33.22	38.8	14.4
0.15	42.62	47.78	10.8
0.20	48.56	56.62	7.7
n=100			
p=0.05	34.26	42.6	19.6
0.10	57.38	65.96	13.0
0.15	69.5	75.58	8.0
0.20	76.98	81.12	5.1
n=120			
p=0.05	48.62	60.14	19.2
0.10	75.26	85.06	11.5
0.15	87.62	95.08	7.8
0.20	95.74	100.68	4.9

table 4.1

4. FEEDBACK VERTEX SETS IN TOURNAMENTS

A **tournament** is a digraph T=(V,A), where any two different vertices of T are joined by exactly one arc.

Theorem 6: Let T=(V,A) be a tournament and t be a positive integer. The problem of deciding whether T has a fvs F of at most t vertices is NP-complete.

Sketch of proof:
We will reduce the **vertex cover** problem (vc-problem) to the fvs-problem. The vc-
-problem, as language problem, is defined as follows. Input is an (undirected)
graph G=(V,E) and a positive integer k. The question to be answered is, whether
there is some subset C of at most k vertices, such that G-C has no edges. (We met
this problem already in the proof of theorem 5.)
Let 1, 2,...,n be some arbitrarily chosen, fixed linear ordering of the vertices
of G. Then define a tournament T as follows. V(T) consists of the vertices of G
and for every vertex iεV, 1≤i≤n-1, V(T) contains two more vertices x_i and y_i. Order
the vertices of V(T) linearly as follows:

$1, x_1, y_1, 2, x_2, y_2, 3, \ldots, n-1, x_{n-1}, y_{n-1}, n$. The arc set $A(T)$ of the tournament T is defined as follows:

Let u, v be two different vertices of T, where u lies left from v in the linear ordering. If $u-v$ is an edge of G, then $(v,u)\varepsilon A(T)$, in all other cases $(u,v)\varepsilon A(T)$. Note that all arcs of T, corresponding to edges of G, are directed from right to left, and all other arcs of T are directed from left to right.

Claim: G has a vc C of at most k vertices iff T has a fvs F of at most k vertices.

The proof of this claim, using the fact, that all cycles in T are destroyed, if all "right to left arcs" are deleted from T, will not be given here.

Because the vc-problem is known to be NP-complete, see [4], the fvs-problem is NP-complete, too, because of the claim. (Membership of the fvs-problem in NP is obvious.) □

Theorem 7: There is a $O(|F| \cdot n^2)$ approximation algorithm for the fvs-problem of tournaments T, which determines fvs's F of T satisfying $|F|/f(T) \leq 3$.

Sketch of proof:

Decompose T first into its s.c.c.'s in time $O(n^2)$. Then process every s.c.c. T' with at least 3 vertices as follows. Determine a triangle D (= cycle of length 3) in T'. Add $V(D)$ into F and process $T'-V(D)$ like T.

A triangle D in T' can be determined as follows. Determine any cycle C in T'. If C' is not a triangle, then take two arbitrary vertices $u, v \varepsilon V(C)$, which are nonadjacent on C. Because either $(u,v)\varepsilon A(T')$ or $(v,u)\varepsilon A(T')$, a shorter cycle C' is obtained, et cetera, until a triangle will be found.

Finding and deleting a triangle D from T' can be performed in $O(n)$ steps (using adjacency representation of T). Therefore the claimed running time follows.

The worst case ratio of 3 comes from the fact, that the approximation algorithm described above, besides the fvs F, also determines $|F|/3$ vertex disjoint triangles of T. □

5. CONCLUDING REMARKS

Introducing the procedure SIMPLIFY as a subroutine of MFVS results into a polynomial time approximation algorithm for the fvs-problem, which often will find optimal or nearly optimal fvs's of digraphs. A challenging aim is to reduce its running time, by introducing parallelism and by approximating the mean return times r_i, needed in MFVS, by an experiment.

The unsolved question in this paper, which we are most interested in at the moment, is to prove or disprove whether the worst case ratio of MFVS is bounded by $O(\log n)$, for arbitrary digraphs.

231

6. ACKNOWLEDGEMENTS

Thanks to P. Heusch and to F. Höfting for several helpful discussions and for sup-
plying the implementations and test results mentioned in this paper.
Thanks also to an unknown referee, who suggested the simplified proof of Theorem 1.

7. REFERENCES

[1] M. Bidjan-Irani. Wissensbasierte Systeme zur Sicherstellung der Entwurfsqua-
lität hochintegrierter Schaltungen und Systeme. PhD-thesis in preparation,
Paderborn (1988)

[2] P. Erdös and J. Spencer. Probabilistic Methods in Combinatorics. Academic Press,
New York, 1974

[3] R.W. Floyd. Assigning meaning to programs. Proc. Symp. Appl. Math. 19 (1967),
19 - 32

[4] M.R. Garey and D.S. Johnson. Computers and Intractability, a guide to the theory of NP-completeness. W.H. Freeman and Comp., San Francisco, 1979

[5] M.R. Garey and R.E. Tarjan. A linear time algorithm for finding all feedback
vertices. Inform. Process. Lett. 7 (1978), 274 - 276 .

[6] P. Heusch. Untersuchungen über das Feedback-Arc-Set in planaren Graphen.(Diplomarbeit) Bericht Nr. 51, Reihe Informatik, Paderborn 1988

[7] D.S. Johnson. Approximation algorithms for combinatorial problems. J. Comput.
System Sci. 9 (1974), 256 - 278

[8] K. Mehlhorn. Data Structures and Algorithms 2: NP-Completeness and Graph Algorithms. Springer-Verlag Berlin, 1984

[9] B. Monien and R. Schulz. Four approximation algorithms for the feedback vertex
set problem. Proc. 7th Conf. on Graph Theoretic Conc. of Comput. Sci., Hanser
Verlag, München, 1981, 315 - 326

[10] K.B. Reid and L.W. Beineke. Tournaments. In: Selected topics in graph theory 1.
L.W. Beineke/R.J. Wilson, eds.. Academic Press, London, 1978, 169 - 204

[11] B.K. Rosen. Robust linear algorithms for cutsets. J. Algorithms 3 (1982),
205 - 217

[12] S.M. Ross. Introduction to Probability Models. Academic Press, Inc., Orlando,
1985

[13] A. Shamir. A linear time algorithm for finding minimum cutsets in reducible
graphs. SIAM J. Comput. 8 (1979), 645 - 655

[14] E. Speckenmeyer. On feedback vertex sets and nonseparating independent sets
in cubic graphs. in: J.Graph Theory 12 (1988), 405-412

[15] C. Wang, E.L. Lloyd, and M.L. Soffa. Feedback vertex sets and cyclically reducible graphs. J. Assoc. Comput. Mach. 32 (1985), 296 - 313

[16] F. Höfting and E. Speckenmeyer. Feedback vertex set simple digraphs.
Technical Report, Paderborn 1989

IMPROVED SELF-REDUCTION ALGORITHMS FOR GRAPHS WITH BOUNDED TREEWIDTH

Hans L. Bodlaender

Department of Computer Science, University of Utrecht

P.O.Box 80.089, 3508 TB Utrecht, the Netherlands

Abstract

Recent results of Robertson and Seymour show, that every class that is closed under taking of minors can be recognized in $\mathcal{O}(n^3)$ time. If there is a fixed upper bound on the treewidth of the graphs in the class, i.e. if there is a planar graph not in the class, then the class can be recognized in $\mathcal{O}(n^2)$ time. However, this result is non-constructive in two ways: the algorithm only decides on membership, but does not construct 'a solution', e.g. a linear ordering, decomposition or embedding; and no method is given to find the algorithms. In many cases, both non-constructive elements can be avoided, using techniques of Fellows and Langston, based on self-reduction. In this paper we introduce two techniques that help to reduce the running time of self-reduction algorithms. With help of these techniques we show that there exist $\mathcal{O}(n^2)$ algorithms, that decide on membership and construct solutions for treewidth, pathwidth, search number, vertex search number, node search number, cutwidth, modified cutwidth, vertex separation number, gate matrix layout, and progressive black-white pebbling, where in each case the parameter k is a fixed constant.

1 Introduction.

A graph G is said to be a minor of a graph H, if G can be obtained from a subgraph of H by a number of edge-contractions. (An edge-contraction is the operation that replaces two adjacent vertices v, w by a new vertex that is adjacent to all vertices, adjacent to v or w.) Robertson and Seymour [25] have shown that for every class of graphs F, that is closed under taking of minors, there is a finite set of graphs $ob(F)$, the obstruction set of F, such that for all graphs G: $G \in F$, if and only if there is no graph $H \in ob(F)$ that is a minor of G. Further, there is an $\mathcal{O}(n^3)$ algorithm for every fixed graph H, that tests whether H is a minor of a given graph G [29]. Thus, one can test membership in F in $\mathcal{O}(n^3)$ time. If the treewidth of graphs in F is bounded by some constant (or, equivalently, if there is at least one planar graph that is not in F [26]), then the minor-tests, and hence the membership in F-test can be done in $\mathcal{O}(n^2)$ time [29]). A similar characterization with an obstruction set exists for classes of graphs that are closed under immersions [24].

G is an immersion of H, if G can be obtained from a subgraph of H by a number of edge-lifts. An edge-lift is the operation that replaces edges (v, w) and (w, x) by an edge (v, x). For fixed H, one can test whether a given graph G contains H as an immersion in polynomial time, and in $\mathcal{O}(n^2)$ time if the treewidth of G is bounded. Many applications of these results were obtained by Fellows and Langston [14, 12, 15, 13].

Note that these results are non-constructive in two ways: the algorithms only decide on membership in the class, but do not construct a solution like a linear ordering, decomposition or embedding, and no method is given to construct the algorithm: to write down this type of algorithm we must know the obstruction set of the class of graphs we want to recognize. However, in many cases, both non-constructive elements can be avoided with techniques of Brown, Fellows and Langston

[10] and Fellows and Langston [16, 13], based on self-reductions. We concentrate in this paper on the problem on finding solutions, for the case that there is a bound on the treewidth of the graphs.

Self-reduction is the technique to consult the decision algorithm a number of times with inputs derived from the original input, in order to construct the 'solution' to the problem. Algorithms of this type are also called 'oracle algorithms', and the decision algorithm is called the 'oracle'. The overhead of an oracle algorithm is the time, required for all operations, except those of the oracle, where each call to the oracle is counted as one unit of time.

This paper is organized as follows. In section 2 we review a number of definitions and results. In sections 3 and 4 we introduce two new techniques, that help to design faster constructive algorithms for immersion and minor closed classes of graphs with a fixed bound on the maximum treewidth. In section 5 we apply these techniques, and show that there exist $\mathcal{O}(n^2)$ algorithms that decide on membership and construct solutions for treewidth, pathwidth, search number, vertex search number, node search number, cutwidth, modified cutwidth, vertex separation number and gate matrix layout, where in each case the parameter k is a fixed constant. For each of these problems, except treewidth, algorithms with running time between $\mathcal{O}(n^3)$ and $\mathcal{O}(n^4)$ were designed by Fellows and Langston [13, 16]. These algorithms are non-constructive, in the sense, that we know that the algorithms exists, but we do not know the algorithms itself, as we do not know the corresponding obstruction sets. However, in section 6 we show that we can avoid this problem, and construct the corresponding algorithms, using a technique of Fellows and Langston [16].

2 Definitions and preliminary results.

In this section we give a number of well-known definitions and results. First we consider the important notion of treewidth, which was introduced by Robertson and Seymour [27].

Definition.
Let $G = (V, E)$ be a graph. A tree-decomposition of G is a pair $(\{X_i | i \in I\}, T = (I, F))$, with $\{X_i | i \in I\}$ a family of subsets of V, and T a tree, with the following properties

- $\bigcup_{i \in I} X_i = V$

- For every edge $e = (v, w) \in E$, there is an $i \in I$, with $v \in X_i$ and $w \in X_i$.

- For all $i, j, k \in I$: if j lies on the path from i to k in T, then $X_i \cap X_k \subseteq X_j$.

The treewidth of a tree-decomposition $(\{X_i | i \in I\}, T)$ is $\max_{i \in I} |X_i| - 1$. The treewidth of G, denoted by treewidth(G), is the minimum treewidth of a tree-decomposition of G, taken over all possible tree-decompositions of G.

There are several alternative ways to characterize the class of graphs with treewidth $\leq k$. See e.g. [1].

Very few NP-hard problems stay NP-hard, when we restrict them to a class of graphs with some fixed upper bound on the treewidth of the graphs in the class (see e.g. [3, 5, 7, 11, 19, 30]). In this paper we consider the approach of Courcelle [11] and Arnborg, Lagergren and Seese [3], as this approach appears to be most suitable for our purposes.

Courcelle [11] showed that every property that can be expressed in monadic second order form, can be tested in linear time for graphs that are given together with a tree-decomposition with constant bounded treewidth. Arnborg, Lagergren and Seese [3] extended the class of problems that can be dealt with. Consider logical formula's, that can use the following ingredients: the usual logical operations ($\land, \lor, \rceil, \Rightarrow$, etc.), quantifications over vertices ($\exists v \in V, \forall v \in V$), edges ($\exists e \in E, \forall e \in E$), sets of vertices ($\exists W \subseteq V, \forall W \subseteq V$), and sets of edges ($\exists F \subseteq E, \forall F \subseteq E$), equality tests $(v = w, e = f, (v,w) = e)$, membership tests $(v \in W, e \in F)$, and incidence tests $((v,w) \in E, (v,w) \in F)$. The formula may be open, where the free variables are given interpretations as pre-specified vertices, edges, sets of vertices, or sets of edges, respectively. Properties, that are expressed in this way are called monadic second order graph properties. We use the following variant of the results of Courcelle [11] and Arnborg, Lagergren and Seese [3].

Theorem 2.1 [11, 3]
Let k be a constant. Let $\exists v \in V \Phi(G, v)$ $(\exists e \in E \Phi(G, e))$ be a monadic second order graph property. Then a *linear time* algorithm can be constructed, that given a graph $G = (V, E)$ together with a tree-decomposition of G with treewidth $\leq k$, either finds a vertex $v \in V$ such that $\Phi(G, v)$ (an edge $e \in E$, such that $\Phi(G, e)$), or decides that such a vertex v (edge e) does not exist.

For our purposes, it is important to note that for fixed graphs H, the properties "H is a minor of G", or "H is an immersion of G" can be expressed as monadic second order graph properties. In order to find tree-decomposition with small treewidth, we can use the following result, which is a direct corollary of results of Robertson and Seymour [28, 29].

Theorem 2.2 (Robertson, Seymour)
For every fixed $k \geq 1$, there is an $\mathcal{O}(n^2)$ algorithm that given a graph $G = (V, E)$, either decides that the treewidth of G is larger that k, or finds a tree-decomposition of G with treewidth $\leq 4\frac{1}{2}k$.

(Robertson and Seymour considered "branchwidth", and obtained a bound of $3k$.) Thus, for graphs with constant bounded treewidth, we can find in $\mathcal{O}(n^2)$ time a tree-decomposition with treewidth still bounded by a constant, although it does not have optimal treewidth. (The algorithm is fully constructive, and uses $\mathcal{O}(p(k) \cdot 3^k \cdot n)$ time, where p is a polynomial.)

Some other definitions we use:

$$
\begin{aligned}
G[W] \quad &: \quad \text{the subgraph of } G, \text{ induced by } W, \text{ i.e. } (W, \{(v,w) \in \\
&\quad\ E | v, w \in W\}) \\
G - \{v\} \quad &: \quad \text{the subgraph of } G, \text{ induced by } V - \{v\}, G[V - \{v\}]. \\
G - \{e\} \quad &: \quad \text{the graph } (V, E - \{e\}) \\
\text{clique } \{v_1, \ldots, v_k\} \quad &: \quad \text{the complete graph on } \{v_1, \ldots, v_k\} : (\{v_1, \ldots, v_k\}, \{(v_i, v_j) \mid \\
&\quad\ 1 \leq i, j \leq k, i \neq j\}) \\
G \cup H \quad &: \quad \text{the (not necessarily disjoint) union of } G \text{ and } H.
\end{aligned}
$$

3 Quiet self-reductions.

In this section we propose the notion of "quiet" self-reductions. This rather simple idea is based on a closer observation of the $\mathcal{O}(n^2)$ minor test algorithm for graphs with bounded treewidth. Basically, this algorithm consists of two phases. In the first phase, the algorithm, indicated in theorem 2.2 is run. Either we decide that the treewidth of G is too large, and we know that G is a "no"-instance, or we find a tree-decomposition with treewidth $\mathcal{O}(1)$. This phase costs $\mathcal{O}(n^2)$ time. In the second phase, the tree-decomposition is used to perform the actual minor test, using dynamic programming, as in [3, 5, 7, 11, 30]. This second phase uses $\mathcal{O}(n)$ time.

An oracle algorithm may call this procedure a number of times, each time for a new graph G' that is obtained from modifications of the original input graph G. The number of such calls is usually at least $\mathcal{O}(n)$. Thus, it may be possible to save time, if we could be able to avoid the first phase for most of the calls to the minor-test algorithm. The following definition expresses a class of such oracle algorithms.

Definition.
An oracle algorithm is *quiet*, if, there are constants c_1, c_2, such that for any graph $G = (V, E)$, when

the algorithm runs with G as input, then every graph $H = (W, F)$, that is input to the oracle, fulfills:

(i) $|W - (V \cap W)| \leq c_1$

(ii) $\exists W' \subseteq V : |W'| \leq c_2 \wedge ((v, w) \in F - (E \cap F) \Rightarrow v \in W' \vee w \in W')$

In other words, every graph that is input to the oracle, can be obtained by taking a subgraph of G, adding a constant number of new vertices, and for a constant number of vertices, adding a number of edges, starting at that vertex.

Theorem 3.1
Let k, l be constants. Let A be a quiet oracle algorithm, such that

(i) A yields the answer "no", if the treewidth of input graph G is larger than k.

(ii) A has overhead $f(n)$, and makes $g(n)$ oracle calls to an oracle $\mathcal{O}$.

(iii) A call to oracle $\mathcal{O}$ costs $\mathcal{O}(n)$ time, if the input-graph H to the oracle, is given together with a tree-decomposition of H with treewidth $\leq l$.

Then A can be implemented with an algorithm, that uses $\mathcal{O}(f(n) + n^2 + g(n) \cdot n)$ time.

Proof.
First run the algorithm of Robertson and Seymour, that either decides that treewidth $(G) \geq k$, or finds a tree-decomposition of G with constant treewidth. In the former case, output "no", and we are done. In the latter case, note that for input graph H to the oracle, one can find in $\mathcal{O}(n)$ time a tree-decomposition of H with constant treewidth: use the tree-decomposition of G, remove all vertices in $G - H$, and then add to each set X_i all vertices in $H - G$, and all vertices in the set W', defined by (v, w) edge in $H - G \Rightarrow v \in W' \vee \in W'$, $|W'|$ bounded by a constant. A tree-decomposition of H with constant bounded treewidth results. Hence, the total time for all oracle calls is bounded by $\mathcal{O}(g(n) \cdot n)$, and the total time for the algorithm is bounded by $\mathcal{O}(f(n) + n^2 + g(n) \cdot n)$. $\qquad \square$

Theorem 3.1. can be applied to several problems, considered in [13]. In many cases, improvements with a factor up to $\mathcal{O}(n)$ can be made. However, we need a second technique in order to obtain $\mathcal{O}(n^2)$ algorithms.

4 Using monadic second order graph properties instead of minor tests.

In this section we describe our second technique. It is based on the observation, that not only minor-tests, but also more complicated questions, if we write them as monadic second order graph properties, can be tested in linear time, given a tree-decomposition of G with constant bounded treewidth.

Lemma 4.1
Let $\varphi(G)$ be a monadic second order graph property. Each of the following properties can be expressed in monadic second order form:

(i) $\varphi(G - \{v\})$ $(G - \{v\} = (V - \{v\}, \{(w, x) | w \neq v, x \neq v\})$

(ii) $\varphi(G - \{e\})$ $(G - \{e\} = (V, E - \{e\}))$

(iii) $\varphi(G' = (V, E \cup \{(v, w)\}))$

(iv) $\varphi(G' = (V, E \cup \{(v, w) | w \in W\}))$

(v) $\varphi(G' = (V, E - \{(v, w) | w \in W\}))$

(v, w, e, W are free variables in the resulting formulas, but not in φ.)

Proof.
We can rewrite φ, inductively. For example, consider (i). We only consider a few cases, the other are similar. If $\varphi = \varphi_1 \vee \varphi_2$, then $\varphi(G - \{v\}) = \varphi_1(G - \{v\}) \vee \varphi_2(G - \{v\})$. If $\varphi = \exists w \in V \varphi_1(W, G)$, then $\varphi(G - \{v\}) = \exists w \in V : v \neq w \wedge \varphi_1(w, G - \{v\})$. If $\varphi = \exists W \subseteq V : \varphi_1(W, G)$, then $\varphi(G - \{v\}) = \exists W \subseteq V : \daleth(v \in w) \wedge \varphi_1(W, G - \{v\})$. If $\varphi = (w = x)$, or $((w, x) \in F)$, then $\varphi(G - \{v\}) = \varphi$. The other cases are similar. $\qquad\square$

This result can often be applied in the following way. Suppose we look for a vertex v (or edge e), such that G, with some local operations applied on v (or e) remains in a minor-closed class F. Lemma 4.1 shows that, if we can write these local operations in a suitable form, then we can find such a vertex v (or edge e), in linear time (supposing G is given with a constant width tree-decomposition). This can save up to a factor of $\mathcal{O}(n)$ time in comparison to algorithms that test for each vertex v the resulting modified graph G separately.

As a first example, consider the "k vertices within F" problem for a minor closed class of graphs F, with a fixed upper bound on the treewidth of graphs in F. (This problem, without assumptions on the treewidth was considered in [10].) I.e., we must find k vertices $v_1, \ldots, v_k$, such that $G - \{v_1, \ldots, v_k\} \in F$. One easily sees that if G is a "yes"-instance to this problem, then the treewidth of G is bounded by the maximum treewidth of a graph in F plus k.

Using the characterization with obstructions, the property $G \in F$ can be written as a monadic second order graph property, hence we can write $\exists v_1 \exists v_2 \ldots \exists v_k \ G - \{v_1, \ldots, v_k\} \in F$ as a monadic second order property. (Apply lemma 4.1(i) k times.) So suppose G is given together with a constant width tree-decomposition . Then we can find v_1 in linear time, using lemma 2.1. Using again lemma 2.1 on the property $\exists v_2 \ldots \exists v_k G - \{v_1, \ldots, v_k\} \in F$ we see that we can find v_2 in linear time. (v_1 is now a free variable with a predetermined value.) So, in k steps, each using $\mathcal{O}(n)$ time, we find $v_1, \ldots, v_k$, such that $G - \{v_1, \ldots, v_k\} \in F$, if they exist.

Theorem 4.2
Let k be a constant, and let F be a minor-closed class of graphs with a fixed upper bound on the treewidth of graphs in F. Then there exists an $\mathcal{O}(n)$ time algorithm, that given a graph $G = (V, E)$, together with a tree-decomposition of G with constant bounded treewidth , finds k vertices $v_1, \ldots, v_k$ (k edges $e_1, \ldots, e_k$), such that $G - \{v_1, \ldots, v_k\} \in F$ ($G - \{e_1, \ldots, e_k\} \in F$), or decides that such collection of vertices (edges) does not exists.

5 Faster constructive algorithms for various problems.

5.1 Treewidth.

We now show that, for fixed k, there exists an $\mathcal{O}(n^2)$ algorithm that constructs a tree-decomposition with treewidth $\leq k$, or decides that such a tree-decomposition does not exist for a given graph G. This improves on an $\mathcal{O}(n^{k+2})$ algorithm by Arnborg, Corneil and Proskurowski [2]. For $k = 1, 2, 3$ there exist linear time algorithms [4, 21]. For variable k, the problem is NP-complete [2].

Lemma 5.1 [6]
Let $G = (V, E)$ be a graph, with $W \subseteq V$ is a clique in G. Then, for any tree-decomposition $(\{X_i | i \in I\}, T = (I, F))$ of G, there exists an $i \in I$ with $W \subseteq X_i$.

Our algorithm is based on the following lemma's, which are slight modifications of lemma's of Arnborg, Corneil and Proskurowski [2].

Lemma 5.2

(i) If $G = (V, E)$ has treewidth $\leq k$, and $|V| \geq k$, then there exist vertices $v_1, \ldots, v_k \in V$, with $G' = (V, E \cup \{(v_i, v_j)|1 \leq i, j \leq k, i \neq j\}) = G \cup$ clique $(v_1, \ldots, v_k)$ has treewidth $\leq k$.

(ii) Suppose treewidth $(G) \leq k$, $v_1, \ldots, v_k$ form a clique in G. Let $V_1, \ldots, V_r$ be the sets of vertices of the connected components of $G[V - \{v_1, \ldots, v_k\}]$. Then, for all $i, 1 \leq i \leq r$: $G[V_i \cup \{v_1, \ldots, v_k\}]$ has treewidth $\leq k$.

(iii) Suppose treewidth $(G) \leq k, v_1, \ldots, v_k$ form a clique in G. Suppose $G[V - \{v_1, \ldots, v_k\}]$ is connected, $|V| \geq k + 2$. Then there exists a vertex $w \in V - \{v_1, \ldots, v_k\}$, such that $G' = (V, E \cup \{(v_i, w)|1 \leq i \leq k\})$ has treewidth $\leq k$. Moreover, for each connected component V_i of $G[V - \{v_1, \ldots, v_k, w\}]$, there is at least one vertex $w_i \in \{v_1, \ldots, v_k, w\}$ with $\forall v \in V_i (v, w_i) \neq E$, and $G'[V_i \cup (\{v_1, \ldots, v_k, w\} - \{w_i\}]$ has treewidth $\leq k$.

We now sketch our algorithm.

1. Run the "approximate tree-decomposition" algorithm of Robertson and Seymour (see lemma 2.2). If it tells us that the treewidth of G is larger than k, then output "no", and stop. Otherwise we have a tree-decomposition of G with treewidth $\mathcal{O}(1)$.

2. Test whether G contains a minor in the obstruction set of the graphs with treewidth $\leq k$. If so, then output "no", and stop. Otherwise, we know that treewidth $(G) \leq k$, and we continue with step 3. (Step 1 and 2 basically form the recognition algorithm of Robertson and Seymour [29]).

3. Find vertices $v_1, \ldots, v_k$ as indicated in lemma 5.2(i). This can be done in linear time, using lemma 2.1, observing that

$$\exists v_1 \in V \exists v_2 \in V \cdots \exists v_k \in V : G' = (V, E \cup \{(v_i, v_j)|1 \leq i, j \leq k, i \neq j\}) \text{has treewidth} \leq k.$$

can be written as a monadic second order graph property, by using lemma 4.1, and the characterization with forbidden minors. With k applications of lemma 2.1 we find vertices $v_1, v_2, \ldots, v_k$.

4. Determine the connected components $V_1, \ldots, V_r$ of $G[V - \{v_1, \ldots, v_k\}]$. Let $G' = G \cup$ clique$\{v_1, \ldots, v_k\}$.

5. Now, for each $i, 1 \leq i \leq r$, find a tree-decomposition with treewidth $\leq k$ of $G'[V_i \cup \{v_1, \ldots, v_k\}]$, with a procedure, described below. Then build a tree-decomposition of G with treewidth $\leq k$ as follows:

   (a) Observe that, for each $i, 1 \leq i \leq r$, there must be a set $X_{\alpha(i)}$ in the tree-decomposition of $G'[V_i \cup \{v_i, \ldots, v_k\}]$ that contains $v_1, \ldots, v_k$, as these vertices form a clique in this graph.

   (b) Now take the disjoint union of all r tree-decompositions, add an extra set $X_o = \{v_1, \ldots, v_k\}$, and add a tree-edge from X_o to $X_{\alpha(i)}$ for all $i, 1 \leq i \leq r$. One can now check that a correct tree-decomposition of G with treewidth $\leq k$ results.

6. Next we describe a recursive procedure, that given a set $W \subseteq V$, with $G[W]$ connected, and vertices $v_1, \ldots, v_k \in V$, finds a tree-decomposition of $G' = G[W \cup \{v_1, \ldots, v_k\}] \cup$ clique$(\{v_1, \ldots, v_k\})$, with treewidth $\leq k$. The set, associated to the root of the resulting tree contains $v_1, \ldots, v_k$.

   (a) If $|W| \leq 1$, then take the tree-decomposition $(\{X_1 = W \cup \{v_1, \ldots, v_k\} \}, (\{1\}, \emptyset))$.

(b) Otherwise, find a vertex $w \in W$, with $G[W \cup \{v_1, \ldots, v_k\}] \cup$ clique $(\{v_1, \ldots, v_k, w\})$ has treewidth $\leq k$. This can be done in linear time, with the methods, exposed in sections 3 and 4. Lemma 5.2 guarantees us, that such a vertex w exists.

(c) Determine the connected components $W_1, \ldots, W_r$, of $G[W - \{w\}]$.

(d) For each of these connected components W_i, find $W_i \in \{v_1, \ldots, v_k, w\}$ with $\forall v \in W_i(v, w_i) \notin E$. (See lemma 5.2). Now call the procedure recursively with set W_i and vertices $\{v_1, \ldots, v_k, w\} - \{w_i\}$.

(e) So now we have tree-decompositions of all graphs $G_i = G[W_i \cup (\{v_1, \ldots, v_k, w\} - \{w_i\})] \cup$ clique$(\{v_1, \ldots, v_k, w\} - \{w_i\})$, with treewidth $\leq k$. The root of such a tree-decomposition contains $\{v_1, \ldots, v_k, w\} - \{w_i\}$. The desired tree-decomposition of G' can be built as follows: take the disjoint union of the tree-decomposition of G_i $(1 \leq i \leq r')$. Take a new set $X_r = \{v_1, \ldots, v_k, w\}$, which is taken as the root of the new tree-decompositions. Connect X_r to each of the roots of the tree-decomposition of graphs G_i. One can check that indeed the resulting structure is a tree-decomposition of G' with treewidth $\leq k$.

This completes the description of the algorithm. The time needed for steps 1,2,3,4 and 5 is bounded by $\mathcal{O}(n^2)$. Each call of the procedure in step 6 costs $\mathcal{O}(n)$ time, and this procedure is called $\mathcal{O}(n)$ times. Thus, the total time of our algorithm is $\mathcal{O}(n^2)$.

Theorem 5.3
For each constant k, there exists an $\mathcal{O}(n^2)$ algorithm, that for a given graph $G = (V, E)$ either decides that the treewidth of G is larger than k, or finds a tree-decomposition of G with treewidth $\leq k$.

(In [8] an (easier) $\mathcal{O}(n^3)$ algorithm is given.)

5.2 Search number

In this section we consider the search number, vertex search number and node search number of a graph. A search strategy of a graph is a sequence of the following types of moves:

1. Place a searcher on a vertex.

2. Delete a searcher from a vertex.

3. Move a searcher over an edge.

All edges are initially *contaminated*. An edge (v, w) can become *cleared* by moving a searcher from v to w, while there is a second searcher on v, or all other edges, adjacent to v are already cleared. An edge can become *recontaminated*, when a move results in a path without searchers from a contaminated edge to the edge. The search number of G is the minimum number of searchers needed to clear all edges. It has been shown by LaPaugh [18] that for every graph G, there exists a search sequence, that uses the optimal number of searchers, and does not allow recontamination. Such a search sequence is called *progressive*. If we let vertices instead of edges be cleared or contaminated, then the vertex search number is the minimum number of searchers, needed to clear all vertices with a progressive search sequence. Determining the minimum number of searchers needed is NP-complete [22]. Note that for a progressive search strategy that clears all vertices, we may assume that never two searchers occupy the same vertex, and that once a searcher has left a vertex, then no searcher will visit that vertex again. Note that, for fixed k, the classes of graphs with search number of vertex search number $\leq k$ are closed under taking of minors. Also, if the (vertex) search number of G is k, then treewidth $(G) \leq k$ (see e.g. [9]). The node search number of a graph was introduced by Kirousis and Papadimitriou [17]. Here an edge is cleared by having a searcher on both its endpoints.

Definition.
$BW(n, k) = (\{v_1, \ldots, v_n\}, \{(v_i, v_j) | 1 \leq i, j \leq n, |i - j| \leq k\}).$

($BW(n, k)$ is the maximal graph on n vertices with bandwidth k.)

Lemma 5.4

(i) If there exists a progressive search strategy that clears all vertices of $G = (V, E)$ with k searchers, then there exists one with the first k moves the placing of a searcher ($|V| \geq k$).

(ii) There exists a progressive search strategy that clears all vertices of G, and that starts with placing a searcher on vertices $w_1, \ldots, w_k$, if and only if the graph G', obtained by taking the disjoint union of G and $BW(2k + 1, k)$ and then identifying vertices v_i and w_i for $1 \leq i \leq k$, has vertex search number $\leq k$.

Proof.
(i) One can obtain the desired search strategy by first executing the first k moves of the type "place a searcher on a vertex", and then executing all other moves in sequence.

(ii) $\Rightarrow$ Use the following search strategy: place searchers on vertices $v_{k+2}, v_{k+3}, \ldots, v_{2k+1}$. Then move a searcher from v_i to v_{i-k}, for $i = 2k + 1, 2k, 2k - 1, \ldots, k + 1$. Now we have searchers on $w_1 = v_1, \ldots, w_k = v_k$. Continue with the given search strategy that clears all vertices of G and starts with searchers on $w_1, \ldots, w_k$.

$\Leftarrow$ Consider a progressive search strategy that clears all vertices of G' with k searchers. Consider the first move that removes a searcher from a vertex $v_i \in \{v_{k+1}, \ldots, v_{2k+1}\}$ or moves a searcher away from a vertex $v_i \in \{v_{k+1}, \ldots, v_{2k+1}\}$. It follows that after this move, all vertices in $\{v_{k+1}, \ldots, v_{2k+1}\} - \{v_i\}$ must contain a searcher. Also, vertices $v_{i-1}, \ldots, v_{i-k}$ must be cleared or contain a searcher. It follows that either all vertices in G are cleared or uncleared. Suppose the latter. It follows that $i = 2k + 1$, and that the next $k + 1$ moves are: move a searcher from v_j to v_{j-k}, for $j = 2k + 1, \ldots, k + 1$. We then have searchers on $v_1 = w_1, \ldots, v_k = w_k$. The remaining search sequence is a progressive search sequence that clears all vertices of G, starting with searchers on $v_1, \ldots, v_k$. In the case that all vertices in G are cleared after moving the searcher from v_i, we use the same argument for the search strategy, that is obtained by "reversing" the original strategy. $\square$

Lemma 5.4 gives us a method to find the first k vertices where a searcher is placed: we must find vertices $v_1, \ldots, v_k$, such that the graph $(V \cup \{w_{k+1}, \ldots, w_{2k+1}\}, E \cup \{(v_i, v_j) | 1 \leq i, j \leq k, i \neq j\} \cup \{(v_i, w_j) | 1 \leq i \leq k < j \leq i + k\} \cup \{(w_i, w_j) | k \leq i, j \leq 2k + 1, 0 < |i - j| \leq k\})$ has vertex search number k. This can be done in linear time, by combining the techniques of section 3 and 4. (Add $w_{k+1}, \ldots, w_{2k+1}$ to G, and all edges between them. Construct a constant width tree-decomposition of the resulting graph G'. Write: "$\exists v_1 \ldots \exists v_k$: if we add edges between $v_i, v_j, 1 \leq i \leq j \leq k$, and v_i, w_j with $j - i \leq k$, then the resulting graph has vertex search number $\leq k$" as a monadic second order graph property, using the characterization with forbidden minors, and lemma 4.1.).

In order to find the next move, we distinguish between two cases.

Case 1: There exists a vertex v_i, containing a searcher, that is not adjacent to a vertex that does not contain a searcher. Then one may assume that the next move in our search strategy is the removal of the searcher from v_i. Now remove v_i from G, and find, recursively a progressive search strategy that clears all vertices of $G - \{v\}$ with k searchers, that starts with placing a vertex on each of the vertices $v_1, \ldots, v_{i-1}, v_{i+1}, \ldots, v_k$. (This can be done, similar as how we found the first k moves. Only let $v_1, \ldots, v_{i-1}, v_{i+1}, \ldots, v_k$ be free variables in the monadic second order graph property that is tested.) The resulting search sequence can easily be extended to the desired search sequence.

<u>Case 2</u>: Such a vertex v_i does not exists. Then necessarily the next move must be the moving of a searcher from a vertex v_i to an adjacent uncleared vertex. Note that for each searcher, there is at most one vertex where it can move to, otherwise the vertex it leaves will become recontaminated. So we have to consider at most k possible moves. For each possible move from v_i to w, test whether there exists a progressive search strategy that clears the vertices of $G - \{v_i\}$ with k searchers, and that starts with placing a searcher on $v_1, \ldots, v_{i-1}, v_{i+1}, v_k$, and w. This can be tested, similar as above, in linear time. Do this for each of the k possible moves, until a good move is found. Then find recursively a progressive search strategy, that clears the vertices of G with k searchers, that starts with placing a searcher on $v_1, \ldots, v_{i-1}, v_{i+1}, \ldots, v_k$ and w.

As in total $\mathcal{O}(n)$ moves are made, the total time of the algorithm is $\mathcal{O}(n^2)$.

Theorem 5.5
For each constant k, there exists an $\mathcal{O}(n^2)$ algorithm, that for a given graph $G = (V, E)$, either decides that the vertex search number of G is larger than k, or finds a progressive search sequence that clears the vertices of G with k searchers.

We can use this result to obtain the following theorem as an easy corollary.

Theorem 5.6
(i) For each constant k, there exists an $\mathcal{O}(n^2)$ algorithm, that for a given graph $G = (v, E)$ either decides that the search number of G is larger than k, or finds a progressive search strategy that clears all edges of G with k searchers.
(ii) For each constant k, there exists an $\mathcal{O}(n^2)$ algorithm, that for a given graph $G = (v, E)$ either decides that the node search number of G is larger than k, or finds a progressive search strategy that clears all edges of G with k searchers.

Proof.
(i) Let G' be the graph, obtained by subdividing each edge in G once, i.e. $G' = (V \cup E, \{(v, e) | v \in V, e \in E, \exists w \in V : (v, w) = e\})$. The vertex search number of G' equals the search number of G, and the corresponding search strategies can be easily transformed into each other.
(ii) Use the transformation of theorem 2.5 of [17]. $\qquad\square$

5.3 Other problems.

Several other problems can be dealt with in the same manner. The techniques are similar to those used for the vertex search number problem, but the details are different.

We give the definitions of the problems, that are considered. A linear ordering of a graph $G = (V, E)$ is a bijection $V \to \{1, 2, \ldots, |V|\}$. The cutwidth of a linear ordering f is $\max_{1 \leq i < n} |\{(v, w) \in E | f(v) \leq i < f(w)\}|$. The cutwidth of G is the minimum cutwidth over all linear orderings of G. The vertex separation of a linear ordering f is $\max_{1 \leq i < n} |\{v \in V | f(v) \leq i \wedge \exists w \in V : (v, w) \in E \cap f(w) > i\}|$. The vertex separation number of G is the minimum vertex separation over all linear orderings of G. The modified cutwidth of a linear ordering f is $\max_{1 \leq i < n} |\{(v, w) \in E | f(v) < i < f(w)\}|$. The modified cutwidth of G is the minimum modified cutwidth over all linear orderings of G. A path-decomposition of G is a tree-decomposition $(\{X_i | i \in I\}, T = (I, F))$ of G, where T is a path, i.e. T is a tree with every node degree 1 or 2. The pathwidth of path-decomposition $(\{X_i | i \in I\}, T = (I, F))$ is $\max_{i \in I} |X_i| - 1$. The pathwidth of G is the minimum pathwidth over all path-decompositions of G.

Determining the cutwidth, modified cutwidth, vertex separation number or pathwidth of a graph is NP-complete (see e.g. [2, 23]). For fixed k, Fellows and Langston obtained $\mathcal{O}(n^3)$ and $\mathcal{O}(n^3 \log n)$ algorithms [13, 16].

Theorem 5.7

(i) For every fixed k, there exists an $\mathcal{O}(n^2)$ algorithm that for a given graph $G = (V, E)$ either decides that the cutwidth of G is larger than k, or finds a linear ordering of G with cutwidth $\leq k$.

(ii) For every fixed k, there exists an $\mathcal{O}(n^2)$ algorithm that for a given graph $G = (V, E)$ either decides that the pathwidth of G is larger than k, or finds a path-decomposition of G with pathwidth $\leq k$.

(iii) For every fixed k, there exists an $\mathcal{O}(n^2)$ algorithm that for a given graph $G = (V, E)$ either decides that the vertex separation number of G is larger than k, or finds a linear ordering of G with vertex separation $\leq k$.

(iv) For every fixed k, there exists an $\mathcal{O}(n^2)$ algorithm either decides that for a given graph $G = (V, E)$, the modified cutwidth of G is larger than k, or finds a linear ordering of G with modified cutwidth $\leq k$.

Proof.
(i) Use the self-reduction of Brown, Fellows and Langston [10], and apply the techniques of section 3 and 4. Parallel edges can be avoided by putting a vertex on the middle of these edges.

(ii) The vertex separation number of a graph equals its node search number minus 1 [17] and the corresponding search strategy can be transformed into a linear ordering with correct vertex separation number, and vice versa, within $\mathcal{O}(n^2)$ time.

(iii) It is folklore that the pathwidth of a graph equals its node search number minus 1 and easy $O(n^2)$ transformations exist from the corresponding search strategy to a path-decomposition and vice-versa.

(iv) Omitted.

$\square$

Corollary 5.8
For every fixed k, there exists an $\mathcal{O}(n^2)$ algorithm, that for a given Boolean matrix M, either finds a column permutation of M such that if in each row every 0 lying between the row's leftmost and rightmost 1 is changed to a *, then no column contains more than k 1's and *s, or decides that such a column permutation does not exist.

Proof.
There is a linear transformation from this problem to the problem of finding path-decompositions with pathwidth $\leq k - 1$ [12, 16].

$\square$

Corollary 5.9
For every fixed k, there exists an $\mathcal{O}(n^2)$ algorithm that, for a given directed acyclic graph $G = (V, E)$, either gives a pebbling strategy for the progressive black-white pebbling game on G, that uses k pebbles (see [20] for a definition) or decides that such pebbling strategy does not exist.

Proof.
There is a linear transformation from this problem to the vertex separation problem and vice versa [20].

$\square$

6 Final remarks.

As discussed in section 1, the algorithms are non-constructive in the sense that one knows that an algorithm exists, but we do not have a concrete algorithm of which we know that it is correct. In order to overcome this problem, Fellows and Langston [13] designed a technique that allows us to actually construct algorithms without knowing the exact obstruction set. The technique does not work always, but only if we have an (efficient) oracle algorithm, producing "solutions", an (efficient) algorithm that checks whether a "candidate-solution" is a correct solution, and an arbitrary (other) algorithm, that decides on membership in the considered class of graphs. The resulting algorithm has the following form. Let T be a minor or immersion closed class of graphs.

1. Let S be a subset of the obstruction set of T.

2. Check whether the input graph G has a graph in S as a minor (or immersion). If so, output "no" ($G \neq T$), and stop. (If T is a class of graphs with bounded treewidth, then we output also "no" here and stop, if in this step it is detemined that the treewidth of G is too large.)

3. Use the oracle algorithm to produce a 'solution', using S as an obstruction set.

4. Check the produced solution. If it is a correct solution, output it ($G \in T$), and stop.

5. (S is not the complete obstruction set of T). Enumerate all graphs, until a graph $G, G \notin S, G \notin T$, all minors of G are in T, are found. (G is a new element of the obstruction set). Put G in S. Go to step 2.

Thus, we have the following theorem, that is a small variant of a result of Fellows and Langston ([13], thrm. 13).

Theorem 6.1 Let Σ be a finite alphabet. Let $\mathcal{G}$ denote the set of all graphs. Let $\Pi \subseteq \mathcal{G} \times \Sigma$ be a relation, such that the set of graphs $T = \{G \in \mathcal{G} \mid \exists x \in \Sigma^* : \Pi(G, x)\}$ is closed under taking of minors. Suppose the following are known:

1. an algorithm, that given a graph $G = (V, E)$ and a string $x \in \Sigma^*$, determines whether $\Pi(G, x)$ holds, and that uses $\mathcal{O}(f_1(|V|))$ time.

2. an algorithm, that given a graph $G = (V, E)$ and a graph $H = (W, F)$ time, either determines whether H is a minor of G or outputs correctly that $G \notin T$, and that uses $\mathcal{O}(f_2(|V|) \cdot f_3(|W|))$ time. (Note that in the former case, G may or may not be a member of T.)

3. an algorithm, that given the obstruction set of T, $ob(T)$, and given a graph $G = (V, E)$, either determines that $G \notin T$, or it outputs a string $x \in \Sigma^*$ with $\Pi(G, x)$, and that uses $\mathcal{O}(f_4(|V|))$ time, when given a graph G, and a subset of the obstruction set of T.

4. an algorithm that determines whether a given graph G is in T.

Then an algorithm is known, that given a graph $G = (V, E)$, either determines that $G \notin T$, or it outputs a string $x \in \Sigma^*$ with $\Pi(G, x)$, and that uses $\mathcal{O}(\max(f_1(|V|), f_2(|V|), f_4(|V|)))$ time.

We can apply this technique to each of the problems considered in section 5. In each case we get not only that an $\mathcal{O}(n^2)$ algorithm exists, but we can also construct such an algorithm. The only non-constructive element is that we do not know the size of the constants, as these depend on the size of the obstruction set. Observe that the method of Arnborg, Lagergren and Seese [3] that obtains a linear time algorithm on bounded treewidth graphs, given an extended monadic second order graph property, is in fact an automatic procedure. As our obstruction sets may grow, our monadic second order graph properties can vary during this algorithm, so we need to implement this procedure in order to combine the techniques, and obtain fully constructive $\mathcal{O}(n^2)$ algorithms for the problems considered in section 5.

An important disadvantage of the algorithms of this paper are the large constant factors: these are usually at least double exponential in k. Similar constant factors appear with the algorithms in [8], [13] and [14]. It should however be noted, that our constant factors are still elementary (in contrast with the extremely large constant factors that appear e.g., in [27, 29]).

References

[1] S. Arnborg. Efficient algorithms for combinatorial problems on graphs with bounded decomposability – A survey. *BIT*, 25:2–23, 1985.

[2] S. Arnborg, D. G. Corneil, and A. Proskurowski. Complexity of finding embeddings in a k-tree. *SIAM J. Alg. Disc. Meth.*, 8:277–284, 1987.

[3] S. Arnborg, J. Lagergren, and D. Seese. Problems easy for tree-decomposable graphs (extended abstract). In *Proc. 15 th ICALP*, pages 38–51. Springer Verlag, Lect. Notes in Comp. Sc. 317, 1988.

[4] S. Arnborg and A. Proskurowski. Characterization and recognition of partial 3-trees. *SIAM J. Alg. Disc. Meth.*, 7:305–314, 1986.

[5] S. Arnborg and A. Proskurowski. Linear time algorithms for NP-hard problems restricted to partial k-trees. *Disc. Appl. Math.*, 23:11–24, 1989.

[6] H. L. Bodlaender. Dynamic programming algorithms on graphs with bounded tree-width. Tech. rep., Lab. for Comp. Science, M.I.T., 1987. Ext. abstract in proceedings ICALP 88.

[7] H. L. Bodlaender. NC-algorithms for graphs with small treewidth. In J. van Leeuwen, editor, *Proc. Workshop on Graph-Theoretic Concepts in Computer Science WG'88*, pages 1–10. Springer Verlag, LNCS 344, 1988.

[8] H. L. Bodlaender. Polynomial algorithms for graph isomorphism and chromatic index on partial k-trees. In *Proc. 1st Scandinavian Workshop on Algorithm Theory*, pages 223–232, 1988.

[9] H. L. Bodlaender. Some classes of graphs with bounded treewidth. *Bulletin of the EATCS*, 1988.

[10] D. Brown, M. Fellows, and M. Langston. Nonconstructive polynomial-time decidability and self-reducibility. In *Princeton Forum on Algorithms and Complexity*, 1987.

[11] B. Courcelle. The monadic second-order logic of graphs I: Recognizable sets of finite graphs. Technical Report I-8837, Dept. Comp. Sc, Univ. Bordeaux 1, 1988.

[12] M. R. Fellows and M. A. Langston. Nonconstructive advances in polynomial-time complexity. *Inform. Proc. Letters*, 26:157–162, 1987.

[13] M. R. Fellows and M. A. Langston. Fast search algorithms for graph layout permutation problems. Technical Report CS-88-189, Dept. of Comp. Sc., Washington State Univ., 1988.

[14] M. R. Fellows and M. A. Langston. Layout permutation problems and well-partially-ordered sets. In *5th MIT Conf. on Advanced Research in VLSI*, pages 315–327, 1988.

[15] M. R. Fellows and M. A. Langston. Nonconstructive tools for proving polynomial-time decidability. *J. ACM*, 35:727–739, 1988.

[16] M. R. Fellows and M. A. Langston. On search, decision and the efficiency of polynomial-time algorithms. In *Proc. STOC*, pages 501–512, 1989.

[17] L. M. Kirousis and C. H. Papadimitriou. Searching and pebbling. *Theor. Comp. Sc.*, 47:205–218, 1986.

[18] A. S. LaPaugh. Recontamination does not help to search a graph. Technical report, Comp. Sci. Dept, Princeton Univ., New Yersey, 1982.

[19] C. Lautemann. Efficient algorithms on context-free graph languages. In *Proc. 15'th ICALP*, pages 362–378. Springer Verlag, Lect. Notes in Comp. Sc. 317, 1988.

[20] T. Lengauer. Black-white pebbles and graph separation. *Acta Inf.*, 16:465–475, 1981.

[21] J. Matoušek and R. Thomas. Algorithms finding tree-decompositions of graphs. Unpublished paper, 1988.

[22] N. Megiddo, S. L. Hakimi, M. R. Garey, D. S. Johnson, and C. H. Papadimitriou. The complexity of searching a graph. *J. ACM*, 35:18–44, 1988.

[23] B. Monien and I. H. Sudborough. Min cut is NP-complete for edge weighted trees. *Theor. Comp. Sc.*, 58:209–229, 1988.

[24] N. Robertson and P. Seymour. Graph minors. IV. Tree-width and well-quadi-ordering. Manuscript.

[25] N. Robertson and P. Seymour. Graph minors. XVI. Wagner's conjecture. To appear.

[26] N. Robertson and P. Seymour. Graph minors. III. Planar tree-width. *J. Comb. Theory Series B*, 36:49–64, 1984.

[27] N. Robertson and P. Seymour. Graph minors. II. Algorithmic aspects of tree-width. *J. Algorithms*, 7:309–322, 1986.

[28] N. Robertson and P. Seymour. Graph minors. X. Obstructions to tree-decompositions. Manuscript., 1986.

[29] N. Robertson and P. Seymour. Graph minors. XIII. The disjoint paths problem. Manuscript., 1986.

[30] P. Scheffler. Linear-time algorithms for NP-complete problems restricted to partial k-trees. Report R-MATH-03/87, Karl-Weierstrass-Institut Für Mathematik, Berlin, GDR, 1987.

Finding a Minimal Transitive Reduction in a Strongly Connected Digraph within Linear Time

Klaus Simon

ETH Zürich, Institut für Theoretische Informatik, CH-8092 Zürich, e-mail : simon@inf.ethz.ch

This paper describes an algorithm for finding a minimal transitive reduction G_{red} of a given directed graph G, where G_{red} means a subgraph of G with the same transitive closure as G but itself not contains a proper subgraph G_1 with the same property too. The algorithm uses depth-first search and two graph transformations preserving the transitive closure to achieve a time bound of $O(n+m)$, where n stands for the number of vertices and m is the number of the edges.

1. Introduction

One of the classic topics in computer science is computing the path information of a directed graph (digraph) $G = G(V, E)$, where $V(E)$ represents the vertex (edge) set. In this connection there are two extreme points of view:

   i) **A minimum query time representation.** Determine the unique digraph $G^*(V, E^*)$ which contains a directed edge from v to w if and only if there is a directed path from v to w in G.

   ii) **A minimum storage representation.** Determine the "edge minimal" graph $G_r(V, E_r)$ with the property $G_r^* = G^*$.

The graph G^* is traditionally called *transitive closure* of G. The appellation for G_r is not unique. In former papers (MOYES AND THOMPSON, HSU) the indication *minimum corresponding graph* is used for G_r but nowadays G_r is normally named *transitive reduction* of G. In order to avoid baffling in the application of the conception transitive reduction we have first of all to differentiate between the adjectives "minimal" and "minimum" related to G_r or E_r and on the other hand between $E_r \subseteq E$ and $E_r \subseteq V \times V$. In terms of a set "minimal" means a set property, whereas "minimum" refers to the size of the set. We, especially, say G_r is a minimal transitive reduction of G if and only if $G_r^* = G^*$ and there is no proper subgraph G_1 of G_r with $G_1^* = G_r^*$. In other words a minimal transitive reduction is a subgraph G_r of G with $G_r^* = G^*$ and no *reducible* edge (v, w) in it, where (v, w) is reducible in G_r if and only if there is a simple path $P = v_0, \ldots, v_s$ from $v = v_1$ to $w = v_s$ in $G_r - (v, w)$. A minimum transitive reduction is a transitive reduction of minimum size.

The second point to be regarded is the condition $E_r \subseteq E$ or $E_r \subseteq V \times V$. AHO, GAREY AND ULLMAN have shown, that this makes no difference on acyclic digraphs.

Moreover, on acyclic digraphs a minimal transitive reduction is also a minimum transitive reduction. For computing the transitive reduction on acyclic digraphs see SIMON, 1988[12]. But in a strongly connected digraph — a digraph $G(V, E)$ with $E^* = V \times V$ — the problem of finding a transitive reduction becomes trivial demanding on condition that $E_r \subseteq V \times V$ since every permutation of the nodes is a minimal and a minimum solution. Otherwise, on condition $E_r \subseteq E$ the decision whether digraph G has a minimum transitive reduction with at most c edges — for a given c — is NP-complete. Note for example: A hamiltonian circuit is always a minimum transitive reduction. This is a result by SAHNI. But back to our problem again. We are describing an linear algorithm for finding a minimal transitive reduction $G_r(V, E_r)$ of a given strongly connected digraph $G(V, E)$, where the transitive reduction is defined on condition $E_r \subseteq E$ and "minimal" stands for containing no reducible edge. The simple idea which the algorithm follows is to delete as many edges as possible without changing the transitive closure. The algorithm is based on an iterative use of depth-first search, TARJAN, 1972, and two kinds of graph transformations which do not change the transitive reduction.

The paper is divided into several sections. Section 2 contains some notations and a short description of depth-first search on directed graphs. In section 3 we are developing an algorithm to delete all reducible edges which are not in the tree constructed by depth-first search. In section 4 we make use of the results of section 3, in order to reduce the problem testing the reducibility of tree arcs to a dominator problem in flowgraphs. Section 5 contains a short conclusion and open problems.

2. Notations and Basic Concepts

As usual we need some notations. First of all the graph-theoretic definitions which are more or less standard. For an advanced reader conversant with digraphs and depth-first search it is not necessary to read the section. A *(directed) graph* $G = G(V, E)$ is an ordered couple consisting of a set of *vertices (nodes)* V, $V = \{ 1, ..., n \}$, and a set of *edges (arcs)* E, $E \subseteq V \times V$. A graph is always a simple graph containing no parallel edges or self-loops — arcs of the form (v, v). The graph $G_r = (V, E_r)$ is said to be the *reversal* or the *reverse graph* of G, where

$$E_r = \{ (x, y) \mid (y, x) \in E \},$$

or in other words,

$$(x, y) \in E_r \iff (y, x) \in E.$$

In the same way (x, y) is called the *reverse edge* of (y, x). We say edge (v, w) leaves v, enters w and joins v with w. Unless we specify otherwise, any subgraph is the subgraph induced by its vertex set. A subgraph G' of G is called nonreducible if g' does not include a reducible edge. A path P in G from vertex v_0 to vertex v_s is a sequence of vertices $v_0, v_1, \ldots, v_s$ such a way that (v_{i-1}, v_i) is an edge for $i \in \{ 1, 2, \ldots, s \}$; s is the length of the path. The path P is simple if all its vertices are pairwise distinct. A path $v_0, \ldots, v_s$ is a cycle if $s > 1$ and $v_0 = v_s$ and a simple cycle if in addition $v_1, \ldots, v_{s-1}$ are pairwise distinct. A graph without cycles is acyclic. Let be $A \subseteq E$ then we write $v \xrightarrow[A]{*} w$ if there is a path from v to w using only arcs of A. If there is $v \xrightarrow[E]{*} w$ (or

$v \xrightarrow{*}_G w$ or simply $v \xrightarrow{*} w$) then w is said *reachable* from v. If $v \xrightarrow{*} w$ is a simple path with length greater than 0 then we write $v \xrightarrow{+} w$. A single edge (v, w) is also written as $v \to w$. A digraph G is strongly connected if it satisfies the conditions $v \xrightarrow{*} w \xrightarrow{*} v$ for all $v, w \in V$. Suppose G is a strongly connected digraph with a singular vertex s. If vertex x lies on every path from vertex s to vertex y then x is called a *dominator* of y. This is equivalent to say x dominates y if and only if y is not reachable from s in $G - \{ x \}$. The node x is an *immediate dominator* of y if x is a dominator of y and x is dominated by every other dominator z of y. It is easy to see that each vertex has a unique immediate dominator if it has any dominators. Furthermore we use the conception of domination for an edge e in the same way as for a vertex. In particular, e is a dominating edge of y iff s does not reach the node y in $G - e$. An edge or a vertex is called *reverse dominating* if it is dominating in the reverse graph G_r.

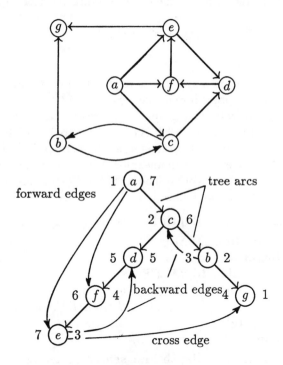

Figure 1

A digraph G and its DFS-tree T. The inscription of a vertex v in T is given by dfsnum|v|compnum. Throughout this paper the nodes in T are arranged from right to left and from top to bottom relative to their dfsnum. Later the nodes will be identified with their dfsnum.

A *(directed, rooted)* tree T is an acyclic digraph with one distinguished vertex r called the *root* r such a way that $r \xrightarrow{+} v$ for all vertices v, $v \neq r$, and no edges enter r. Please note: In a tree exactly one arc enters every other vertex as the root. A tree vertex with no existing edges is a *leaf*. Let (v, w) be a tree edge, then v is the father of w and w is a son of v. If $v \xrightarrow[T]{*} w$, then v is an ancestor of w and w is a descendant of v. A tree node z is called a common ancestor of the node v and w if and only if $z \xrightarrow[T]{*} v$ and $z \xrightarrow[T]{*} w$. The vertex z is the *nearest common ancestor* of v and w iff there is no other common ancestor x of v and w with $x \neq z$ and $z \xrightarrow[T]{*} x$. A subtree T_1 of a tree T with root v is the subgraph of T induced by the nodes $\{ w \in V \mid v \xrightarrow[T]{*} w \}$. If a tree T contains all the vertices of G, then T is called a *spanning* tree of G. The *height* of a node v in T is the length of path with maximum length from v to a leaf in the subtree of v. A *sink-tree* S is a tree in which the edges are directed from a son to its father or in other words it is

$$w \xrightarrow[S]{*} r \quad \forall w \in V$$

where r is the root of S.

Our algorithm is based on a systematic exploration of a graph. Especially we are using the depth-first search procedure in the version of TARJAN, 1972, to recognize the strongly connected components of a digraph G. Therefore, we need a short description of this version of depth-first search (DFS). It is convenient to formulate DFS as a recursive procedure $DFS(v)$ with vertex v as parameter, see TARJAN, 1972. In general we search for unexplored vertices by traversing an unexplored edge from the most recently reached vertex which still has unexplored edges. The set $REACH$ contains the explored vertices. On these conditions DFS has the following main structure

────────────────────── **depth-first search** ──────────────────────

```
procedure DFS(v : V)
begin
        add v to REACH;
        for ∀ w with (v, w) ∈ E do
                if w ∉ REACH then DFS(w) fi
        od
end
```

────────────────────── **depth-first search** ──────────────────────

The procedure starts with

$$REACH \leftarrow \emptyset;$$
$$DFS(r);$$

and marks all vertices reachable from start vertex r. But DFS gives some further information about the digraph G. In particular, DFS computes a spanning tree DFS-tree with root r, a numbering *dfsnum* with respect to calling time of procedure DFS and a second numbering *compnum* with respect to completion time of DFS. With these numberings we partition the edges of the graph into 4 classes: The tree edges T, the forward edges F, the backward edges B and the cross edges C. This partition is defined by

(i) $\qquad$ $(v, w) \in T \quad \Longleftrightarrow \quad w$ unmarked when (v, w) is explored.

(ii) $\qquad$ $(v, w) \in F \quad \Longleftrightarrow \quad dfsnum(v) < dfsnum(w)$

$\qquad\qquad\qquad\qquad\qquad$ and $compnum(w) < compnum(v)$

$\qquad\qquad\qquad\qquad\qquad$ and $(v, w) \notin T$.

(iii) $\qquad$ $(v, w) \in B \quad \Longleftrightarrow \quad dfsnum(w) < dfsnum(v)$

$\qquad\qquad\qquad\qquad\qquad$ and $compnum(v) < compnum(w)$.

(iv) $\qquad$ $(v, w) \in C \quad \Longleftrightarrow \quad dfsnum(v) > dfsnum(w)$

$\qquad\qquad\qquad\qquad\qquad$ and $compnum(w) < compnum(v)$.

Figure 1 shows an example for such a partition. For the proof of this and some further properties of DFS see MEHLHORN, 1984. This is also a good reference for a detailed description of DFS to compute $dfsnum$, DFS-tree, B and so on, in linear time.

In order to recognize the strongly connected components of a given digraph G, the procedure DFS uses a map

$$lowpoint : V \to V$$

defined by

$$lowpoint(v) = \min\{\ dfsnum(w)\ |\ \exists\ (u, w) \in E \text{ with } v \xrightarrow[T]{*} u\ \}.$$

Let v be the root of DFS-tree. We then have the following well-known result: The digraph $G(V, E)$ is strongly connected if and only if

$$lowpoint(v) < dfsnum(v) \quad \forall\, v \in V,\ v \neq r.$$

For proof see MEHLHORN, 1984. Let now $lowpoint(v) = w$ then $lowpoint(v)$ is defined by an edge $(u, w) \in E$, where u is a node in the subtree of v with respect to DFS-tree. This edge we indicate as $lowedge(v)$. Now we obtain a special subgraph $G_{low}(V, E_{low})$ of G by

$$G_{low} = \text{DFS-tree} + \{\ lowedge(v)\ |\ v \in V, v \neq r\ \}.$$

For the subgraph G_{low} it is easy to see: G is strongly connected if and only if G_{low} is strongly connected. Note: G_{low} has at most $2 \cdot n$ edges. Naturally the question arises

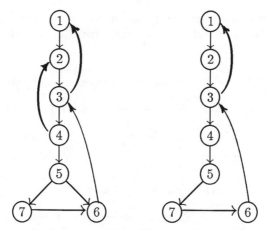

Figure 2: A low-graph and its transitive reduction.

whether perhaps G_{low} is a minimal transitive reduction too? In general this is incorrect, see figure 2.

After this preparation we can start with our analysis of finding a minimal transitive reduction.

3. Deleting Reducible Nontree Arcs

We begin by exploring G, starting DFS at any vertex v. Then DFS partitions the edges of G in tree arcs T, forward edges F, the backward edges B and cross edges C. In order to simplify discussion we shall identify vertices using the *dfsnum* assigned them by DFS. Now a simple idea to obtain a minimal transitive reduction is to delete as many reducible edges as possible. For getting a linear time complexity we have to check in time $O(1)$ for each and every selected edge e: Is e reducible? For some edges it is not very difficult to answer, i.e. see lemma 1. Whereas for other edges e it is not easy, see lemma 3. In these cases we give no answer but we replace the edge e by an edge e' of a more simple type. The new edge e' has the property: e' is reducible in $G - e + e'$ if and only if e is reducible in G. This replacement allows us to postpone the check to a later date. Moreover, if e'' is not a tree arc then e'' is reducible in G if and only if e'' is reducible in $G - e + e'$. This implies our general strategy for this chapter. At each step select one edge e, $e \in B \cup C \cup F$, and in time $O(1)$ either test e to be reducible or replace e by another edge e'. Let's now come to a detailed treatment.

Note: In the following we assume that the DFS-tree is always a subgraph to G. In this section we do not delete any tree arcs from G. On this condition we get

Lemma 1. *Each forward edge* $e = (v, w)$ *is reducible.*

Proof. By definition of forward edges there is a simple path $v \xrightarrow[T]{+} w$ with length greater or equal than two. But this is a reducing path for the edge (v, w). ∎

The next lemma is giving us an analogous proposition for backward edges.

Lemma 2. *Let $\{ (v,w_1),\dots,(v,w_s) \}$ be the set of backward edges emanating from node v such a way that $w_1 < w_2 < \dots < w_s$ then all the edges $(v,w_2),\dots,(v,w_s)$ are reducible.*

Proof. Remember we have $dfsnum(v) = v$ for all vertices. Now all vertices $w_1,\dots,w_s$ are nodes on the path $r \xrightarrow{*}_{T} v$. For this reason there is a reducing path $P = v \to w_1 \xrightarrow{+}_{T} w_i$ for each backward edge (v,w_i), $2 \le i \le s$. ∎

To check the cross edges we need a more complicated proposition which is given in the following lemma.

Lemma 3. *Let $e = (v,z)$ be a cross edge and w is the nearest common ancestor of v and w in the DFS-tree. If $(v,w) \in E$ then (v,z) is reducible in G. If (v,w) is not an edge of G we obtain:*

1. *The edge (v,z) is reducible in G if and only if the edge $e' = (v,w)$ is reducible in the digraph $G_1 = G - (v,z) + (v,w)$.*

2. *$G_1^* = G^*$.*

3. *Let $e'' \in C \cup B \cup F$ be an edge with $e \ne e'' \ne e'$ then e'' is reducible in G if and only if e'' is reducible in G_1.*

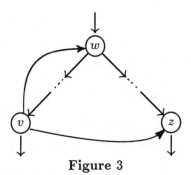

Figure 3

Proof. Let's have a look at figure 3. First of all we see if $(v,w) \in E$ then the reducing path to (v,z) is given by $v \to w \xrightarrow{+}_{T} z$.

1. On the other hand let now (v,w) be not an edge of G. Further let G_2 be the graph formed by G by deleting edge (v,z). Assuming (v,z) is reducible in G there is a reducing path P_1 from v to z in G. Since G is strongly connected there is a simple path P_2 from z to w. Joining P_1 with P_2 we get a path P_3 from v to w which does not contain (v,z). But P_3 is a path in G_2 and, therefore, there is also a simple path $P_4 = v \xrightarrow{+}_{G_2} w$ in G_2. Since $G_1 = G_2 + (v,w)$, P_4 must be a reducing to (v,w) in G_1. Conversely, suppose P_1 is a reducing path to (v,w) in G_1. Extending P_1 by $w \xrightarrow{+}_{T} z$ we observe a path P_2 in G from v to z. By shrinking cycles we find a reducing path for (v,z) in G.

2. In every path P in G which contains the arc (v,z) we can replace this arc by the path $v \to w \xrightarrow{+}_{T} z$ in G.

3. Let now $e'' = (x, y)$ be an edge in G, with $e'' \notin T$, $(v, w) \neq e'' \neq (v, z)$. Suppose e'' is reducible in G and P_1 is the corresponding reducing path. If (v, z) is not an edge of P_1 then there is nothing to show. Therefore, we assume (v, z) is an edge of P_1. Replacing (v, z) in P_1 by $v \to w \xrightarrow[T]{+} z$ produces a path P_2 in G_1 which connects the same endpoints as P_1. By construction e'' is not an edge of P_2. Hence, shrinking cycles gets the reducing path to e'' in G_1. Conversely, suppose $e'' = (x, y)$ is reducible in G_1. Then we consider two cases. The trivial case is given if there is a reducing path P_1 for the edge e'' which does not contain the edge (v, w). This path P_1 exists in G too. In the notrivial case every reducing path P_1 contains the edge (v, w). Let u be the first node after the starting node w on the tree path from w to v, thus $w \xrightarrow[T]{+} u \xrightarrow[T]{+} v$. By our presumption about P_1 we find that the node x cannot be less than u. This is shown by a short view on the structure of DFS. Let $w = w_1, \ldots, w_s$, with $w_1 = r$ and $w_s = w$, be the tree path $v \xrightarrow[T]{+} w$ from the root r to the vertex w. We infer from the definition of depth-first search: If (w_{s-1}, w_s) is a tree edge then it exists no edge (a, w_s) with $a < w_s$ and $a \neq w_i$, $1 \leq i \leq s - 1$. In other words, the vertices $w_1, \ldots, w_s$ builds a vertex separator for all nodes a and b with $a < w_s \leq b$. That means every path connecting a and b passes through at least one vertex $w_i, 1 \leq i \leq s - 1$. Now we can show our proposition. If we assume $x < u$ conversely to our proposition then we observe with $a = x$ and $b = v$: The reducing path P_1 from x over v and w to y must pass through a vertex $w_i, 1 \leq i \leq s - 1$, before reaching v. But the part $w_1 \xrightarrow{+} v \to w$ in the path P_1 can be replaced by $w_i \xrightarrow[T]{+} w$ to get a reducing path which does not contain (v, w) in contradiction to our presumption that no such path exists. Therefore, it must hold $x \geq u$. The same argument as above shows further that there is a path W in G from z to w which does not contain a node h with $h > u$. Since the edge $e'' = (x, y)$ cannot be an edge of this path W. Consequently we can replace the arc (v, w) in path P_1 by (v, z) and W to get a path P_2 from x to y in G. Shrinking cycles in P_2 gives us the reducing path P_3 for the edge (x, y) in G and the lemma follows. ∎

Now lemma 3 allows us the following strategy to find and delete reducible edge. First delete all forward edges by lemma 1. Then delete or replace all cross edges by lemma 3. We use a map *standfor* to store up the origin (v, z) of a new edge (v, w). Now the graph contains only backward edges others than tree arcs. In the third step we delete all backward edges except one for each node v by lemma 2. In the following we call this resulting digraph G'. Note that the DFS-tree for G is also a DFS-tree for G' and G' is strongly connected by induction on the construction process.

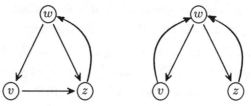

Figure 4: (w, z) is reducible in G but not in G_1.

Remark: Figure 4 shows an easy example that Lemma 3.3 is not correct for tree arcs. Therefore, we need a separate method to find reducible tree arcs, see chapter 4.

The digraph G' contains the DFS-tree and at most one backward edge (v, w) for every node v. Now we examine this edge $e = (v, w)$ to its reducibility. We shall give a bottom up procedure to make this. Therefore, we start in a leaf v and this case is easy, because it is clear that the only edge emanating v cannot be reducible. The next lemma describes the situation in which v is an inner node.

Lemma 4. Let (v, w) be a backward edge in G, x be a descendant of v with $v \neq x$ and z be a vertex with $v \xrightarrow[T]{+} x \to z$ and $(x, z) \in E$. (In other words, (x, z) is an edge from the subtree of v to a node z on the tree path from w to v.) If $(z, w) \in E$ then (v, w) is reducible. Otherwise, if $(z, w) \notin E$ then we obtain:

1. The edge (v, w) is reducible in G if and only if the edge (z, w) is reducible in the digraph $G_1 = G - (v, w) + (z, w)$.
2. $G^* = G_1^*$.
3. Let $e = (a, b)$ be an edge in G with $(v, w) \neq e \neq (z, w)$, $e \in B$, and there is no path $v \xrightarrow[T]{*} a$. Then e is reducible in G if and only if e is reducible G_1.

Proof. Clearly, if $(z, w) \in E$ then the path $v \xrightarrow[T]{*} x \to z \to w$ is a reducing path relating to (v, w), see figure 5. On the other hand, let now (z, w) not be an edge in G.

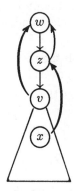

Figure 5

1. First we suppose (v, w) is reducible in G. Then there exists a reducing path P_1 from v to w in G. P_1 is also a path in G_1. Now it can be enlarged through $z \xrightarrow[T]{+} v$ getting a path $P_2 = z \xrightarrow[T]{+} w$. By shrinking cycles in P_2 we find a reducing path for (z, w) in G_1. Conversely, suppose P_1 is a reducing path for (z, w) in G_1. By our presumption there is a path $v \xrightarrow[T]{+} x \to z$ in G. Joining this path with P_1 we obtain a $P_2 = v \xrightarrow[T]{+} x \to z \xrightarrow[T]{+} w$ from v to w in G. Therefore, there also exists a simple path in P_3 of length at least two from v to w in G.

2. All nodes stay reachable in G_1 by replacing the edge (v,w) through $v \xrightarrow[T]{+} x \to z \to w$. Otherwise we use the path $z \xrightarrow[T]{+} v \to w$ for the edge (z,w) in G.

3. Let's first have the assumption: The edge $e = (a,b)$ is reducible in G and the reducing path P_1 contains the edge (v,w). The case $(v,w) \notin P_1$ is trivial. In order to observe a path P_2 connecting a with b in G_1 we apply replacement of edge (v,w) by $v \xrightarrow[T]{+} x \to z \to w$. By our presumption this new path P_2 cannot embody the edge (a,b). Hence shrinking cycles in P_2 gives a reducing path P_3 for (a,b) in G_1. Inversely, suppose (a,b) has a reducing path P_1 in G_1 holding within itself the edge (z,w). The replacement of (z,w) by $z \xrightarrow[T]{+} v \to w$ obtains a path P_2 from a to b in G. The lemma follows by same argumentation as above. ∎

Now lemma 4 allows us the following method to test the reducibility of a backward edge (v,w). By induction about the height of v we assume that no backward edge (x,y), $x \neq v$, emanating from the subtree of v is reducible. Then we calculate the node $z = z(v)$ defined by

$$(*) \qquad z = \min\{ \, y \in V \mid \exists \, x \in V \text{ with } v \xrightarrow[T]{+} x \text{ and } (x,y) \in E \, \}.$$

Note, since G is strongly connected it is always $z \leq v$. If $z = v$ then (v,w) is the only edge leaving the subtree of v and $G - (v,w)$ would not be longer a strongly connected digraph. (Remember the observation about $lowpoint(v)$.) For that reason the edge (v,w) is not reducible. Otherwise we delete and/or replace the edge (v,w) by lemma 4. This completes the induction step.

After this preparations we have all the results needed to build an efficient reducing algorithm for nontree arcs, the algorithm A.

This algorithm works correctly by the results in preceding discussion. Let $G'' = (V, E'')$ be the digraph given by the algorithm A before step 5 is executed. This digraph G'' is constructed via the sequence of digraphs

$$G = G_0, G_1, \ldots, G_s = G'',$$

where G_{i+1} arises from G_i by deletion of a reducible edge e_i, or by replacement of an edge e_i' by another e_i'' using lemma 3 or 4. Therefore, we have $G_{i+1}^* = G_i^*$ and by induction on i the fact $G_0^* = G_s^*$. If we replace e_i' by e_i'' then we keep the origin e_i' of e_i'' in the variable $standfor$, with $standfor(e_i'') = e_i'$. With a replacement we do not change the reducibility of any edge e, $e \notin T$ and $e_i' \neq e \neq e_i''$, see Lemma 3.3 or Lemma 4.3. In Lemma 4.3 the proof is only given for the case if $e = (a,b)$ is not an edge emanating from the subtree of v. But we see by induction on the height (v): If e is emanating from the subtree of v then e is not reducible. Note, that the node $z = z(v)$ computed in $R-test$ satisfies the definition in expression $(*)$. This completes our above proposition, since an edge which is not reducible in a digraph G is also not reducible in any subgraph of G. So we obtain first that every remaining nontree arc e in G'' is not reducible in G''. Secondly we have by induction on i with starting node $i = s$: If a nontree arc e is

_____ **algorithm A** _____

input: A strongly connected digraph $G(V, E)$; the partition of the edges T, B, C, F; the DFS-tree of G.

output: A subgraph $G' = (V, E')$ of G, which does not contain a reducible nontree arc.

(0)   **for** $\forall\ e \in E$ **do** $standfor(e) \leftarrow e$ **od;**
        for $\forall\ v \in V$ **do** $backpoint(v) \leftarrow v$ **od;**

(1)   **for** $\forall\ e \in F$ **do** delete e from G **od;**

(2)   **for** $\forall\ e = (v, w) \in B$ **do**
        $backpoint(v) \leftarrow \min(backpoint(v), w)$ **od;**
        for $\forall\ e = (v, w \in B$ with $w \neq backpoint(v)$ **do**
            delete (v, w) from G **od;**

(3)   **for** $\forall\ e = (v, z) \in C$ **do**
        let w be the nearest common ancestor of v and z;
        if $w < backpoint(v)$
        then
            add (v, w) to G;
            delete $(v, backpoint(v))$ from G;
            $standfor(v, w) \leftarrow (v, z)$;
            $backpoint(v) \leftarrow w$;
        fi
        delete (v, z) from G;
    od;

(4) $R - test($ "root of DFS-tree" $)$;

(5)   replace all edges e in G by their origin $standfor(e)$

with

(6) $R - test(v : V)$
        for $\forall\ w$ with $(v, w) \in T$ **do** $R - test(w)$ **od;**
        if v is not a leaf
        then
            $z \leftarrow \min(\{\ backpoint(w) \mid (v, w) \in T\ \})$;
            if $v \neq z$ $\quad (*z < v*)$
            then
               if $backpoint(v) < backpoint(z)$
               then
                  delete $(z, backpoint(z))$ from G;
                  add $(z, backpoint(v))$ to G;
                  $standfor((z, backpoint(v))) \leftarrow standfor((v, backpoint(v)))$;
                  $backpoint(z) \leftarrow backpoint(v)$;
               fi;
               delete $(v, backpoint(v))$ from G;
               $backpoint(v) \leftarrow z$
            fi
        fi

_____ **algorithm A** _____

not reducible in G''' then its origin in G is not reducible in G too. Therefore, we can summarize our analysis in the following Theorem.

Theorem 1. *Algorithm A computes a strongly connected subgraph G' of a digraph G where G' does not contain any reducible nontree arc relating to a given DFS-tree. The algorithm A has a linear time and storage complexity $O(|V| + |E|)$.*

Proof. The correctness of theorem 1 is clear by the preceding discussion. So we only have to show the complexity bound. First of all we use depth-first search to get the input data structures of the algorithm in particular the DFS-tree and the partition of the edges into T, B, F, C. It is well-known that this can be done with linear time and storage complexity. Now we come to algorithm A. It is trivial to see that step 0, 1, 2 and 5 can be implemented in time $O(1)$ per considered edge. This is also true for the step 3 without calculation of the nearest ancestor w for the node v and z. Using a data structure by HAREL and TARJAN we can compute w in time $O(1)$ and storage $O(n)$. Now, for any fixed vertex v the procedure R-test takes time linear in the number of edges emanating from v. So the total costs of R-test are linear in the number of edges referring to the calling time of R-test. Because we have linear complexity in each of our 6 steps we reach a linear time and storage complexity for the whole algorithm A too. ∎

4. The Reducibility of Tree Arcs

For the resulting subgraph of algorithm A we have to test its tree arcs on reducibility. We do not solve this problem directly but we use algorithm A as a fundamental step to obtain a spanning tree respectively, a spanning sink-tree which does not contain any reducible edges. To reach this aim we further need a result from HAREL, 1985, about finding immediate dominators in linear time.

Let us remember our definition of the statement "$G = (V, E)$ is strongly connected" which stands for

$$v \xrightarrow[E]{*} w \quad \text{and} \quad w \xrightarrow[E]{*} v \quad \forall\, v, w \in V.$$

For this we infer that the digraph G contains for every vertex z_1 a spanning sink-tree T_1 with root z_1 such that

$$z_1 \xrightarrow[T_1]{*} w \quad \forall\, w \in V,$$

and further that G includes for every vertex z_2 a spanning sink-tree T_2 with root z_2 such that

$$w \xrightarrow[T_2]{*} z_2 \quad \forall\, w \in V.$$

For a simple argumentation we select $z_1 = z_2 = r = 1$ throughout this section. Now we can start our analysis. The main result for this section is given in theorem 2.

Theorem 2. *For a given strongly connected digraph $G = (V, E)$ a strongly connected subgraph G'' is computable in linear time such that G'' contains a nonreducible spanning sink-tree.*

Proof. We start the construction of G'' with the computation of a spanning sink-tree T_s of G with root r. Of course, T_s is not strongly connected. Therefore we add some edges to T_s to get a spanning tree T'. But we want guarantee that every edge e added to T_s is not reducible in the subgraph $T_s \cup T'$. For that reason we construct T_s and T' in the following way. We reverse the direction of every edge in $G = (V, E)$ getting the reverse graph $G_r = (V, E_r)$ of G. Next we start depth-first search on G_r in the node r. Then the algorithm A is applied to G_r resulting a subgraph G'_r which does not embody a reducible nontree arc. Now, the spanning sink-tree T_s of G is given by the reversal of the DFS-tree of G_r and the edges in $T' - T_s$ are defined through the reversals of the nontree arcs of G_r. The subgraph of G formed by $T_s \cup T'$ is called G'. Of course, G' is the reverse graph of G'_r, G' is strongly connected, and every edge (v, w) in G' is reducible iff (w, v) is reducible in G'_r. Now repeat the procedure above for the graph G'. Especially we compute a new DFS-tree T for G' and use again algorithm A to find and delete reducible nontree acrs related to this new DFS-tree T. After this last step the resulting graph is called G''. For G'' we observe that every edge which can be reducible is both an edge of T_s and an edge of T. Therefore we call such edges *critical* and by this notation we find G'':

(a) There is at most one critical edge (x, y) for every fixed node x since in T_s every node has at most one emanating edge.

(b) Because a critical edge (x, y) is an arc of T we infer that y is a son of x in T.

Now we want construct a nonreducible spanning sink-tree. This leads us to the following question for a critical edge (x, y): Is there existing a vertex z in G'' such that z does not reach the root r in the graph $G - (x, y)$? Obviously, if there is such a node z then the critical edge (x, y) is not reducible. On the other hand, it is not necessary to test every possible combination of z and (x, y). Since if z does not reach r in $G'' - (x, y)$ then it infers directly that x does not reach r in $G'' - (x, y)$, or in other words: If (x, y) is a reverse dominating edge for z then (x, y) is a reverse dominating edge for x, too. Therefore we have only to check: Is the critical edge (x, y) also a reverse dominating edge for x? If the answer is "yes" then we mark (x, y) as nonreducible and in this way we get a final digraph G''' which satifies:

(c) No critical edge in G''' is reverse dominating for any vertex of G'''.

By induction on the height of a vertex x in T it is now easy to see that G''' includes a nonreducible spanning sink-tree. We use the induction hypothesis that no node u from the subtree of x in T must use a critical edge from a subtree of a son of x to reach the root r. If x is a leaf then it is nothing to show. Otherwise, let now x be a inner node of T and (x, y) its critical edge. Since (x, y) is not reverse dominating for x we infer (x, y) is not reverse dominating for any vertex u in the subtree of x. For that reason and the induction hypothesis we get: There is a path from u to r containing no critical edge from the subtree of x. Applying this proposition to the root r we get a spanning sink-tree including no reducible edges. Note that the argumentation above is only correct if node x has at most one critical edge outgoing from it. We come to the running time. First we use depth-first search and algorithm A to the reverse graph G_r of G. This takes time $O(|V| + |E|)$. Then we repeat this step on G' with cost $O(n)$ getting G''. In G'' we have to decide for every critical edge (x, y): Is (x, y) reverse dominating for x? But finding dominating edges looks very similar to find dominating vertices. Indeed, it is easy to formulate our problem as a dominating problem for vertices. To make this we reverse the direction of all edges in G'' to get its reversal G''_r. Then we split up the edge (x, y) into two edges (x, z) and (z, y). The new vertex $z = z(x, y)$ is not joined with any other vertices. Now the new vertex z is a immediate dominator of x in G''_r if and only if the edge (x, y) is reverse dominating for x in G''. But computing immediate dominators for a rooted digraph is a well-known problem in graph theory. We use a result from HAREL, 1985, to solve it in linear time. Therefore, we reach also time $O(|V|)$ to compute G'''. Sum up the three steps we obtain total costs

$$O(n + e) + O(n) + O(n) = O(n + e)$$

to find G''' and the theorem follows. ∎

In the rest of section we describe how to use theorem 2 in order to find a minimal transitive reduction of $G = (V, E)$. But now this easy. We apply theorem 2 to calculate a strongly connected subgraph of G which includes a nonreducible spanning sink-tree T_1. Next we apply again theorem 2 to the reversal of G_1 getting a subgraph G_2 which embodies a nonreducible spanning tree T_2. Since all edges of T_1 are not reducible they are also edges of G_2. Now we observe the final subgraph G_3 of G by deleting all remaining critical edges of G_2. We summarize in theorem 3.

Theorem 3. *The final subgraph G''' of G is a transitive reduction of G. It can be computed in linear time.*

Proof. By construction G_3 embodies a spanning sink-tree T_1 and a spanning tree T_2. Further all edges in G_3 are not reducible. Hence we have

$$v \xrightarrow[T_1]{*} r \xrightarrow[T_2]{*} w \xrightarrow[T_1]{*} r \xrightarrow[T_2]{*} v \quad \forall\, v, w \in V,$$

and this shows that G_3 is strongly connected. Second we find for every edge (x, y) of G not in G_3 a reducing path

$$x \xrightarrow[T_1]{*} r \xrightarrow[T_2]{*} y$$

by shrinking cycles. The time to calculate G_3 is clear linear, since we use a combination of linear algorithms. This completes the theorem. ∎

5. Conclusions and Open Problems

This paper presents an algorithm for finding a minimal transitive reduction of a given strongly connected digraph G. The algorithm is based on depth-first search partly as an algorithm and partly as induced data structure. Some relations to find dominators in flowgraphs are shown. We think that the following questions are of some interest in view of the minimal transitive reduction. Is it possible to decide whether digraph G is Bellman-Ford-Orderable on its minimal transitive reduction? Is it necessary to compute dominators in section 4? What is the connection between transitive reduction and maximum flow, see SIMON, 1988[13]?

6. References

[1] A.V. AHO, M.R. GAREY AND J.D. ULLMAN: *The Transitive Reduction of a Directed Graph,* SIAM J. Computing, 1 (1972), 131-137.

[2] S. EVEN: *Graph Algorithms,* Computer Science Press, Potomac, MD, 1979.

[3] D. HAREL: *A linear time algorithm for finding dominators in flow graphs and related problems,* seventeenth annual ACM Symposiumon theory of computing, Providence, 1985, 185-194.

[4] D. HAREL AND R.E. TARJAN: *Fast Algorithms for Finding Nearest Common Ancestors,* SIAM J. Computing 13 (1984), 339-355.

[5] H.T. HSU: *An Algorithm for Finding a Minimal Equivalent Graph of a Digraph,* J. ACM 22 (1975), 11-16.

[6] T.LENGAUER A. R.E. TARJAN: *A fast algorithm for finding dominators in a flowgraph.* ACM transactions on programming languages a. systems, 1 (1979), 121-141.

[7] K. MEHLHORN: *Data Structures and Algorithms, Vol 2: Graph Algorithms and NP-Completeness,* Springer, EATCS Monographs in Computer Science, 1984.

[8] K. MEHLHORN AND B. H. SCHMIDT: *On BF-Orderable Graphs* Discrete Applied Mathematics 15(1986), 315-327

[9] D.M. MOYLES AND G.L. THOMPSON: *Finding a minimum equivalent graph of a digraph,* J. ACM 16 (1969), 455-460.

[10] R. HADDAD, A. SCHÄFFER: *Recognizing Bellman-Ford-Orderable Graphs,* Computer Science Department Stanford University, Standford, California 94305-2140.

[11] S. SAHNI: *Computationally Related Problems,* SIAM J. Computing 3 (1974), 262-279.

[12] K. SIMON: *An Improved Algorithm for Transitive Closure on Acyclic Digraphs,* Theoretical Computer Science 58 (1988), 325-346.

[13] K. SIMON: *On Minimum Flow and Transitive Reduction,* Proceedings ICALP'88, Tampere, Lecture Notes in Computer Science, Springer-Verlag, 317, 535-560.

[14] R.E. TARJAN: *Depth first search and linear graph algorithms,* SIAM J. Computing 1 (1972), 146-160.

[15] R.E. TARJAN: *Finding Dominators In Directed Graphs,* SIAM J. Computing 3 (1974), 62-89.

Paging binary trees with external balancing *

Andreas Henrich

FernUniversität Hagen
5800 Hagen
West Germany

Hans-Werner Six

FernUniversität Hagen
5800 Hagen
West Germany

Peter Widmayer

Universität Freiburg
7800 Freiburg
West Germany

Abstract

We propose the partially paged binary tree principle (PPbin tree principle, for short) for maintaining binary trees which do not fit into core and hence must be (at least partially) paged on secondary storage. The PPbin tree principle can be applied to balanced as well as unbalanced binary trees. Paging a balanced binary tree results in a balanced external binary tree. However, main advantage of the new principle is that even for unbalanced binary trees it is very unlikely that long external access paths will arise. As an example, we describe the partially paged k-d tree which is used as directory in a spatial data structure. The analysis of the expected storage utilization and the expected external height proves the efficiency of the new data structure derived from the application of the PPbin tree principle.

1. Introduction

Binary search trees are useful and easy to handle data structures used in many applications. In the one-dimensional case, several balanced tree classes like AVL trees (see e.g. [AVL62], [Wir75]), brother trees [OW81] or weight-balanced trees [NR73] have been introduced. In the multidimensional case, the unbalanced k-d trees [Ben75] have been proposed.

Unfortunately, binary search trees in general suffer from the need to be kept in main memory. In this paper, we propose a new principle for maintaining binary trees which do not fit into main memory and hence must be paged on secondary storage. We call it the **Partially Paged binary tree principle** (PPbin tree principle, for short). Here, "partially paged" means that the binary tree can be maintained in core as usually as far as its size does not exceed the main memory capacity. If its size grows such that the whole tree cannot be kept in core any longer, a prefix tree whose size depends on the core size is maintained internally and only subtrees outside the prefix tree must be paged. The PPbin tree principle can be applied to many different binary tree classes. Main advantage of the PPbin tree principle is that even for unbalanced binary trees it is very unlikely that long external access paths will arise, i.e. many external pages must be traversed during a search. Hence, main application area of the PPbin tree principle are unbalanced binary trees occurring, for instance, in multidimensional geometric applications. As an example, we describe the partially paged k-d tree which is used as directory in a spatial data structure accommodating point- and non-point geometric objects. The analysis of the expected storage utilization and the expected external height proves the efficiency of the new data structure.

In [IKO87], binary priority search trees which support three sided range queries and updates with optimal worst case complexity are adapted to secondary memory. The resulting data structure is derived from B-trees [BM72] and from a generalized version of red-black trees [GS78] and therefore balanced. In contrast to our spatial data structure based on the application of the PPbin tree principle to k-d trees, however, the external priority search tree does not efficiently support regular (i.e. four sided) range queries and cannot be extended to more than two dimensions and to arbitrary (i.e. non-point) geometric objects.

Robinson [Rob81] proposes the K-D-B tree as an extension of the B-tree for storing k-dimensional points. However, the performance and storage utilization is outperformed by several other spatial data

---

* This work has been supported by DFG grants Si 374/1 and Wi 810/2.

structures like [NHS84], [KS88], [Free87], and our k-d PPbin tree based data structure. Furthermore, it is not suitable for non-point geometric objects.

Although unbalanced binary trees are the natural application area of the PPbin tree principle, it can also be carried over to balanced binary trees with advantage. Applying the PPbin tree principle to an AVL-tree, for example, results in a PPbin tree with logarithmic bound on the external height [HSW89].

In the next section, we explain the PPbin tree principle for one-dimensional unbalanced binary trees. In section 3, an application of the principle to the (unbalanced, multidimensional) k-d tree serving as directory in a spatial data structure is provided. In section 4, the expected storage utilization and the expected external height of the spatial data structure based on the k-d PPbin tree are analyzed.

2. The PPbin tree principle

2.1 Basic ideas and properties

We are facing the problem that the number of objects to be stored in a binary search tree exceeds the main memory capacity. Here, the whole tree or at least subtrees must be paged on secondary storage. This is not an easy task, especially if the pages should be reasonably filled and access paths should not contain too many pages in order to get the number of disk accesses for maintaining data objects small. We propose to apply the PPbin tree principle resulting in a *PPbin tree* which consists of two basic parts:

1. the internal (prefix-)tree T_i containing nodes near the root, and
2. several external subtrees stored in pages.

Furthermore, for a PPbin tree the *external balancing property* holds:

The number of external pages which are traversed on any two paths from the root to a leaf differs by at most 1.

Defining the *external height* $h_{ext}(T)$ of a PPbin tree T to be the maximal number of external pages occurring on a path in T, the external balancing property ensures that each path in T traverses either $h_{ext}(T)$ or $h_{ext}(T)-1$ external pages.

When objects are inserted into an initially empty PPbin tree, the tree grows up to a size when it cannot be kept in the dedicated part of the main memory any longer. Then a paging algorithm determines a subtree T_s to be paged on secondary storage such that the external balancing property is preserved. If T_s consists of n_s nodes, the main memory is now able to receive additional n_s nodes until a further invocation of the paging algorithm must take place.

If a paged subtree T_P grows up to a size where it cannot be kept in one page, the page is split into two. To this end, the left and right subtree of the root of T_P are stored in distinct pages. If the split page has been referenced by a node in the internal prefix tree T_i, the root of T_P is inserted into T_i. If the split page has been referenced by a node in a page, the root of T_P is inserted into this page.

Figure 2.1 shows a PPbin tree.

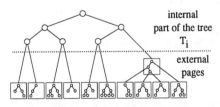

internal
part of the tree
T_i

external
pages

Figure 2.1: Overall structure of a PPbin tree

2.2 A closer look

We now discuss the PPbin tree in more detail by explaining the insertion of a new object into the structure. We formulate the procedures needed to perform an insertion operation in a pseudo programming language.

First, we explain procedure *Insert*, which inserts a new leaf q into a PPbin tree T.

PROCEDURE **Insert (q, T);**
{searches for the node p in the PPbin tree T which will be the father of the new leaf q and calls *TreeInsert*, if T is not empty, and creates a new root for T, otherwise.}
BEGIN
    IF T is empty THEN
        let q be the root of T;
    ELSE
        search for the node p which will be the father of q;
        IF q will be the left son of p THEN dir := left ELSE dir := right END IF;
        TreeInsert(T, p, dir, q);
    END IF;
END **Insert**;

The procedure *TreeInsert* which we explain afterwards stores the new leaf q as the dir son of p (dir $\in$ {left, right}) in the PPbin tree T while preserving the external balancing property for T. For this purpose, *TreeInsert* calls the procedures *Page* and *PageSplit* which we describe next.

The procedure *Page* is called when after the insertion of an additional node the size of the internal prefix tree T_i reaches the maximal possible number n_i of internal nodes. Then *Page* searches for a subtree T_s in T_i such that paging T_s preserves the external balancing property and stores T_s in a page.

The external balancing property is preserved if T_s is a *paging candidate*, i.e. T_s fulfills the following conditions:

1. Any path from the root of T_s down to a leaf contains the minimal number of external pages (of all paths in T).
2. The height of T_s is at most h_P.

The second condition stems from the fact that a page can hold a tree up to a height of h_P.

In order to direct the search for a paging candidate in T_i the following numbers are attached to each node v in T_i:

$nep_{minl}(v)$, resp. $nep_{minr}(v)$,: the minimal number of external pages occurring on any path in T traversing the left, resp. right, son of v.

$nep_{maxl}(v)$, resp. $nep_{maxr}(v)$,: the maximal number of external pages occurring on any path in T traversing the left, resp. right, son of v.

$s(v)$: the number of nodes of the biggest paging candidate which can be reached from v.

$h(v)$: The height of the subtree with root v in T_i.

Now we can define procedure *Page*:

PROCEDURE **Page (T);**
{pages a subtree T_s of the internal prefix tree T_i.}
BEGIN
    r := root(T_i);
    WHILE r $\neq$ NIL AND r is a node in T_i AND
        ($nep_{minl}(r) \neq nep_{maxl}(r)$ OR $nep_{minr}(r) \neq nep_{maxr}(r)$ OR $nep_{minl}(r) \neq nep_{minr}(r)$ OR
        $h(r) > h_P$) DO

```
        IF nep_minl(r) < nep_minr(r) THEN
            r := the left son of r;
        ELSIF nep_minl(r) > nep_minr(r) THEN
            r := the right son of r;
        ELSE
            r := the son of r with greater s(son);
            {if s(left son) = s(right son) each son might be chosen}
        END IF;
    END WHILE;
    f := r;
    allocate a new (empty) page P;
    IF r is a node in T_i THEN
```

```
        store the subtree T_s with root r of T_i in P;
        make the pointer to r pointing to P;
    ELSIF r = NIL THEN
```

```
        replace r by the page P;
        make P pointing to NIL;
        Page(T);
    ELSE {r is a page}
```

```
        make the pointer to r pointing to P;
        make P pointing to r
        Page(T);
    END IF;
    REPEAT
        f := the father of f in T_i;
        recompute nep_minl(f), nep_maxl(f), nep_minr(f), nep_maxr(f), s(f) and h(f) from the correspond-
            ing values of the direct sons of f;
    UNTIL f = root(T_i);
END Page;
```

If more than one paging candidate occurs in T_i, the paging algorithm chooses a candidate with the maximal possible number of nodes. If the search for a paging candidate ends with NIL or in a page, an empty page P is attached and *Page*(T) is invoked again. The recursive call of *Page* may happen $\lceil \frac{n_i}{2} \rceil$ times in the worst case. However, in our simulations, where 100,000 skew distributed objects have been inserted into an LSD tree (a spatial data structure with a PPbin tree as directory; see section 3) with bucket capacity 5, no empty page was created. The reason is that, if after the insertion of a new node the size of T_i reaches the main memory capacity n_i, *Page* searches for a paging candidate in T_i independently of the actual insertion which has caused the call of *Page*. Hence, a paging candidate is determined with one *Page* call if it exists at all. The performance complexity of one *Page* call is O(height(T_i)).

The procedure *PageSplit* is called when an additional node has to be stored in a page which is not able to take it without splitting. In a page a subtree is organized as a sequential heap of fixed height h_P. We choose the heap organization because the PPbin tree is "height-balanced" w.r.t external pages. Hence, it seems natural to use a height-balanced criterion inside a page, too: when the insertion of an additional node would cause the height of a paged subtree to exceed h_P, the page is split into two. After splitting the new node can be inserted.

PROCEDURE **PageSplit (p, dir, q, T_P, T)**;
{The left and right subtree of the root of T_P are stored in two distinct pages P_l and P_u, and the root of T_P is inserted into T by calling *TreeInsert*. After the page split the new node q is the dir son of p (dir $\in$ {left, right}).}
BEGIN
    P := the page containing T_P;
    r := root(T_P);
    allocate two new pages P_l and P_u;
    store the left subtree of r in P_l and make P_l be the left son of r;
    store the right subtree of r in P_u and make P_u be the right son of r;
    IF p is stored in P_l THEN
        store q as the dir son of p in P_l;
    ELSE
        store q as the dir son of p in P_u;
    END IF;
    f := the node in T referencing P;
    IF P is the left son of f THEN dirf := left ELSE dirf := right END IF;
    deallocate P;
    TreeInsert(T, f, dirf, r);
END **PageSplit**;

Figure 2.2 depicts the effect of a page split.

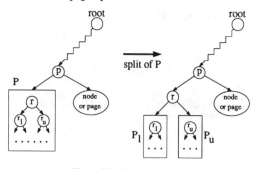

Figure 2.2: Split of page P

We are now in a position to define *TreeInsert*:

PROCEDURE **TreeInsert (T, p, dir, q)**;
{stores the node q as the dir son of p (dir $\in$ {left, right}) in the PPbin tree T. q is the new leaf to be inserted or the root of a subtree whose page has been split.}
BEGIN
    IF p is a node in T_i THEN
        make q the dir son of p in T_i;
        f := q;
        REPEAT
            f := the node in T_i referencing f;
            recompute $nep_{minl}(f)$, $nep_{maxl}(f)$, $nep_{minr}(f)$, $nep_{maxr}(f)$, s(f) and h(f) from the corre-
                sponding values of the direct sons of f;
        UNTIL f = root(T_i);
        IF (number of nodes in T_i) = n_i THEN *Page*(T) END IF;
    ELSE {p is stored in a page}
        T_P := the paged subtree containing node p;
        IF the height of T_P will be greater than h_P after the insertion of p THEN
            PageSplit(p, dir, q, T_P, T);
        ELSE
            Store q as the dir son of p in T_P;
        END IF;
    END IF;
END **TreeInsert**;

Note that before and after the execution of *TreeInsert* the number of internal nodes is always smaller than n_i. Hence, T_i is always able to take a new node before an eventual call of the procedure *Page* takes place.

Figure 2.3 depicts the overall structure of the insertion algorithm for the PPbin tree.

The heart of the insertion algorithm for PPbin trees is the procedure *Page* which determines a paging candidate and stores it on secondary storage. Since *Page* always determines a paging candidate with the maximal number of nodes among all paging candidates, the page utilization is satisfactory and a degeneration of the external height is rather unlikely. Hence, a degeneration of the (original) binary tree is mainly reflected by a degeneration of the internal prefix tree and less by a degeneration of the external height of the PPbin tree.

The analysis of the application of the PPbin tree principle to arbitrary binary trees is extremely difficult and still open. For the class of k-d trees, the analysis provided in section 4 demonstrates that the expected external height of k-d PPbin trees is near the minimal possible external height.

2.3 The operations

Search

A search is performed similar to a search in unpaged binary trees.

Insertion

The insertion algorithm has been explained in detail in the previous section.

Deletion

In a PPbin tree T, a node is deleted similar to a node deletion in unpaged binary trees, i.e. we can restrict the discussion to the deletion of a node v with at most one son. If v is an internal node, i.e. in T_i, the deletion is trivial. If v is a node in a paged subtree T_P the height of T_P may decrease by one.

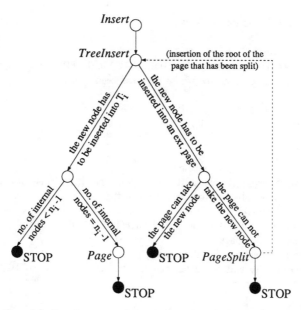

Figure 2.3: Overall structure of the insertion algorithm for the PPbin tree

If the page P storing T_P has a brother page which contains a subtree of height less than h_P, both pages can be merged. Such a merge works just inversely to a page split.

If the page P becomes empty, it can be removed from T if the external balancing property will not be violated. According to this property a path in T contains either $h_{ext}(T)$ or $h_{ext}(T)$-1 external pages. If paths of both external lengths exist, P can only be removed, if all paths containing P contain $h_{ext}(T)$ external pages. Otherwise, P has to remain in T. If all paths in T contain equally many external pages, P can also be removed.

If deletions cause the external height $h_{ext}(T)$ to decrease by one, empty pages, which could not be removed so far, can now be removed as far as the external balancing property is not violated. In any case, removing an empty page can be performed by the search procedure on the fly.

3. The LSD tree: an application of the PPbin tree principle in the multidimensional case

In this section, we present a new spatial data structure as an application of the PPbin tree principle in the multidimensional case. The data structure efficiently supports spatial queries like the retrieval of an object by its coordinates (exact match) and range queries where all objects geometrically intersecting the query region are selected for further processing or presentation on the screen. Since the set of objects varies over time, insertions and deletions of objects are supported as well.

The new structure using a partially paged k-d tree as directory has several advantages over other spatial data structures which have been proposed (see e.g. [Free87], [HSW88a], [HSW88b], [KS86], [KS88], [KW85], [NHS84], [Otoo86]):

1. It provides a great freedom for using the split strategy best suited for the actual application.
2. It is extremely insensitive to skew distributed objects.
3. The order of insertion does not influence the structure (as far as distribution dependent split strategies (see section 3.2) are applied).
4. For B data buckets the size (number of nodes) of the directory is B-1.
5. The structure can be extended to arbitrary geometric objects, i.e. non-point objects using the transformation technique ([Hin85], [SK88]). This technique, for instance, stores two-dimensional

rectangles as four-dimensional points (in a four-dimensional data structure). Unfortunately, the distribution of the four-dimensional points in general is extremely skew and therefore only data structures which are insensitive to such distributions perform well in this environment.

3.1 Basic ideas and properties

For sake of simplicity, we restrict the discussion to two-dimensional points. A generalization to arbitrary dimensions is straightforward.

Like most spatial data structures, the new structure partitions the data space into pairwise disjoint cells and stores all objects located in a cell in an associated data bucket (bucket, for short). In contrast to the grid file [NHS84], however, it is not grid oriented, i.e. a cell boundary may occur at an arbitrary position. The free choice of cell boundaries, i.e. split positions, is the basis of the graceful adaptation to arbitrary skew object distributions. Since a new split position can be chosen locally optimal, i.e. optimal with respect only to the cell to be split and independent from other existing cell boundaries, we call the structure **Local Split Decision tree** (LSD tree, for short).

Figure 3.1 shows a possible partition of the data space.

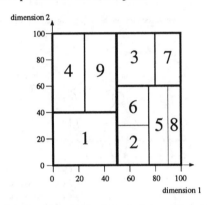

Figure 3.1: Possible partition of the data space for an LSD tree

This flexible data space partition has to be maintained by a directory. For this purpose we use a binary tree similar to a k-d tree [Ben75], which is embedded into a PPbin tree. Each node of the k-d PPbin tree represents one split decision by storing the split dimension and the split position (in that dimension). Figure 3.2 illustrates the LSD tree associated with the data space partition of Figure 3.1.

The first split is done in dimension 1 at position 50. All points whose first coordinates are smaller than, resp. greater than or equal to, 50 are stored in the left, resp. right, branch of the tree. Note that the tree levels are not necessarily alternatingly associated with data space dimensions.

Figure 3.3 shows the overall structure of the LSD tree with several external directory layers.

3.2 A closer look

In this section, we discuss the LSD tree in more detail by explaining the insertion of a new geometric object o into the structure. This insertion is performed by the procedure *LSDInsert*.

PROCEDURE **LSDInsert** (o, T);
{The search for the bucket B which will receive the new object o is guided by the directory T as in k-d trees. If B does not overflow, o is stored in B and the insertion is finished. Otherwise the procedure *BucketSplit* is called.}

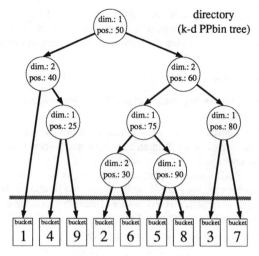

Figure 3.2: LSD tree associated with the data space partition of Figure 3.1

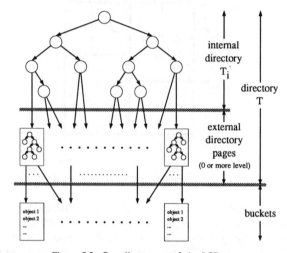

Figure 3.3: Overall structure of the LSD tree

```
BEGIN
    w := root(T);
    WHILE w is not a bucket DO
        s_dim := split dimension stored in w;
        s_pos := split position stored in w;
        IF o[s_dim] ≤ s_pos THEN w := left son of w ELSE w := right son of w END IF;
    END WHILE;
    B := w;
    IF B is not full THEN B := B ∪{o} ELSE BucketSplit(o, B, T) END IF;
END LSDInsert;
```

In *LSDInsert*, an overflowing bucket is handled by *BucketSplit*. The procedure *TreeInsert* which is used in *BucketSplit* is exactly the one described in section 2.2.

PROCEDURE **BucketSplit (o, B, T)**;
{distributes object o and all objects in bucket B over two new buckets.}
BEGIN
    p := the node in T referencing B;
    IF B is the left son of p THEN dir := left ELSE dir := right END IF;
    dim := split dimension stored in p;
    ComputeSplitLine(o, B, dim, s_{pos}, s_{dim});
        {s_{pos} and s_{dim} needed to split B are determined by *ComputeSplitLine*}
    allocate two new buckets B_l and B_u;
    create a new node q, containing s_{pos} and s_{dim}, and referencing B_l and B_u;
    TreeInsert(T, p, dir, q);
    FOR EACH $o' \in B \cup \{o\}$ DO
        LSDInsert(o', T);
        {naturally, an efficient implementation will use locally available information and not
            carry out an *LSDInsert* for each o'}
    END FOR;
    deallocate B;
END **BucketSplit**;

Figure 3.4 depicts the effect of a bucket split.

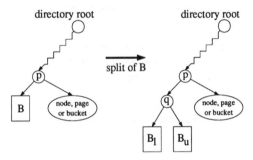

Figure 3.4: Split of data bucket B

It remains to define procedure *ComputeSplitLine*, which determines the split dimension s_{dim} and the split position s_{pos} for the overflowing bucket B. s_{dim} and s_{pos} depend on the split strategy used. We distinguish between two inherently distinct types of **split strategies**:

1. *Data dependent split strategies*
These strategies depend only on the objects stored in the bucket to be split. A typical example is to choose for the split position the mean of all object coordinates with respect to a certain dimension.
2. *Distribution dependent split strategies*
These strategies choose the split dimension and the split position independently of the actual objects stored in the bucket to be split (and of all other existing objects in the structure, as well). A typical example is to split a cell into two cells of equal areas. Note that this "halving split strategy" relies on the assumption of a uniform distribution of the objects.

PROCEDURE **ComputeSplitLine (o, B, dim, s_{pos}, s_{dim})**;
BEGIN
    determine s_{dim} and s_{pos}; {according to the split strategy used}
END **ComputeSplitLine**;

It should be obvious that the use of a binary tree as directory provides the freedom for using the split strategy best suitable for the actual application. This is an important advantage over other structures (see e.g. [Free87], [HSW88a], [KS86], [KS88], [NHS84]) where split decisions are more or less influenced by previous split decisions even if completely different data regions are concerned. Furthermore, the size of the directory is directly related to the number of buckets, i.e. for B buckets the directory contains B-1 nodes. This is in contrast to the grid file, for instance, where several entries in the directory may point to the same bucket.

3.3 The operations

Exact match

A search is guided by the k-d PPbin tree and ends in a bucket. This bucket is scanned until the object is determined or the search ends unsuccessfully.

Insertion

The insertion algorithm has been explained in detail in the previous section.

Deletion

An exact match locates the bucket B containing the object to be deleted and the object is removed from B. If the brother of B is a bucket, too, i.e. both buckets stem from the same bucket split, and the number of objects stored in both buckets is not greater than the bucket capacity, the two buckets are merged and the corresponding node is deleted from the directory as described for PPbin trees in Section 2.3.

Range query

In a range query all points located in a query region Q are reported. The function *RangeQuery* computes the set of requested objects. D(w) denotes the data region which is the union of all data cells whose corresponding buckets can be reached from the node w.

```
FUNCTION RangeQuery (Q, T) : SET OF objects;
{returns all objects which are located in the query region Q.}
BEGIN
    RangeQuery := ∅;
    stack := EmptyStack;
    w := root(T);
    PUSH(stack, w);
    WHILE NOT IsEmpty(stack) DO
        w := POP(stack);
        WHILE w is not a bucket DO
            IF Q ∩ D(right son of w) = ∅ THEN
                w := left son of w;
            ELSIF Q ∩ D(left son of w) = ∅ THEN
                w := right son of w;
            ELSE
                PUSH(stack, right son of w);
                w := left son of w;
            END IF;
        END WHILE;
        B := w;
        RangeQuery := RangeQuery ∪ {o | o is stored in B ∧ o is located in Q};
    END WHILE;
END RangeQuery;
```

Note that $Q \cap D(w) \neq \emptyset$ is the invariant condition of the inner WHILE-loop. Hence, if $Q \cap D(\text{one son of } w) = \emptyset$ then $Q \cap D(\text{other son of } w) \neq \emptyset$.

4. Analysis of the LSD tree

In this section, we provide an analysis of the LSD tree. We estimate the expected utilization of data buckets and directory pages, the expected number of external directory layers, and the distribution of directory pages over the external layers. Since the directory of the LSD tree is a k-d PPbin tree, the analysis comprises the k-d PPbin tree, too. The analysis is based on the following assumptions:

1. The split decisions are independent of each other.
2. For each node v in T the probability that a point in $D(v)$ is located to the left or to the right of the split line associated with v is $\frac{1}{2}$ each.

The first assumption is clearly ensured by the Local Split Decision tree. The second assumption is ensured if the split position is chosen as the mean of all object coordinates w.r.t. a certain dimension and the objects are inserted in random order. Hence, independent of the distribution of objects the assumptions are ensured in an LSD tree based on an appropriate data dependent split strategy, as far as the objects are inserted in random order.

The *level of a node* v in an LSD tree is as usual the length of the path from the root to v. The level of the root is 0. If v is a node in a page the level of v is defined as above assuming that a pointer to a page is a pointer to the root of the subtree stored in that page.

We regard the k-d PPbin tree used as directory T as an infinite binary tree with certain nodes active and others inactive. Of course, only the active nodes are actually stored in T. The number of nodes on level k is 2^k, $k \geq 0$.

The probability that a random point is located in the data region $D(v)$ of a node v on level k in the directory T is $\frac{1}{2^k}$. If n points have been inserted into the LSD tree with directory T, the random variable X, denoting the number of points in the data region $D(v)$ of a node v on level k is $B\left(n, \frac{1}{2^k}\right)$ distributed, i.e.

$$P(X = i) = \binom{n}{i} \cdot \left(\frac{1}{2^k}\right)^i \cdot \left(1 - \frac{1}{2^k}\right)^{n-i} .$$

Since in all interesting cases $\frac{1}{2^k} < 0.1$ and $n > 30$ the binomial distribution can be approximated by a Poisson distribution:

$$P(X = i) \simeq \frac{\left(\frac{n}{2^k}\right)^i}{i!} \cdot e^{-\left(\frac{n}{2^k}\right)} .$$

The probability that a certain node v in the directory T of an LSD tree is active equals the probability that more than b (= bucket capacity) points are located in the data region $D(v)$ because for storing more than b points a bucket split must have been performed activating v. Hence, the probability that a node v on level k in T is active is

$$P(X > b) = 1 - P(X \leq b)$$

$$\simeq 1 - \sum_{i=0}^{b} \frac{\left(\frac{n}{2^k}\right)^i}{i!} \cdot e^{-\left(\frac{n}{2^k}\right)} .$$

Denoting

$$P(X \leq b) \simeq \xi_T(n, b, k) = \sum_{i=0}^{b} \frac{\left(\frac{n}{2^k}\right)^i}{i!} \cdot e^{-\left(\frac{n}{2^k}\right)}$$

the expected number of active nodes on level k in the directory T is

$$NLev_T(n, b, k) = (1 - \xi_T(n, b, k)) \cdot 2^k .$$

The expected number of active nodes in the directory T is

$$N_T(n, b) = \sum_{k=0}^{\infty} NLev_T(n, b, k)$$

and the expected number of buckets in the LSD tree is

$$B(n, b) = N_T(n, b) + 1 .$$

Figure 4.1 shows the expected bucket utilization

$$\beta(n, b) \stackrel{\text{def}}{=} \frac{n}{b \cdot B(n, b)}$$

for several bucket capacities.

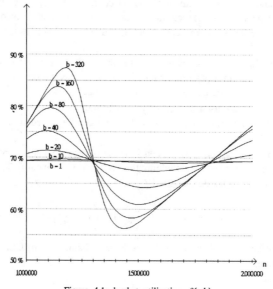

Figure 4.1: bucket utilization $\beta(n,b)$

Note that we assume that after a bucket split b instead of (the actual) b+1 objects are distributed over the two resulting buckets.

$\beta(n,b)$ is a periodic function of n if b is fixed and $n \gg b$. Each doubling of n corresponds to one period.

Besides the bucket utilization we are interested in the directory page utilization. A directory page can accommodate a subtree up to a height of h_P. We start the analysis assuming that only one external layer is used.

A node v on level k in T_i is active (w.r.t. T_i) if and only if at least one node on level $k+h_P$ in T which can be reached from v is active (w.r.t. T) (otherwise v is stored in a page). Hence, a node v on level k in T_i is inactive if and only if each of the 2^{h_P} nodes on level $k+h_P$ in T is which can be reached from v is inactive. The probability that v is inactive is therefore

$$\xi_{T_i}(n, b, k, h_P) = [\xi_T(n, b, k + h_P)]^{2^{h_P}} .$$

Then the probability that v is active on level k in T_i is $1 - \xi_{T_i}(n,b,k,h_P)$.

For T with one external directory layer the expected number of active nodes on level k in T_i is

$$NLev_{T_i}(n, b, k, h_P) = (1 - \xi_{T_i}(n, b, k, h_P)) \cdot 2^k ,$$

and the expected number of active nodes in T_i is

$$N_{T_i}(n, b, h_P) = \sum_{k=0}^{\infty} NLev_{T_i}(n, b, k, h_P) .$$

The expected number of directory pages in T with one external directory layer is

$$PLay_1(n, b, h_P) = N_{T_i}(n, b, h_P) + 1 .$$

Defining the expected directory page utilization as the quotient of the expected number of nodes stored in directory pages and the product of the expected number of directory pages and the page capacity, we get

$$\gamma_1(n, b, h_P) \stackrel{\text{def}}{=} \frac{N_T(n, b) - N_{T_i}(n, b, h_P)}{PLay_1(n, b, h_P) \cdot (2^{h_P} - 1)} .$$

Figure 4.2 shows the directory page utilization $\gamma_1(n,b,h_P)$ for T with one external directory layer for several bucket capacities.

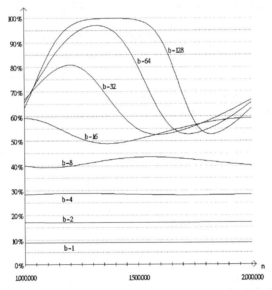

Figure 4.2: $\gamma_1(n,b,h_P)$ for T with one external directory layer as a function of n ($h_P = 6$)

Like $\beta(n,b)$, $\gamma_1(n,b,h_P)$ is a periodic function of n if b is fixed and $n \gg b$. Again, each doubling of n corresponds to one period.

We now extend the analysis of the directory page utilization if $x \geq 1$ external directory layers are used. Note that layer 1 is the layer closest to the buckets and layer x contains the pages referenced from nodes in T_i.

Let T_i^x denote the internal part of the directory T when x external layers are used. A node v on level k in T_i^x is inactive (w.r.t T_i^x) if and only if each of the 2^{h_P} nodes on level k+h_P in T_i^{x-1} which can be reached from v is inactive (w.r.t T_i^{x-1}). The probability that v (in T_i^x) is inactive is therefore given by the recurrence relation

$$\xi_{T_i^x}(n,b,k,h_P) = \begin{cases} \left[\xi_T(n,b,k+h_P)\right]^{2^{h_P}} & x=1 \\ \left[\xi_{T_i^{x-1}}(n,b,k+h_P,h_P)\right]^{2^{h_P}} & x>1 \end{cases}$$

The expected number of active nodes in T_i^x on level k is

$$NLev_{T_i^x}(n,b,k,h_P) = \left(1 - \xi_{T_i^x}(n,b,k,h_P)\right) \cdot 2^k .$$

The expected number of active nodes in T_i^x is

$$N_{T_i^x}(n,b,h_P) = \sum_{k=0}^{\infty} NLev_{T_i^x}(n,b,k,h_P) .$$

Since for each external layer $z \in \{1, ..., x\}$

$$N_{T_i^z}(n,b,h_P) = \sum_{k=0}^{\infty} NLev_{T_i^z}(n,b,k,h_P)$$

the expected number of directory pages on each external layer z is

$$PLay_z(n,b,h_P) = N_{T_i^z}(n,b,h_P) + 1 .$$

The expected directory page utilization γ_z of each external layer z is given by

$$\gamma_z(n,b,h_P) \stackrel{def}{=} \begin{cases} \dfrac{N_T(n,b)-N_{T_i^z}(n,b,h_P)}{PLay_z(n,b,h_P)\cdot\left(2^{h_P}-1\right)} & z=1 \\ \dfrac{N_{T_i^{z-1}}(n,b,h_P)-N_{T_i^z}(n,b,h_P)}{PLay_z(n,b,h_P)\cdot\left(2^{h_P}-1\right)} & z>1 \end{cases}$$

Figure 4.3 shows the directory page utilization $\gamma_z(n,b,h_P)$ for z = 1, 2, 3, i.e. for layer 1, layer 2 and layer 3. The bucket capacity is 16.

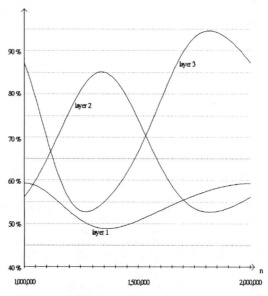

Figure 4.3: $\gamma_z(n,b,h_P)$ for z = 1, 2, 3, i.e. for layer 1, layer 2 and layer 3 (b = 16 and h_P = 6)

We are now in a position to derive the expected number of external layers and the expected number of directory pages if an upper bound n_i for the number of internal nodes is given.

The expected number χ of external layers can be calculated as follows:

1. If $N_T(n, b) \leq n_i$ then $\chi = 0$.
2. If $N_{T_i^1}(n, b, h_P) \leq n_i < N_T(n, b)$ then $\chi = 1$.
3. If $n_i < N_{T_i^1}(n, b, h_P)$ then choose χ such that $N_{T_i^\chi}(n, b, h_P) \leq n_i < N_{T_i^{\chi-1}}(n, b, h_P)$.

Note that χ is the expected value of the external height $h_{ext}(T)$ of T.

The expected number of directory pages on layer z is given by

$$
PLay_z^*(n, b, h_P, n_i) = \begin{cases} \left\lceil \frac{N_{T_i^{z-1}}(n,b,h_P)-n_i}{\gamma_z(n,b,h_P)\cdot(2^{h_P}-1)} \right\rceil & z = \chi \\[2ex] PLay_z(n, b, h_P) & z < \chi \\[2ex] 0 & z > \chi \end{cases}
$$

where $N_{T_i^0}(n, b, h_P) := N_T(n, b)$.

Table 4.1 shows, for instance, that on the average two external layers suffice for 10 million objects for realistic values of b, h_P and n_i.

n	pages on layer 1	pages on layer 2	pages on layer 3
10.000	0	0	0
100.000	238	0	0
500.000	1166	5	0
1.000.000	2332	38	0
5.000.000	13886	253	0
10.000.000	27771	526	0
20.000.000	55543	1070	3

Table 4.1: Expected number of directory pages needed on each layer if b = 16, h_P = 6 and n_i = 1000

At the end of this section, we sketch the analysis of the situation when the second assumption is relaxed. We assume that the probability that a point is located to the left, resp. to the right, of a split line is α, resp. $1-\alpha$, for $0 < \alpha < 1$.

For $\alpha > 0.5$ the probability that the leftmost path p_l in T (whose expected length is maximal among all paths in T) is shorter than k, i.e. the number of active nodes on p_l is less than k+1, is

$$
\xi_{T,\alpha}(n, b, k) = \sum_{i=0}^{b} \frac{(n \cdot \alpha^k)^i}{i!} \cdot e^{-(n\cdot\alpha^k)}.
$$

Hence, the probability that the length of p_l equals k is

$$
\delta_\alpha(n, b, k) = \xi_{T,\alpha}(n, b, k+1) - \xi_{T,\alpha}(n, b, k).
$$

Figure 4.3 depicts δ_α as a function of k for multiple values of α. It turns out that the height of the directory of an LSD tree does not degenerate even if a clumsy split strategy is chosen.

The analysis demonstrates that under realistic assumptions the expected bucket utilization of the LSD tree as well as the expected page utilization of the k-d PPbin tree used as directory is competitive to other spatial data structures (see e.g. [Free87], [HSW88a], [KS86], [KS88], [KW85], [NHS84], [Otoo86]). Furthermore, the external height of the LSD tree does not degenerate even if a clumsy split strategy is used. Remember, however, that one of the main advantages of the LSD tree over other structures is the freedom of choosing the best suitable split strategy.

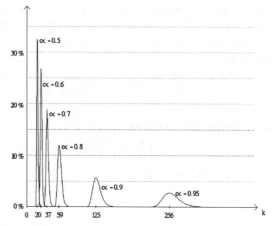

Figure 4.4: δ_α(n,b,k) as a function of k (b = 1 and n = 1,000,000)

References

[AVL62] Adelson-Velskii, G.M., Landis, E.M.: 'An algorithm for the organization of information', Soviet Math. Dokl. 3 (1962), 1259–1263

[BM72] Bayer, R., McCreight, E.: 'Organization and maintenance of large ordered indexes', Acta Informatika 1, 3, 173–189, 1972

[Ben75] Bentley, J.L.: 'Multidimensional Binary Search Trees Used in Database Applications', Communications of the ACM, Vol. 18, 9, 509-517, 1975

[Free87] Freeston, M.: 'The BANG file: a new kind of grid file', Proc. ACM SIGMOD Int. Conf. on Management of Data, 260-269, 1987

[GS78] Guibas, L.J., Sedgewick, R.: 'A Dichromatic Framework for Balanced trees', 19th Annual IEEE Symposium on Foundations of Computer Science, 8–21, 1978

[Hin85] Hinrichs, K.: 'The Grid File System: Implementation and Case Studies of Applications', Doctoral Thesis No. 7734, ETH Zürich, 1985

[HSW88a] Hutflez, A., Six, H.-W., Widmayer, P.: 'Globally Order Preserving Multidimensional Linear Hashing', Proc. IEEE 4th Int. Conf. on Data Engineering, 572-579, 1988

[HSW88b] Hutflez, A., Six, H.-W., Widmayer, P.: 'Twin Grid Files: Space Optimizing Access Schemes', Proc. ACM SIGMOD Int. Conf. on Management of Data, 183-190, 1988

[HSW89] Henrich, A., Six, H.-W., Widmayer, P.: unpublished manuscript

[IKO87] Icking, Ch., Klein, R., Ottmann, Th.: 'Priority Search Trees in Secondary Memory', Universität Freiburg Institut für Informatik, Bericht 4, November 1987

[KS86] Kriegel, H.-P., Seeger, B.: 'Multidimensional Order Preserving Linear Hashing with Partial Expansions', Proc. Int. Conf. on Database Theory, 203-220, 1986

[KS88] Kriegel, H.-P., Seeger, B.: 'PLOP-Hashing: A Grid File without Directory', Proc. IEEE 4th Int. Conf. on Data Engineering, 369-376, 1988

[KW85] Krishnamurthy, R., Whang, K.-Y.: 'Multilevel Grid Files', IBM Research Report, Yorktown Heights, 1985

[NHS84] Nievergelt, J., Hinterberger, H., Sevcik, K.C.: 'The Grid File: An Adaptable Symmetric Multikey File Structure', ACM Transactions on Database Systems, Vol. 9, 1, 38-71, 1984

[NR73] Nievergelt, J., Reingold, E.M.: 'Binary Search Trees of Bounded Balance', SIAM J. Computing 2 (1973) 33-43

[Otoo86] Otoo, E.J.: 'Balanced Multidimensional Extendible Hash Tree', Proc. 5th ACM SIGACT / SIGMOD Symposium on Principles of Database Systems, 100-113, 1986

[OW81] Ottmann, Th., Wood, D.: '1-2 brother trees or AVL trees revisited', Comput. J., 23 (1981), 248-255

[Rob81] Robinson, J.T.: 'The K-D-B-Tree: A Search Structure for Large Multidimensional Dynamic Indexes', Proc. ACM SIGMOD Int. Conf. on Management of Data, 10-18, 1981

[SK88] Seeger, B., Kriegel, H.-P.: 'Techniques for Design and Implementation of Efficient Spatial Access Methods', Proc. 14th Int. Conf. on VLDB, 360–371, 1988

[Wir75] Wirth, N.: 'Algorithmen und Datenstrukturen', B.G. Teubner, Stuttgart 1975

THE COMPLEXITY OF GRAPH PROBLEMS
FOR SUCCINCTLY REPRESENTED GRAPHS

Antonio Lozano and José L. Balcázar
Department of Software (Llenguatges i Sistemes Informàtics)
Universitat Politècnica de Catalunya
08028 Barcelona, SPAIN
e-mail: eabalqui@ebrupc51.bitnet

Abstract. Highly regular graphs can be represented advantageously by some kind of description shorter than the full adjacency matrix; a natural succinct representation is by means of a boolean circuit computing the adjacency matrix as a boolean function. The complexity of the decision problems for several graph-theoretic properties changes drastically when this succinct representation is used to present the input. We close up substantially the gaps between the known lower and upper bounds for these succinct problems, in most cases matching optimally the lower and the upper bound.

1. Introduction

Representing graphs with regularities by means of data structures smaller than the adjacency matrix is a very natural idea; for instance, data structures for sparse matrices, with a small number of nonzero entries, have been studied since long time. Several other possibilities have been proposed to obtain shorter representations of regular graphs ([4], [16]). In particular, we will continue the study initiated in [4], where the adjacency matrix is represented by a hopefully small boolean circuit, which on input the binary representations of i and j computes the (i, j) entry of the adjacency matrix.

In particular, we are interested in using this representation to describe the inputs to algorithms for various decisional graph-theoretic problems. On the one hand, since the size of the input may be much smaller, the complexity of the algorithms must be expected to grow. On the other hand, since only very regular graphs can be succinctly described, one might hope that some trade-off exists, making the problem easier for compactly described instances due to the regularity of the represented graph.

In [4] and subsequently in [12] these hopes are destroyed for many problems. Indeed, it is shown in [4] that quite simple graph problems, such as simply checking nonemptiness of the edge set, become *NP*-hard if the input is given by a circuit. A sufficient condition for *NP*-completeness of succinct problems is given there. The interested reader can find a small, natural correction (and a construction showing the necessity of that correction) in [11]. It was shown in [12] that sets that are *NP*-complete for projection reducibility (to be defined below) become *NEXPTIME*-complete when the input is given by a circuit representation. This last reference states that similar techniques can be applied to projection-complete sets in other complexity classes, such as *P* or *NL*, and asserts that from such a construction for *NL* other lower bounds of [4] can be improved to optimal.

However, two objections can be made to their statements. First, the proof is complete only for the case of *NP*, and for the other complexity classes the constructions themselves are not given; although they are similar in spirit, each one has its own peculiarities and small technical difficulties. We present here a much simpler proof of the main result of [12], as corollary of a couple of lemmas whose statements regard just decisional problems and their succinct versions. This presentation is both easier and better, since the independence of the complexity classes in which the problems lie allows us to carry over immediately the result to other classes.

Second, to improve the other lower bounds they mention, neither of the constructions indicated is appropriate, since the closest one is directed graph accesibility but the graphs of [4] are undirected. Thus, the proof in [12] does not apply to connectivity, bipartiteness, nor acyclicity in undirected graphs, nor to undirected acessibility itself, unless $NL = SL$ (where SL denotes symmetric logarithmic space [10]). We prove here these results by a different proof. Also, we study one of the cases to which the results of [12] do not apply, namely planarity, and using an argument of Reif [13] we settle the particular case of bounded degree graphs.

The table in figure 1 shows the improvements to the results of [4] that stem from [12] and from our work. It can be seen that 6 out of the 15 problems considered in [4] were given optimal bounds there, and that for two more of them optimal bounds were obtained in [12], where also three more were claimed. We confirm the claims of [12] (using a different proof), present a technique that unifies many previous results into a much simpler construction, settle two more problems by this technique, and close up substantially the gap for the remaining two problems (perfect matching and planarity). Finally we obtain optimal bounds for a subproblem of planarity, when the input is restricted to bounded degree graphs.

2. Graphs, reducibilities, and succinctness

This section proposes a different, more general way of presenting the results of [12], so that both the results there, and the previous ones in [4], follow from the same technique. Thus, this section provides a method to recast all previous results in a uniform way. No new contributions to the table in figure 1 will be proved in this section, but the theorems will be used repeatedly as main tools in the next section.

Throughout this paper we use the term *graph* to mean an undirected finite graph, with no multiple edges. A graph can be described by a boolean square matrix (its adjacency matrix). We mention several complexity classes, i.e. classes of sets that can be decided within a given resource bound by a sequential computation model such as, e.g., multitape Turing machines. Results regarding complexity classes may be found, among others, in [1] and [5]. These complexity classes are: deterministic logarithmic space, denoted L, nondeterministic logarithmic space, denoted NL, deterministic polynomial time, denoted P, nondeterministic polynomial time, denoted NP, polynomial space, for which deterministic and nondeterministic classes coincide and therefore denoted just $PSPACE$, deterministic exponential (i.e. $O(2^{n^k})$ for constant k) time, denoted $EXPTIME$, and nondeterministic exponential time, denoted $NEXPTIME$. Given a reducibility and a complexity class, a problem is *hard* for the class under the reducibility if every problem in the class is reducible to it, and is *complete* for the class if it is hard for it and also belongs to it.

We consider inputs to decisional problems encoded as words over the two letter alphabet $\{0, 1\}$; in particular the standard encoding of a graph is its adjacency matrix. The length (i.e. number of bits) of a binary string x is denoted $|x|$, and the i-th bit of x is denoted x_i. We identify each decisional problem with the set of encodings of inputs on which the answer is YES. A succinct representation of a binary word x is a boolean circuit that on input (the binary representation of) i outputs two boolean values, one indicating whether $i \leq |x|$ and another one indicating, in that case, the i-th bit of x. The succinct version sA of a decisional problem A is: given a boolean circuit describing a word x, decide whether $x \in A$.

The succinctly represented problem is at most exponentially more difficult than the problem on standard encodings. The following observation is proved for graphs in [4], and can be stated similarly for any binary encoding.

1. **Lemma.** Let $f(n) \geq \log n$ be a nondecreasing bound, and let A be a decisional problem. Assume that $A \in DTIME(f(n))$ (resp. $NTIME(f(n))$, $DSPACE(f(n))$, or $NSPACE(f(n))$). Then, for the succinct version of A, we have $sA \in DTIME(f(2^n)n^2)$ (resp. $NTIME(f(2^n)n^2)$, $DSPACE(f(2^n))$, or $NSPACE(f(2^n))$).

We will use two standard reducibilities. (Polynomial time) m-reducibility is defined as follows: problem A is m-reducible to problem B, denoted $A \leq_m B$, if and only if there is a polynomial time computable function f such that $x \in A \iff f(x) \in B$ for every x.

In order to define the other reducibility, we define *projections*. A projection is a function π from $\{0,1\}^*$ into $\{0,1\}^*$ such that $|\pi(x)| = |x|^k$ for some fixed k, and for all the words x of length n the i-th bit of $\pi(x)$ depends on at most one fixed bit (say the j-th) of x. Thus, for each $i \leq n^k$ there is a fixed $j \leq n$, computable from i in time polynomial in $\log n$, such that $\pi(x)_i$ is either 0, 1, x_j, or $\bar{x}_j$ (i.e. x_j negated), and it is possible to compute within the same time bound which is the case. Finally, problem A is projection-reducible (short, π-reducible) to problem B, denoted $A \leq_\pi B$, if and only if there is a projection π such that $x \in A \iff \pi(x) \in B$ for every x. A nonuniform version of projection reducibility was proposed in [14].

The following lemma, which we name *Conversion Lemma*, shows how to obtain m-reducibilities among succinctly represented problems from projection reducibilities among the standard problems. Its proof is based on a fragment of the proof of the main theorem of [12], and the main advantage is that it is now independent of the fact that the problems be in any particular complexity class.

2. **Lemma.** (*Conversion Lemma*) If $A \leq_\pi B$ then $sA \leq_m sB$.

Proof. Let w be an instance of A. Assume as input an instance of sA, given by a boolean circuit C_w describing w. Observe that the size of C_w is at least $\log |w|$, which is the number of inputs, and therefore $|w| \leq 2^{|C_w|}$.

Consider the projection π that reduces w to $\pi(w)$. The i-th bit of $\pi(w)$ can be computed in time logarithmic on $|w|$, thus polynomial in $|C_w|$, and it is either 0, 1, or the j-th bit of w possibly negated. By a standard translation of Turing machines into circuits (see [1], section 5.4), we find a circuit that computes the i-th bit by combining this computation with the input circuit which finds the j-th bit of w. The resulting circuit is a succinct representation of $\pi(w)$, and therefore is in sB if and only if $\pi(w) \in B$, if and only if $w \in A$, if and only if $C_w \in sA$. Therefore we have a polynomial time reduction from sA to sB. ∎

In order to prove completeness of a succinctly represented problem sA in a complexity class, we need a way of translating arbitrary problems in the class into a region where we can use hypothesis about A, which is exponentially below. To do this, we will encode problems differently, in a form similar to tally sets, which will be of much lower complexity due to the exponential blow-up of the input size. For a problem A, define $\mathrm{long}(A)$ as follows: $x \in \mathrm{long}(A)$ if and only if the binary expression of the length of x is in A. Thus, for each length n, either all the words of that length are in $\mathrm{long}(A)$, if the binary expression of n is in A, or all the words of that length are out of $\mathrm{long}(A)$, otherwise. We have:

3. *Lemma.* Let $f(n) \geq \log n$ be a nondecreasing bound. Then

$$A \in DTIME(f(2^n)) \iff \mathrm{long}(A) \in DTIME(f(n))$$

and similarly for the classes $NTIME(f(n))$, $DSPACE(f(n))$, and $NSPACE(f(n))$.

The proof is quite immediate and therefore omitted. We prove next that the succinct version of $\mathrm{long}(A)$ jumps up again at least to the difficulty of A.

4. *Lemma.* For every A, $A \leq_m s(\mathrm{long}(A))$.

Proof. Denote by $\tilde{x}$ the integer denoted by x in binary notation. Given x, construct a circuit C representing the word $0^{\tilde{x}}$, which outputs zero for all inputs up to $\tilde{x}$. The time required to construct this circuit is easily seen to be polynomial in $|x|$, and by the definitions we have $x \in A \iff 0^{\tilde{x}} \in \mathrm{long}(A) \iff C \in s(\mathrm{long}(A))$. Thus we have the required polynomial time m-reducibility. ∎

Now we can state and easily prove the following result:

5. *Theorem.* Let $f(n) \geq \log n$ be a nondecreasing bound. Then, if A is hard for $DTIME(f(n))$, $NTIME(f(n))$, $DSPACE(f(n))$, or $NSPACE(f(n))$ under projection reducibility, then sA is m-hard for $DTIME(f(2^n))$, $NTIME(f(2^n))$, $DSPACE(f(2^n))$, or $NSPACE(f(2^n))$ respectively.

Proof. Let A be hard for $DTIME(f(n))$, the other three cases being identical. Let $B \in DTIME(f(2^n))$; by lemma 3, $\mathrm{long}(B) \in DTIME(f(n))$, and by the hardness of A under projection reducibility we have $\mathrm{long}(B) \leq_\pi A$. Applying lemmas 2 and 4 we obtain $B \leq_m s(\mathrm{long}(B)) \leq_m sA$, which proves the hardness of sA. ∎

The same result also holds for classes defined by alternating Turing machines. Hamiltonian path and k-colorability are known to be NP-complete, and it can be easily seen that the known reductions to them are projection reductions [5]. We obtain immediately the results from [12]:

6. *Corollary.* The succinct versions of Hamiltonian path and k-colorability are m-complete in $NEXPTIME$.

Moreover, as indicated above, the optimal results about NP-completeness from [4] can be obtained from the same theorems. The following results either appear in [15], or can be proved in an extremely easy fashion using the same methods, employing the logarithmic time computation model of [2].

7. *Fact.* The problems of nonemptiness, triangle, k-cycle, k-path, and $\Delta(G) \geq k$ are projection-complete for nondeterministic logarithmic time; the problem $\delta(G) \geq k$ is projection-complete for Σ_2-alternating logarithmic time.

Therefore, from theorem 5, or essentially from the Conversion Lemma, the first six optimal bounds in the table of figure 1, plus the two *NEXPTIME* optimal bounds for Hamiltonian path and k-colorability follow uniformly. In the next section we will present the classification of the succinct version of additional problems, repeatedly using the Conversion Lemma for many of them.

3. Undirected accesibility and related problems

The main result in this section is a construction of a graph associated to a given quantified boolean formula, which has the following property: for two selected nodes in the graph, there is a path joining them if and only if the quantified boolean formula is true. Although the size of the graph is exponential in the formula, we will see that it can be described by a considerably shorter boolean circuit which can be constructed from the formula in polynomial time.

We employ this construction to present ad-hoc proofs of the *PSPACE*-hardness of the succinct versions of the undirected accesibility and planarity problems. From them, combining the Conversion Lemma with known π-reducibilities among graph problems, we will infer our contributions to the table of figure 1. We end the section by discussing the interesting particular case of bounded degree planarity.

Before presenting the construction, let us define *quantified boolean formulas.* A boolean formula is either a boolean variable, its negation, a boolean constant "true" or "false", the conjunction of two boolean formulas, or the disjunction of two boolean formulas. An assignment of boolean values to the variables *satisfies* the formula if and only if once the substitution is made, the formula evaluates to "true". Quantified boolean formulas are boolean formulas preceded by a string of quantifiers, each refering to one of the variables of the formula, so that no free variables remain. The allowed quantifiers are $\forall$ and $\exists$. Such a formula is evaluated as follows: $\forall v\, F$ is true if, when substituting "true" and "false" for v, both resulting formulas are true; and $\exists v\, F$ is true if, when substituting "true" and "false" for v, at least one of the resulting formulas is true. Otherwise, the formula evaluates to "false".

The problem QBF is the following: given a quantified boolean formula, decide whether it evaluates to "false". More information about quantified boolean formulas and the problem QBF can be found in [1], including the following fact.

8. *Theorem.* QBF is *PSPACE*-complete.

We present now the construction announced above. We describe it inductively on the form of the formula, by induction on the number of quantifiers. The nodes are identified by sequences of binary numbers. Without loss of generality, we assume that the variables are numbered according to their position in the quantifier string, the innermost being smallest.

Induction Basis. The formula has no quantifiers. Since no free variables are allowed, it evaluates either to "true" or to "false". The associated graph consists of two vertices, labeled 0 and 1, joined by an edge in case the fromula is true, or disconnected if the formula is false. Arbitrarily declare *source node* the node 0 and *target node* the node 1.

Induction Step. We have two cases depending on the outermost quantifier. Let F be the given formula, and assume first that F is $\exists v_i \, F'$. Construct F_0 and F_1 substituting "false" and "true", respectively, for v_i, and inductively consider the graphs associated to them, G_0 and G_1. Re-label all nodes in these graphs by concatenating "$i00$," and "$i10$," respectively to their labels. Create two more vertices with labels $i01$ and $i11$. Declare the first *source node* and join it to the source nodes of G_0 and G_1. Declare the other *target node* and join it to the target nodes of G_0 and G_1. The case of F being $\forall v_i \, F'$ is similar, but the graphs are connected in series instead of in parallel, i.e. the new source node is linked to the source of G_0, the target of G_0 is linked to the source of G_1, and the target of G_1 is linked to the new target node.

It is easy to see by induction that the source and target nodes of the obtained graph are connected if and only if the formula evaluates to false. The graph requires exponential time to be constructed, and is itself of size exponential in the size of the formula. However, given the formula, it is possible to decide whether an edge exists between two nodes in time polynomial on the labels of the nodes. Assume that the formula has n variables. If both labels consist of n numbers then the edge corresponds to a nonquantified formula, and the labels provide the value of each variable, so that it only remains to evaluate the formula. This can be done in polynomial time [9]. Else, only sources and targets of the same stage or two consecutive stages can be linked, and this is easy to check by looking at one of the quantifiers in the prefix of the formula.

This algorithm can be presented by a boolean circuit depending only on the formula F, and the circuit can be constructed in polynomial time from F using the techniques of [9]. Thus, we have a circuit C_F which can be constructed in time polynomial in the size of F, which represents a graph in which a source and a target node are selected, such that F evaluates to true if and only if there is a path from the source node to the target node in the graph. This is a polynomial time m-reducibility from QBF to succinct graph accessibility, and therefore we obtain the following theorem:

9. *Theorem.* The succinct version of the graph accessibility problem for undirected graphs is *PSPACE*-hard for the m-reducibility.

A similar reduction works for planarity testing:

10. *Theorem.* The succinct version of the planarity problem for undirected graphs is *PSPACE*-hard for the m-reducibility, even when restricted to graphs of bounded degree.

Proof. Given F, construct a circuit describing a complete graph with five nodes, and then substitute the previously constructed graph for one of the edges, identifying the endpoints with the selected source and target nodes. The resulting graph is nonplanar if and only if there is a path between the source and target nodes, if and only if the formula evaluates to true. Observe that the degree of the graph is bounded. ∎

Some more results will be derived from known relationships between graph problems. In [3] many problems are compared according to projection and constant depth reducibility. Although their projections are more general than ours, since no uniformity is required, it is easily seen that all the reductions required here are polynomial time computable. We will use also a reduction from [8], where it is stated for logarithmic space reducibility; from the proof it is easily seen that it can be computed by a projection.

11. *Theorem.* [8] The undirected graph accesibility problem is π-reducible to the problem of deciding whether an undirected graph is bipartite.

12. *Theorem.* [3] The undirected graph accesibility problem is π-reducible to the following problems on undirected graphs: connectivity, having a cycle, having an Eulerian path, and having a perfect matching (the last, for bipartite graphs).

In fact, that reference shows that all these problems except perfect matching are equivalent under projection reducibility. From these two results and the Conversion Lemma, we obtain many lower bounds announced in the table. Some upper bounds follow from lemma 1, using the facts that acyclicity is in L [6], and connectivity and Eulerian path [3] and bipartiteness [8] are in NL.

13. *Theorem.* The succinct versions of acyclicity, connectivity, bipartiteness, and Eulerian path are $PSPACE$-complete.

For perfect matching we are left with:

14. *Theorem.* The succinct version of perfect matching is $PSPACE$-hard.

Since perfect matching and planarity are in P, their succinct versions are in EXP-$TIME$, and therefore are the only ones left in which the upper and lower bounds are still far apart. However, we can show that a combination of known results proves that planarity of succinctly represented graphs of bounded degree is in $PSPACE$, and therefore $PSPACE$-complete by theorem 10.

15. *Theorem.* Testing planarity of bounded degree graphs is in NL, and therefore its succinct version is $PSPACE$-complete.

Proof. In [13] it is shown that this problem is in the third level of the symmetric complementation logarithmic space hierarchy, which is included by definition in the logarithmic space hierarchy. However, Immerman [7] has shown that NL is closed under complementation, and therefore the logarithmic space hierarchy coincides with NL. Thus, planarity of bounded degree graphs is in NL. ∎

4. Conclusions

We have considered decisional graph-theoretic problems, in which the input is represented in a non-standard form: since the adjacency matrix of the graph can be viewed as a boolean function, it can be represented by a possibly small boolean circuit. In case the graphs are particularly regular, such a representation is shorter than the full adjacency matrix. Following the research initiated in [4], we have studied how this change in the presentation of the input affects the complexity of the decisional problems.

It is known that the succinctly represented problem is at most exponentially harder than the problem on standard encodings. In [4] it was shown that quite simple succinct graph problems are already *NP*-hard, and in [12] it was shown that *NP*-complete standard problems (for projection reducibility) become *NEXPTIME*-complete for succinct inputs.

The table of figure 1 shows the upper and lower bounds on the complexity of several well-known graph-theoretic decisional problems, in which all but two of the problems considered in [4] are shown to be completely classified by matching upper and lower bounds. The contributions of this paper to the table are of two kinds: finding new proofs of lower bounds claimed in [12], where the proof was incomplete, and finding new lower bounds. An additional contribution which is not reflected in the table is a generalization of the main result in [12], with a simpler proof, which carries over to other classes and yields in a uniform manner many of the known lower bounds.

More precisely, our Conversion Lemma shows that projection reducibilities among the standard problems translate into *m*-reducibilities among succinctly represented problems, independently of any membership or completeness property of the sets in any particular complexity class. The usefulness of that lemma is clear, since we have obtained all the lower bounds indicated in the table from it, coupled with a new specific construction presented in section 3 (but which was inspired in [4]), and previous results from [3], [8], and [15].

5. References

[1] J.L. Balcázar, J. Díaz, J. Gabarró: *Structural Complexity I*. EATCS Monographs on Theoretical Computer Science, vol. 11, Springer-Verlag (1988).

[2] A. Chandra, D. Kozen, L. Stockmeyer: "Alternation". *Journal ACM* **28** (1981), 114–133.

[3] A. Chandra, L. Stockmeyer, U. Vishkin: "Constant depth reducibility". *SIAM Journal on Computing* **13** (1984), 423–439.

[4] H. Galperin, A. Wigderson: "Succinct representations of graphs". *Information and Control* **56** (1983), 183–198.

[5] M. Garey, D. Johnson: *Computers and Intractability: A Guide to the Theory of NP-completeness*. Freeman, San Francisco (1978).

[6] J. Hong: "On some deterministic space complexity problems". *SIAM Journal on Computing* **11** (1982), 591–601.

[7] N. Immerman: "Nondeterministic space is closed under complementation". *SIAM Journal on Computing* **17** (1988), 935–938.

[8] N.D. Jones, E. Lien, W.T. Laaser: "New problems complete for nondeterministic log space". *Mathematical Systems Theory* **10** (1976), 1–17.

[9] R.E. Ladner: "The circuit value problem is log space complete for P". *SIGACT News* **7** (1975), 18–20.

[10] H.R. Lewis, Ch. Papadimitriou: "Symmetric space-bounded computation". *Theoretical Computer Science* **19** (1982), 161–187.

[11] A. Lozano: "*NP*-hardness on succinct representations of graphs". *Bulletin of the EATCS* **35** (1988), 158–163.

[12] Ch. Papadimitriou, M. Yannakakis: "A note on succinct representations of graphs". *Information and Control* **71** (1986), 181–185.

[13] J.H. Reif: "Symmetric complementation". *Journal ACM* **31** (1984), 401–421.

[14] S. Skyum, L.G. Valiant: "A complexity theory based on boolean algebra". *Journal ACM* **32** (1985), 484–502.

[15] J. Torán: "Succinct representations of counting problems". 6th Int. Conf. on Applied Algebra, Algebraic Algorithms, and Error Correcting Codes, Rome, Italy (1988).

[16] K. Wagner: "The complexity of combinatorial problems with succinct input representation". *Acta Informatica* **23** (1986), 325–356.

Problem	Upper bound	Lower bound
Nonempty edge set	[4]: NP	[4]: NP-hard
Having a triangle	[4]: NP	[4]: NP-hard
Having a k-cycle	[4]: NP	[4]: NP-hard
Having a k-path	[4]: NP	[4]: NP-hard
$\Delta(G) \geq k$	[4]: NP	[4]: NP-hard
$\delta(G) \leq k$	[4]: Σ_2^p	[4]: Σ_2^p-hard
Acyclicity	[4]: $DSPACE(n)$	[4]: co-NP-hard; [12]+[Here]: $PSPACE$-hard
Eulerian circuit	[4]: $NSPACE(n)$	[4]: co-NP-hard; [Here]: $PSPACE$-hard
s-t Accessibility	[4]: $NSPACE(n)$	[4]: Π_2^p-hard; [Here]: $PSPACE$-hard
Connectivity	[4]: $NSPACE(n)$	[4]: Π_2^p-hard; [12]+[Here]: $PSPACE$-hard
Bipartiteness	[4]: $EXPTIME$; [Here]: $PSPACE$	[4]: Σ_2^p-hard; [12]+[Here]: $PSPACE$-hard
Perfect matching	[4]: $EXPTIME$	[4]: Π_2^p-hard; [Here]: $PSPACE$-hard
Planarity	[4]: $EXPTIME$	[4]: Σ_2^p-hard; [Here]: $PSPACE$-hard
Hamiltonian path	[4]: $NEXPTIME$	[4]: Π_2^p-hard; [12]: $NEXPTIME$-hard
k-Colorability	[4]: $NEXPTIME$	[4]: Σ_2^p-hard; [12]: $NEXPTIME$-hard
b-d-Planarity	[Here]: $PSPACE$	[Here]: $PSPACE$-hard

Figure 1. Table of currently known upper and lower bounds

An $O(n \log n)$ Algorithm for 1-D Tile Compaction

Richard Anderson & Simon Kahan[1]
University of Washington
Dept. of Computer Science, FR-35
Seattle, WA 98195

Martine Schlag[1]
University of California, Santa Cruz
225 Applied Sciences
Santa Cruz, CA 95064

Abstract

A puzzle whose solution has applications in VLSI layout compaction of memories and fine grained parallel processors can be phrased as follows:

You are given a set of n rectangles arranged in a coordinate plane such that no two overlap and each rectangle has sides parallel to the coordinate axes. The width of such an arrangement is the length of a longest horizontal line segment having each of its endpoints located within the rectangles. You may slide the rectangles only in the direction of the horizontal axis and may not slide any rectangle over another. Find a minimal width arrangement reachable by sliding from the original arrangement.

The fastest previously known algorithm solving this problem is the iterative approach of Mehlhorn and Rülling [6] requiring $O(n^2 \log n)$ time. This paper develops and proves correct a simple $O(n \log n)$ time algorithm which exploits the geometric structure of the constraints between the rectangles.

1 Introduction

The goal of compaction in VLSI layouts is to squeeze out any extra space in the layout by allowing components to slide in both dimensions [7]. Though this problem is no easier than bin packing, and hence NP-hard [8], a restricted version in which components may slide in only one dimension can be solved quickly. The 1-dimensional compaction used to solve this simpler problem involves construction of a constraint network whose n nodes generally correspond to components and m edges to distance constraints between adjacent components [Fig 1b]. Since the positions of components may be constrained by interconnections and it is generally not desirable to radically alter the layout, performing two such compactions one in each dimension is often a reasonable approximation to allowing complete freedom of movement in both dimensions.

Better approximations to 2-dimensional compaction can be based on the 1-dimensional problem by maintaining the two constraint networks and using heuristics to introduce, remove and exchange constraints in the two constraint networks [3, 8]. These approaches repeatedly solve the two constraint networks using 1-dimensional compaction algorithms, so it is important to be able to perform the 1-dimensional compaction quickly. Since the constraint network has a single origin node corresponding to the left vertical boundary of the layout arena, the 1-dimensional compaction problem can be solved in time $O(m)$ by computing the length of the longest path from the origin to a component as the position of the component [5]. An alternative method uses a min-cost flow algorithm to obtain the positions of the components [1]. Using longest paths to obtain positions in effect flattens the layout against the origin, while the min-cost flow approach tends to distribute any slack in the constraints more equitably throughout the network. The min-cost flow produces a more desirable

---

[1]This research was supported by NSF Presidential Young Investigator Grant MIP-8657693

result in terms of the distances between components – and hence wire lengths – at the expense of additional computation: the best bound known for min-cost flow in a network of m edges and n nodes is $O(n^2(m + n \log n) \log n)$ [2]. However because of its efficiency the longest path method is preferable for solving constraint networks in 2-dimensional compaction.

A special case of interest occurs when an array composed of a single cell is to be compacted, as in a systolic-array parallel processor or a memory. In the optimal solution each instance of the cell may take on a different configuration [4]. However, if all cells in the array must be identical – and this is usually preferable to insure similar electrical characteristics and compact representations – then the entire array can be compacted by compacting a single instance of the cell against itself. In essence the cell is compacted on a torus. Another way to view the problem is as that of computing the smallest area tile which encloses the cell's layout and will 4-tile the plane. The associated 1-dimensional compaction problem can be viewed as cylindrical compaction or computing a tile of minimum width. It can be solved using a constraint network with separation constraints added between the components which are now adjacent across the boundary of the cell on the cylinder. Eichenberger and Horrowitz use the minimum-cost flow approach to assign positions to vertices of a constraint network to which *soft* separation constraints have been added between the components which are now adjacent across the boundary of the cell embedded on a cylinder [1]. Mehlhorn and Rülling also present an $O(n^2 \log n)$ algorithm for this problem when the network is planar[6]. Unfortunately, the more efficient longest path approach is not applicable since there is no origin against which to compact.

In this paper the longest path approach is adapted to solve the 1-dimensional tile compaction problem providing an $O(n \log n)$ time algorithm for planar constraint networks. We begin by defining the problem formally as the *[Single-Layer] Cylindrical Compaction Problem* in terms of finding a minimal width configuration of a set of rectangles. Secondly, because the problem appears at first trivial, we discuss a few of the obvious approaches that fail and motivate the development of our solution. We then define the *Constraint Network*, and state some of its properties. The fourth section reformulates our problem as finding a minimal width configuration of the Constraint Network. In the fifth section we actually solve the problem by constructing a zig-zag shaped tree subgraph of the Constraint Network, the *Skeleton*, which determines the relative positions of a set of origins, the *Left[i]*, analagous to the single origin in the ordinary 1-dimensional problem; much of the fifth section is devoted to proving that these positions do indeed provide a minimal width configuration of the Constraint Network. An algorithm computing the minimal width configuration implied by the Skeleton is given in the sixth section: the bottleneck is computation of the Skeleton itself which, if done in the most straightforward manner, requires time $O(n^2)$; we describe a divide-and-conquer algorithm which reduces this time to $O(n \log n)$. Finally, we discuss what we consider the next steps in making this algorithm practical, and some of the open problems that remain.

(1a) A Rectangle Configuration

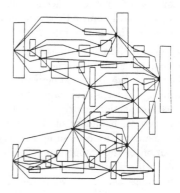
(1b) The Constraint Network

2 Problem Formulation

We represent a circuit as a collection of rectangles [Fig. 1a]. Each rectangle represents a component or wire of the circuit. The size of each rectangle is determined by the space occupied by the corresponding part as well as the constraints separating it from its neighbours. The relative positions of the rectangles are those of the corresponding parts in the initial circuit layout. More formally:

Definitions: Let $R = \{r_1, r_2, \ldots, r_n\}$ be a set of rectangles where rectangle r has height $H(r) > 0$ and width $W(r) > 0$.

- A *Configuration*, $\mathcal{E}$, of R is an embedding of R in the coordinate plane assigning a coordinate to each rectangle such that

  1. the rectangles' edges are parallel to the coordinate axes,
  2. no two rectangles overlap (their interiors are disjoint) and,
  3. no two rectangles abut opposite sides of the same horizontal line.

  The third requirement is to prevent ambiguity when we define visibility; it is an inessential limitation in practice. In fact, for the sake of simplicity we ignore this restriction in our illustrations.

  The mapping $\mathcal{E}$ assigns a coordinate to the lower left corner of each rectangle,

  $$\mathcal{E}(r) \rightarrow (X_{\mathcal{E}}(r), Y_{\mathcal{E}}(r)).$$

- The *vertical extent* of r in $\mathcal{E}$, denoted by $V_{\mathcal{E}}(r)$, is the interval $[Y_{\mathcal{E}}(r), Y_{\mathcal{E}}(r) + H(r)]$.

- The *Width* of a configuration of rectangles is the length of the longest horizontal line segment having each endpoint in any rectangles; formally,

  $$Width(R, \mathcal{E}) = \max\{X_{\mathcal{E}}(r_2) - X_{\mathcal{E}}(r_1) + W(r_2) | r_1, r_2 \in R, \ V_{\mathcal{E}}(r_1) \cap V_{\mathcal{E}}(r_2) \neq \emptyset\}.$$

- A rectangle r_1 is *visible from the right by* rectangle r_2 (or equvalently r_2 is *visible from the left by r_1*) if $X_{\mathcal{E}}(r_1) < X_{\mathcal{E}}(r_2)$ and

  $$V_{\mathcal{E}}(r_1) \cap V_{\mathcal{E}}(r_2) - \bigcup_{X_{\mathcal{E}}(r_1) < X_{\mathcal{E}}(r) < X_{\mathcal{E}}(r_2)} V_{\mathcal{E}}(r) \neq \emptyset.$$

  That is, the union of the vertical extents of all rectangles between r_1 and r_2 does not cover the intersection of their extents. Furthermore, r_2 is *visible by r_1 at y* if

  $$y \in V_{\mathcal{E}}(r_1) \cap V_{\mathcal{E}}(r_2) - \bigcup_{X_{\mathcal{E}}(r_1) < X_{\mathcal{E}}(r) < X_{\mathcal{E}}(r_2)} V_{\mathcal{E}}(r).$$

- The rectangles on the left and right frontiers of the configuration are identified as those which are visible on the right and left respectively. In formally defining these rectangles, it is convenient to introduce two special rectangles r_L and r_R, which bound any set of rectangles on the left and right respectively. The lower left coordinate of r_L is $(-\infty, -\infty)$, $H(r_L) = +\infty$ and $W(r_L) = 1$. The lower left coordinate of r_R is $(+\infty, -\infty)$, $H(r_R) = +\infty$ and $W(r_R) = 1$. Although these rectangles are not in R the notion of visibility can be extended to them.

- A rectangle r is *visible from the left* if it is visible from the left by r_L.

- A rectangle r is *visible from the right* if it is visible from the right by r_R.

- A configuration $\mathcal{E}'$ of R is *reachable directly* from $\mathcal{E}$ if 1) $\mathcal{E}'$ and $\mathcal{E}$ agree on all but one rectangle, say r, such that $Y_{\mathcal{E}'}(r) = Y_{\mathcal{E}}(r)$ and 2) a hypothetical rectangle at $(\min\{X_{\mathcal{E}'}(r), X_{\mathcal{E}}(r)\}, Y_{\mathcal{E}'}(r))$ of height $H(r)$ and width $|X_{\mathcal{E}'}(r) - X_{\mathcal{E}}(r)|$ does not intersect any other rectangles in either configuration. That is, r can slide from its position in $\mathcal{E}$ to its position in $\mathcal{E}'$ without encountering any other rectangle.

  $\mathcal{E}'$ of R is *reachable* from $\mathcal{E}$ if there is a sequence of configurations starting with $\mathcal{E}'$ and ending with $\mathcal{E}$ such that each configuration is reachable directly from the preceding one.

  An instance of the *[Single-Layer] Cylindrical Compaction Problem* consists of a set of rectangles, R, and an initial configuration $\mathcal{E}$. The goal is to obtain a configuration of minimal width by sliding the rectangles horizontally. Hence, a solution to the instance $(R, \mathcal{E})$ is a configuration of R which has the minimal width over all configurations reachable from $\mathcal{E}$.

A more practical characterization of the reachable configurations is given by the following proposition which is stated without proof.

Proposition 1 Given two configurations of R, $\mathcal{E}$ and $\mathcal{E}'$, $\mathcal{E}'$ is reachable from $\mathcal{E}$ if and only if

1. for every r in R, $Y_{\mathcal{E}}(r) = Y_{\mathcal{E}'}(r)$, and

2. for all r_1, r_2 in R, if $X_{\mathcal{E}}(r_1) \leq X_{\mathcal{E}}(r_2)$ and $V_{\mathcal{E}}(r_1) \cap V_{\mathcal{E}}(r_2) \neq \emptyset$ then $X_{\mathcal{E}'}(r_1) \leq X_{\mathcal{E}'}(r_2)$.

We will assume that no horizontal line can be drawn which separates the rectangles into two non-empty sets without intersecting some rectangle. If such a line could be drawn, the rectangles could be split into these two sets of rectangles and the problem solved independently on the two sets. Another way of stating our assumption is:

Assumption 2 The set $I = \bigcup_{r \in R} V_{\mathcal{E}}(r)$ is an interval.

Note that this assumption together with the third constraint in the definition of a configuration implies that any horizontal line $\{(x, y)|y = b, b \in I\}$ intersects the interior of at least one rectangle.

3   Intuition from Failures

In searching for a solution we considered many approaches that looked promising at first but turned out to be flawed. Solving the problem was difficult not because of the complexity of the solution, but because we followed a number of ill fated paths before gaining an intuition that now seems obvious in hindsight. We illustrate one such path here.

Our first approaches were based on the idea that by applying 1-D compaction twice first to the left and then to the right, the configuration would be squeezed together so as to achieve minimal width. Assuming just for the moment that $X(r) > 0$ for each rectangle r, "1-D compaction to the left" slides each rectangle to the left until either it is blocked by another rectangle or has $X(r) = 0$. "1-D compaction to the right" works analagously. Unfortunately, this first approach doesn't achieve a minimal width tiling from the following initial configuration because the upper square ends up too far to the right:

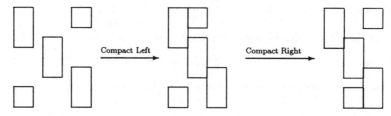

The second application of 1-D compaction appears to undo the desirable effects of the first: what we want is for the upper square to stay put, rather than move back to the right. Our second approach attempts to solve this problem by freezing the positions of all rectangles visible from the right after the first compaction. Then the second compaction moves the non-frozen rectangles as far as possible to the right. This approach works on the example above, but fails on the following because a smaller width would be obtained were the lower left rectangle positioned more to the right:

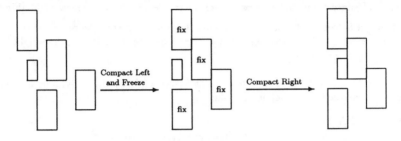

So maybe what we want instead is for the rectangles to stay stuck to those with which they collide during the first compaction. Then in the first example, the upper square would stay stuck to the other rectangles rather than move freely back to the right. This motivates yet another variation on the same theme: after the first compaction, glue together all rectangles which touch; then perform 1-D compaction to the right but treat those rectangles which are glued together as monolithic. The example presented above is now handled properly, but the following is not:

Moreover, the following example shows that what we really need to do is to get rid of the "vertical wall" against which 1-D compaction pushes the rectangles and instead compact the rectangles at different heights separately:

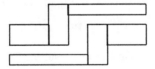

In fact the intuition behind our algorithm is based on the observation that there is a minimum width configuration in which at every height the distance between the leftmost rectangle and the rightmost rectangle is as small as could be obtained in *any* reachable configuration. That this is not obvious is evident from examples in which pushing rectangles at one height closer together may force rectangles at another height further apart. Here, moving r_2 towards r_3 forces r_5 further from r_4.

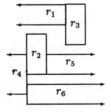

4 The Constraint Network

A natural representation for our rectangle problem is a constraint network. In practice, the network is the starting point for the cylindrical compaction problem: the formation of the network from an initial layout is straightforward.

The Constraint Network corresponding to a cylindrical compaction instance contains a vertex for each rectangle and edges from that vertex to each rectangle visible to it on its right [Fig 1b]. An edge $\overrightarrow{(u,v)}$ of weight w from vertex u to v represents the constraint that the rectangle corresponding to v be located at least w units to the right of that corresponding to u. Hence, the weight is just the width of the rectangle corresponding to u.

Definitions: The *Constraint Network* $N(R, \mathcal{E})$ of an instance $(R, \mathcal{E})$ of the Cylindrical Compaction Problem is a directed graph defined as follows:

- Vertices of $N(R, \mathcal{E})$ are
$$N_V(R, \mathcal{E}) = \{r_1, r_2, \ldots, r_{|R|}\}.$$

- Edges of $N(R, \mathcal{E})$ are
$$N_E(R, \mathcal{E}) = \{\overrightarrow{(r_i, r_j)}|\ rectangle\ r_j\ is\ visible\ from\ the\ left\ by\ r_i\}.$$

- The weight of an edge, $\overrightarrow{(r_i, r_j)}$, is $\|\overrightarrow{(r_i, r_j)}\| = W(r_i)$.

A *Configuration* of a Constraint Network $N(R, \mathcal{E})$ is a mapping $\mathbf{P}$ from the vertices $N_V(R, \mathcal{E})$ of the Network to horizontal positions $\mathbf{P}(r_i)$ such that for all $\overrightarrow{(r_i, r_j)}$ in $N_E(R, \mathcal{E})$:

$$\mathbf{P}(r_j) - \mathbf{P}(r_i) \geq \|\overrightarrow{(r_i, r_j)}\|.$$

A *path* from vertex u_1 to vertex u_i is denoted $< u_1, \ldots, u_i >$. The length of the path, $\sum_{1 \leq j < i} \|\overrightarrow{(u_j, u_{j+1})}\|$, is denoted $\| < u_1, \ldots, u_i > \|$.

A configuration $\mathbf{P}$ of a Constraint Network $N(R, \mathcal{E})$ induces a configuration $\mathcal{P}$ on the rectangles

$$\mathcal{P}(r) \rightarrow (\mathbf{P}(r), Y_{\mathcal{E}}(r)).$$

A few obvious facts about the Constraint Network resulting from our formulation of the Cylindrical Compaction Problem are:

Proposition 3 If $V_{\mathcal{E}}(r_1) \cap V_{\mathcal{E}}(r_2) \neq \emptyset$ and $X_{\mathcal{E}}(r_1) < X_{\mathcal{E}}(r_2)$ then there is a path in $N(R, \mathcal{E})$ from r_1 to r_2.

Along with the assumption that the set of rectangles is not separable by a horizontal line, it follows from Proposition 3 that the Constraint Network is connected:

Proposition 4 The underlying undirected graph of $N(R, \mathcal{E})$ is connected.

Proposition 5 asserts the straightforward correspondence between configurations of $N(R, \mathcal{E})$ and configurations of R reachable from $\mathcal{E}$:

Proposition 5 If $\mathbf{P}$ is a configuration of the Constraint Network $N(R, \mathcal{E})$ then the configuration induced by $\mathbf{P}$ is reachable from $\mathcal{E}$. If $\mathcal{M}$ is a configuration of R reachable from $\mathcal{E}$, then $X_{\mathcal{M}}$ is a configuration of the Constraint Network $N(R, \mathcal{E})$.

Proposition 6 $N(R, \mathcal{E})$ is acyclic and planar.

That $N(R, \mathcal{E})$ is acyclic follows from the observation that a rectangle can't be both to the right and to the left of another rectangle. That it is planar follows from the observation that the construction of the network from the rectangles yields a natural planar embedding of the network.

In fact, the algorithm we present does not require that the network be acyclic; only that it contain no positive cycles. In practice, there may be a need for negatively weighted back-edges in the constraint network for a circuit, so it is important that our algorithm function correctly when non-positive cycles are present. However, our algorithm does depend on the planarity of $N(R, \mathcal{E})$.

5 Zones

Before we can reformulate the Cylindrical Compaction Problem in terms of constraint networks, we need to define the width of a Network Configuration. The projections of the visible portions of edges of rectangles visible from the left form a partitioning of the interval $(-\infty, +\infty)$, as do those of rectangles which are visible from the right. Combining the two partitions, we can form a third partition which consists of all shortest intervals having two endpoints from either partition. This third partition we call the *Zones* of the network [Fig. 2a].

Definitions:

- Let $Endpoints = \{Y_{\mathcal{E}}(r_i) | r_i \text{ is visible by } r_L \text{ or } r_R \text{ at } Y_{\mathcal{E}}(r_i)\}$
  $\bigcup \{Y_{\mathcal{E}}(r_i) + H(r_i) | r_i \text{ is visible by } r_L \text{ or } r_R \text{ at } Y_{\mathcal{E}}(r_i) + H(r_i)\}$.

- Let $e_1, e_2, \ldots, e_{k+1}$ where $k = |Endpoints| - 1$ be $Endpoints$ in increasing order.

- The *Zones* of a Constraint Network $N(R, \mathcal{E})$ are the intervals:

$$\mathcal{Z} = \{z_i = [e_i, e_{i+1}] | 1 \leq i \leq k\}.$$

- For each zone z_i define $Left[i]$ to be the rectangle visible from the left within the interval $[e_i, e_{i+1}]$. Existence of such a rectangle is guaranteed by Assumption 2. The rectangle is unique since for each z_i there is at most one rectangle r which is visible from the left by r_L at y for all $y \in z_i$; otherwise there would be an endpoint of some rectangle visible from the left which would fall in between e_i and e_{i+1}.

- By the same argument we can define $Right[i]$ to be the rectangle visible from the right within the interval $[e_i, e_{i+1}]$.

- The *Width* of z_i in a Network Configuration $\mathbf{P}$ is the distance between $Left[i]$ and $Right[i]$ given by $W_{\mathbf{P}}(z_i) = \mathbf{P}(Right[i]) - \mathbf{P}(Left[i])$.

Note that the $Left[i]$ and $Right[i]$ are not necessarily distinct since the same rectangle may be visible from both the left and right. $Left[i]$ and $Left[i + 1]$ (as well as the pair: $Right[i]$ and $Right[i + 1]$) are also not necessarily distinct since the same rectangle may cover more than one zone; although at least one of the two pairs is distinct. $|\mathcal{Z}| = k$ is bounded above by the number of rectangles in R, but is typically closer to the square root of the number of rectangles, since only rectangles along the left and right frontiers of the configuration can be responsible for zones.

Note that the width of a configuration of rectangles as defined earlier will in fact be greater than the width of the corresponding zone of the network configuration since the width of z_i does not include the width of the rectangle corresponding to $Right[i]$. The width of $Right[i]$ is included in the definition of the width of the Constraint Network:

Definition: The *Width* of a Configuration **P** of a Constraint Network is:

$$\max_{1 \le i \le k} \{W_{\mathbf{P}}(z_i) + W(Right[i])\}.$$

Lemma 7 If **P** is a Configuration of the Constraint Network $N(R, \mathcal{E})$ of minimal width, then the configuration induced by **P** is a solution to the Cylindrical Compaction Problem, $(R, \mathcal{E})$, generating that network.

Proof: The proof follows from Proposition 5 and the definitions of width for configurations of rectangles and constraint networks. □

The sum of the weights along any path from $Left[i]$ to $Right[i]$ is a lower bound on the width of z_i. The longest such path is the minimal width of this zone. A configuration in which each zone achieves its minimal width is clearly a minimal width configuration. That such a configuration exists is not immediate since moving two rectangles closer may force others apart. In fact, there are non-planar networks which do not have such a configuration. Our algorithm will construct such a network configuration for planar networks.

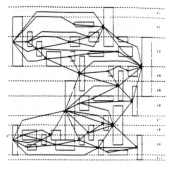 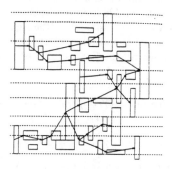

(2a) Zones of a Network   (2b) The Skeleton

6   The Skeleton

Now that we have formulated the Cylindrical Compaction Problem in terms of a Constraint Network, we need only find a minimal width configuration of the Network.

Because $Left[i]$ lies to the left of all other rectangles r_j whose vertical extents intersect $[e_i, e_{i+1}]$ including the rectangle $Right[i]$, there must be a path (possibly of zero length if $Left[i] = Right[i]$) in N from $Left[i]$ to $Right[i]$; for each i.

Lemma 8 For each $1 \le i \le |\mathcal{Z}|$ there is a path from $Left[i]$ to $Right[i]$ in N.

Proof: Since $Left[i]$ and $Right[i]$ are the only rectangles visible by r_L and r_R respectively within the interval $[e_i, e_{i+1}]$, the vertical extents of $Left[i]$ and $Right[i]$ both include the interval $[e_i, e_{i+1}]$. By Proposition 3 there is a path from $Left[i]$ to $Right[i]$. □

There exist paths joining the paths of Lemma 8:

Lemma 9 For each $1 \le i < |\mathcal{Z}|$ there is a path from $Left[i]$ to $Right[i + 1]$ in N.

Proof: The zones of $Left[i]$ and $Right[i + 1]$ are $[e_i, e_{i+1}]$ and $[e_{i+1}, e_{i+2}]$, respectively. By Assumption 2, the horizontal line at $y = e_{i+1}$ must intersect the interior of some rectangle r_j. Then the vertical extent of r_j overlaps with those of $Left[i]$ and $Right[i+1]$. Either $r_j = Left[i]$ or by Proposition 3 there is a path from $Left[i]$ to r_j. Similarly, since either $Right[i + 1]$ is r_j or their vertical extents overlap, there is a path from r_j to $Right[i + 1]$. Putting these paths together gives us the required path from $Left[i]$ to $Right[i + 1]$. □

The heart of our algorithm is the formation of a *Skeleton* which determines the relative displacements of all $Left[i]$'s in a configuration in which each zone attains its minimum width. The Skeleton is an undirected tree constructed from N by including the edges corresponding to a longest path between each $Left[i]$ and $Right[i]$, as well as longest paths joining those paths [Fig. 2b]. Its existence follows from Lemmas 8 and 9. More formally:

Definition: Let $S = (S_V, S_E)$ be the undirected graph underlying any subgraph of N formed by the union of:

- a longest path from $Left[i]$ to $Right[i]$, for $1 \leq i \leq |\mathcal{Z}|$, and
- a longest path from $Left[i]$ to $Right[i + 1]$, for $1 \leq i < |\mathcal{Z}|$

constructed in order:

$< Left[1], \ldots Right[1] >,$
$< Left[1], \ldots Right[2] >,$
$< Left[2], \ldots Right[2] >,$
etc.

such that each path shares with the previous path as many edges as possible. Each of these longest paths we call a *Rib*. S is a *Skeleton* of N.

The reason for each path sharing as many edges as possible with the previous path is to guarantee that there are no cycles in S, as affirmed by Lemma 10. In practice, these cycles are harmless; their absence merely simplifies the exposition.

Lemma 10 S is a tree.

Proof: Suppose at some stage during the construction of S a simple cycle was formed by the addition of the ith path. This cycle must be composed of two edge disjoint paths in S corresponding to two paths in N: a subpath A of the $(i - 1)$st path and a subpath B of the ith path both reaching from some vertex u to some vertex v. Since every subpath used in the construction of S is a longest subpath between the vertices it joins in N, the sum of the weights along A must equal that along B. But the construction requires that the ith path have as many edges in common with the $(i - 1)$st as possible so that B should be identical to A. Hence, there can be no cycle in S. That S is connected is a trivial consequence of its construction. □

So we've shown that S is a tree. Hence, there is exactly one path in S between any given pair of vertices of S. We'll compute displacement in S just as in the directed graph it underlies; that is, direction of edge traversal will matter.

Definitions: The *displacement* D from u to v, $u, v \in S_V$ is defined by:

- $D(u, u) = 0$,

- $D(u, v) = w$ if there is an edge $\in S_E$ between u and v and an edge in N, $\overrightarrow{(u, v)}$ with $\|\overrightarrow{(u, v)}\| = w$,

- $D(u, v) = -w$ if there is an edge in S_E between u and v and an edge in N, $\overrightarrow{(v, u)}$ with $\|\overrightarrow{(v, u)}\| = w$,

- $D(u, v) = D(u, u') + D(u', v)$ if the unique path in S between u and v includes the edge (u, u').

By computing the displacements between vertices along paths in S, we compute relative positions of the vertices of S which satisfy all the constraints represented by the edges of S. We could construct a configuration of the Constraint Network by assigning to each $Left[i]$ its displacement from $Left[1]$ and computing all longest paths from the $Left[i]$'s to obtain the positions of the remaining nodes.

We claim that a minimal width configuration is given by the Network Configuration:

$$\mathbf{M}(v) = \max_{1 \leq i \leq |Z|} \{D(Left[1], Left[i]) + \|P\| \mid P \text{ is a longest path from } Left[i] \text{ to } v \text{ in } N.\}.$$

These positions satisfy the constraints in N as well as those implied by S. Moreover for vertices in S_V, we will show that these positions correspond exactly to the displacements with respect to $Left[1]$. Since for any $v \in S_V$, S includes a path P from some $Left[j]$ to v such that

$$D(Left[1], v) = D(Left[1], Left[j]) + \|P\|,$$

it follows that

$$\mathbf{M}(v) \geq D(Left[1], v), \forall v \in S_V.$$

Lemma 11 will show that in fact $\mathbf{M}(v) = D(Left[1], v)$ for any $v \in S_V$.

Lemma 11 Between any two vertices u and v in S_V there is no path in N of length greater than the displacement $D(u, v)$ from u to v in S.

Proof: Assume that u and v are not identical; otherwise, the statement is trivial. Suppose there exists some path, P, joining u to v in N of length greater than the displacement along the unique path from u to v in S. Let the sequence of paths $\{P^1, P^2, ..., P^m\}$ denote the decomposition of P into paths between vertices in S_V; that is, each P^i has endpoints in S_V and no other vertices in S_V, with the composition of the sequence P^i yielding P. Because N is planar each path added to S during its construction effectively separates the nodes above and below the path. Since each P^i has no vertices in S_V other than its endpoints, its endpoints must fall on the same or consecutive paths added to S during its construction. Each of the P_j's falls into one of the following three classes:

1. Paths consisting of a single edge in S_E.

2. Paths $< x, \ldots y >$, with x, y in S_V, for which there exist a vertex z and paths $< z, \ldots x >$ and $< z, \ldots y >$ in S which are directed paths in N.

3. Paths $< x, \ldots y >$, with x, y in S_V, for which there exist a vertex z and paths $< x, \ldots z >$ and $< y, \ldots z >$ in S which are directed paths in N.

For $\|P\|$ to be greater than the displacement $D(u, v)$ in S, some subpath P^j must be of greater length than the displacement along the path in S joining the endpoints of P^j. Certainly no P^j in the first class is a candidate. That there are no candidates in either the second or third classes follows from two similar proofs; we present the proof only for the second class:

Suppose for some $P^j = < x, \ldots y >$ there is a z in S_V and paths $< z, \ldots x >$ and $< z, \ldots y >$ in S. The displacement from x to y in S is then $D(x, y) = \| < z, \ldots y > \| - \| < z, \ldots x > \|$. Suppose $\|P^j\| > \| < z, \ldots y > \| - \| < z, \ldots x > \|$. Then $\|P^j\| + \| < z, \ldots x > \| > \| < z, \ldots y > \|$, so the path from z to y which is the composition of P^j and $< z, \ldots x >$ has greater length than the path $< z, \ldots y >$ in S. But by the construction of S the path $< z, \ldots y >$ in S must be the longest path from z to y in all of N. This contradiction leads us to believe that there is no path P^j of the second class which is of greater displacement than the path in S joining P^j's endpoints.

Since no subpath of P joining vertices of S is of greater length than the displacement along the path in S, P must have length no greater than the displacement along the path $< u, \ldots v >$ in S. $\square$

The importance of the Skeleton is that it specifies positions for the $Left[i]$ which are consistent with some minimal width configuration of N. The consistency was demonstrated by the previous lemma: if there are no paths in N between vertices of S longer than those in S, there's no constraint forcing any two vertices to be further apart than they are in S. The minimal width follows from the fact that in S the $Left[i]$ are as close as possible to the $Right[i]$: the width of a configuration is minimized when the distance between $Left[i]$ and $Right[i]$ is the closest that these two vertices may be located. Since a longest path between any $Left[i]$ and $Right[i]$ in N is the corresponding Rib in S, the closest $Left[i]$ and $Right[i]$ can be in any configuration is their displacement in S.

Theorem 12 M is a minimal width configuration of the Constraint Network.

Proof: By its construction **M** is a configuration for the Constraint Network. The width of z_i in **M** is given by

$$W_{\mathbf{M}}(z_i) = \mathbf{M}(Right[i]) - \mathbf{M}(Left[i]).$$

Since $Left[i]$ and $Right[i]$ are both in S_V, and by Lemma 11 we had $\mathbf{M}(v) = D(Left[1], v) \forall v \in S_V$,

$$W_{\mathbf{M}}(z_i) = D(Left[1], Right[i]) - D(Left[1], Left[i]) = D(Left[i], Right[i]).$$

To show that **M** has minimal width consider any configuration **P**. Since there is a path of length $D(Left[i], Right[i])$ for each $1 \leq i \leq |\mathcal{Z}|$, this is a lower bound on the $W_{\mathbf{P}}(z_i)$. This gives

$$W_{\mathbf{M}}(z_i) = D(Left[i], Right[i]) \leq W_{\mathbf{P}}(z_i) \text{ for all } 1 \leq i \leq |\mathcal{Z}|,$$

from which we obtain

$$\max_{1 \leq i \leq |\mathcal{Z}|} \{W_{\mathbf{M}}(z_i) + W(Right[i])\} \leq \max_{1 \leq i \leq |\mathcal{Z}|} \{W_{\mathbf{P}}(z_i) + W(Right[i])\}.$$

Therefore the width of **M** is no greater than that of **P**. Since **P** was an arbitrary configuration, **M** must have minimal width. □

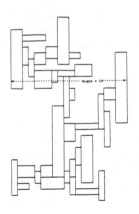
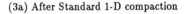

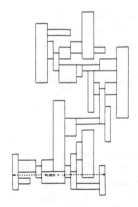

(3a) After Standard 1-D compaction   (3b) After Standard Tile Compaction

7 Algorithm

Our algorithm for solving an instance of the Single-Layer Cylindrical Compaction Problem, $(R, \mathcal{E})$, consists of the following steps:

1. Form the Constraint Network $N(R, \mathcal{E})$ [Fig. 1b].

2. Determine the Zones, $\mathcal{Z}$, from $\mathcal{E}$ [Fig. 2a].

3. Compute the Skeleton, S, from $\mathcal{Z}$ and N [Fig. 2b].

4. Compute the minimal width configuration, $\mathbf{M}$ of N using the displacements of the $Left[i]$'s given by S [Fig. 3b].

From Lemma 7, the minimal width configuration of N yields positions for the rectangles, R, that form a solution to $(R, \mathcal{E})$.

Depending on the system used to generate initial layouts, step 1 may or may not be previously computed. If not, a straightforward method is to determine for each rectangle which other rectangles are visible and creating an edge for each one. Because it requires sorting the rectangles' coordinates this takes $O(n \log n)$ time using a straightforward sweepline method.

The Zones can be computed in $O(n \log n)$ time as well. In fact, they can be determined at the same time as the network is being constructed by noting which rectangles are visible from the left and from the right during the sweeping.

The Skeleton can be computed by finding the longest paths from each $Left[i]$ to $Right[i]$ and $Right[i+1]$. Ensuring that each successive path added in the Skeleton's construction has as many edges as possible in common with the previous path is unnecessary: it is simpler to prove things with an acyclic Skeleton, but the displacement can still be defined and Lemma 11 will still hold if there is a cycle, so long as the two directed subpaths forming the cycle are of equal length. This is guaranteed since only longest paths are used in forming S. The linear time longest path algorithm used in 1-D compaction could be used repeatedly for each of the $|\mathcal{Z}|$ values of i. The total time to compute the Skeleton would then be $O(|\mathcal{Z}| \, n)$ which is bounded by $O(n^2)$.

The final step requires just one more application of 1-D compaction's longest path algorithm: first, one additional node is added to N along with $|\mathcal{Z}|$ edges, each to one of the $Left[i]$'s and of length $D(Left[1], Left[i])$. Then the longest path distance from the origin to each node is assigned as the position of that node in a configuration of N.

So the total time is $O(n^2)$, the bottleneck being the computation of the Skeleton from the Constraint Network. In fact, we can do better: computing the Skeleton from the Constraint Network can be performed in just $O(n \log |\mathcal{Z}|)$ time so that the total time of our algorithm is just $O(n \log n)$.

7.1 $O(n \log |\mathcal{Z}|)$ Algorithm for Computing the Skeleton

While it is true that the only information we need from the Skeleton is the position of the leftmost node for each region, we know of no other way to compute these positions than by computing the whole Skeleton. At first glance, it appears impossible in general to compute the $2|\mathcal{Z}| - 1$ Ribs from which the Skeleton is formed in any less than $(2|\mathcal{Z}| - 1) \in O(n)$ time: simply writing down a longest path between two vertices in a graph of $O(n)$ edges requires $\Omega(n)$ time. Fortunately the formation of the Skeleton doesn't require represention of each of the Ribs explicitly but instead requires knowing only the union of the Ribs. Hence, we need only mark each edge as to whether it is on some Rib without considering which Ribs include it.

We describe a *Divide and Conquer* approach that determines all edges belonging to any of the $2|\mathcal{Z}|-1$ Ribs during two independent recursive computations. The first computation marks all edges that are on any of the Ribs $< Left[i], \dots Right[i] >$, $i \in \{1, \dots |Z|\}$; the second marks edges on Ribs $< Left[i], \dots Right[i+1] >$. Since the second computation is analogous to the first were we simply to shift the contents of the array $Right[]$, we focus on only the first of the two computations in the discussion that follows.

Each recursive computation takes advantage of the fact that Ribs are separators of the planar network: because none of the edges below a given Rib can be on a Rib above it, we need not examine any of the edges below when marking a Rib above. Thus we eliminate much of the redundant work involved were each Rib treated as an independent computation.[1] As a shorthand we refer to the Rib $< Left[i], \ldots Right[i] >$ as Rib i. In order to determine whether edges of a node are "above" or "below" one another, we assume the edgelist for each node is ordered by its position – clockwise starting from π radians – in the initial configuration. For the sake of convenience, we also mark any vertex found to lie on some Rib, though given the edges it is easy enough to infer the vertices.

The routine that solves the problem of marking all edges on Ribs $i+1$ through $k-1$, $Mark(i, k)$. So that we can eliminate special cases from our discussion, we introduce a 0'th Rib and an $(|\mathcal{Z}| + 1)$'st Rib defined as the topmost and bottommost paths in a planar embedding of the network, respectively. Both of these Ribs we mark in advance. Then the single call $Mark(0, |\mathcal{Z}| + 1)$ solves our problem.

$Mark(i, k)$
begin
    if $i \geq k - 1$ then return
    else begin
        $j \leftarrow \lfloor \frac{i+k}{2} \rfloor$
        $Mark(i, j)$
        $Mark(j, k)$
    end
end

As a subroutine of $Mark(i, k)$ we need a procedure that marks all edges on a single Rib j: we'll call this procedure $MarkRib(i, j, k)$. $MarkRib(i, j, k)$ might work by performing a single source longest path search from source $Left[j]$. Edges on the longest path found to $Right[j]$ would then be marked.

This version of $MarkRib$ doesn't, however, achieve our $O(\log n)$ time bound. Suppose the network is shaped like a double ended broom: if the network consists of edges from a number, say $r = \frac{n}{4}$, of sources all directed to the same node, a, an equal number of edges to the corresponding sinks from another node, b, and a path of length $\frac{n}{2} - 1$ from a to b, then every Rib would include the path from a to b of length $\frac{n}{2} - 1$. Moreover, each call to $MarkRib$ would traverse that path. Since there are $|\mathcal{Z}| \in \Omega(n)$ Ribs to mark, each requiring traversal of a path of length $\frac{n}{2} - 1 \in \Omega(n)$ the time spent on this example is $\Omega(n^2)$!

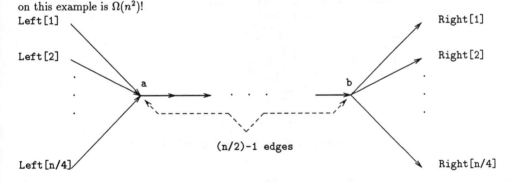

(4) The Difficult Double Ended Broom

However, there is a way to modify $Mark$ to eliminate this problem by taking advantage of the following observation: if ever two Ribs i and k are found to pass thorough the same vertex v, then all Ribs j with $i \mathrel{<} j \mathrel{<} k$ must all pass through v as well. In this case, we use a routine $SolveBroom(v; i, k)$ which solves the single source shortest path problem with v as the source to mark longest paths to all $Right[j]$ and then solves the single source longest path problem with all edges in N reversed to

---
[1]Two similar solutions to this problem were reported independently on the same day. One is due to Tarjan [9]. The simpler one, due to Anderson, we present here.

mark longest paths from v to all $Left[j]$. $SolveBroom$ requires time linear in the number of edges inbetween Ribs i and k. $Mark(i, k)$ becomes:

$Mark(i, k)$
begin

        if $i \geq k - 1$ then return

        else begin

                $j \leftarrow \lfloor \frac{i+k}{2} \rfloor$

                $Mark(i, j, k)$

                If Rib j meets Rib i, $SolveBroom(v; i, j)$ else $Mark(i, j)$

                If Rib j meets Rib k, $SolveBroom(v; j, k)$ else $Mark(j, k)$

        end

end

This version of the algorithm is in fact $O(n \log n)$ time since each of the $O(n)$ edges is traversed at most $O(\log n)$ times.

Once we have marked all edges on the Skeleton by forming the union of edges marked during either of the two recursive computations, we can traverse the Skeleton as before to determine the relative displacements of the $Left[i]$.

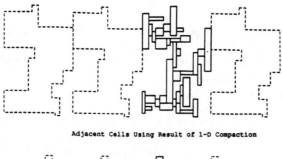

Adjacent Cells Using Result of 1-D Compaction

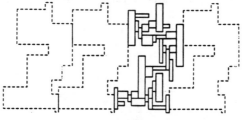

Adjacent Cells Using Result of Tile Compaction

(5) Comparison of Two Compaction Methods

8   Conclusions and Future Work

We have presented a simple $O(n \log n)$ algorithm for performing 1-D tile compaction on a planar constraint network. This is significantly faster than the algorithm described in [6] taking time $O(n^2 \log n)$. Speed is crucial if the algorithm is to be used repeatedly in obtaining an approximation of the optimal 2-D tile compaction as discussed in the introduction.

We believe that the algorithm is easy to understand and implement: we have implemented the $O(n^2)$ version of the algorithm in less than 500 lines of C.

Observe that even on our simple example we achieve a significant savings in space when the circuit is is replicated, the more of a savings tile compaction provides over 1-D compaction.

Before becoming useful as a VLSI design aid, we must confront at least the following issues:

Extension to multi-planar graphs. The algorithm presented works only on planar constraint networks: those resulting from one-layer VLSI circuits. Real VLSI circuits result in non-planar constraint networks. Our algorithm could be used to obtain an initial guess at the width for the more expensive minimum-cost flow method – which does work on non-planar networks – by unnecessarily constraining components on independent layers to maintain their initial relative positions. It remains an open problem whether longest path approaches can be extended to multi-layer circuits. The difficulty is that in non-planar networks it may be impossible to minimize all zone widths simultaneously, so that finding a minimal width configuration does not simply reduce to the problem of minimizing each zone width independently of the others.

Designer interaction Minimizing the overall area of an array of circuits may not aid minimization of the circuit area as a whole. If the circuit has a space available for the array which is rectangular, a heavily skewed parallelogram layout for the array, while possibly the smallest in area, might be less desirable than a slightly larger but less skewed layout. Hence, it may be desirable to feed the Tile Compaction algorithm parameters which bound the allowable skew.

2-D Toroidal Compaction It's not difficult to run the algorithm along the vertical axis after compacting along the horizontal. However, if this is done in the straightforward way, the resulting configuration may actually increase in tiling width, undoing the work done by the initial horizontal tile compaction. Compacting along the skew angle so that rectangles slide not along their vertical axis, but instead along a direction parallel to the skew of the configuration, eliminates this problem but results in an undesirable increased number of jogs in the horizontal wires between rectangles.

References

[1] P. Eichenberger and M. Horowitz, *Toroidal Compaction of Symbolic Layouts for Regular Structures*, 1987 ICCAD, pp. 142-145.

[2] Z. Galil and E. Tardos, *An $O(n^2(m + n \log n) \log n)$ Min-Cost Flow Algorithm*, 1986 FOCS, pp. 1-9.

[3] G. Kedem and H. Watanabe, *Graph-Optimization Techniques for IC layout and Compaction*, 1983 DAC, pp. 113-120.

[4] T. Lengauer, *The Complexity of Compacting Hierachically Specified Layouts of Integrated Circuits*, 1982 FOCS, pp. 358-368.

[5] Y.Z. Liao and C. K. Wong, *An Algorithm to Compact a VLSI Symbolic Layout with Mixed Constraints*, IEEE Transactions on CAD of Integrated Circuits and Systems, Vol. CAD-2, No.2, 1983.

[6] K. Mehlhorn and W. Rülling, *Compaction on the Torus*, 1988 AWOC, pp 212-225.

[7] T. Ohtsuki, gen. ed., <u>Layout Design and Verification</u>, Elsevier Science Pub. B.V. (North Holland), 1986, Chapter 6: *Layout Compaction*, D. Mlynski and C-H. Sung.

[8] M. Schlag, Y. Z. Liao, and C. K. Wong, *An Algorithm for Optimal Two-Dimensional Compaction of VLSI Layouts*, Research Report RC 9739, IBM T. J. Watson Research Center, 1982.

[9] R. E. Tarjan, Personal communication, 1988.

Weighted Parallel Triangulation of Simple Polygons

K. Menzel   B. Monien

University of Paderborn
Paderborn, West Germany

Abstract

This paper presents a sequential and a parallel algorithm for computing an inner triangulation of a simple polygon which is optimal with respect to some weight function. This weight function can be chosen rather arbitrarily. The sequential algorithm runs in $O(n^3)$ time and the parallel algorithm runs in $O(log^2 n)$ time with $O(n^6)$ processors on a Concurrent-Read, Exclusive-Write Parallel RAM model (CREW PRAM).

1   Introduction

The problem of triangulating a simple polygon in the plane is a fundamental problem in computational geometry. Given a weight function, in this paper we will show how to compute an inner triangulation with optimal weight. Applications can be found in Computer Aided Design (CAD) to prevent special color effects on graphic images.

To our knowledge, this is the first paper about weighted triangulations with an arbitrary weight-function. One special case is known as the Delauney triangulation. In this particular case, a triangulation is computed, wherin all triangles are nearly equilateral. The Delauney Triangulation can be constructed sequentially in $O(n \, log \, n)$ time [1]. Algorithms concerning efficient solutions to the ordinary triangulation problem can be found in [2][4][5].

Our algorithmic methods are similar to the methods known for recognition of context-free languages. To any inner triangulation of a simple polygon P, one can associate a tree. This tree is defined by associating a "node" to every triangle of the triangulation and connecting two "nodes" by an edge iff the corresponding triangles have a common line. The graph constructed this way is a tree. Thus our problem is equivalent of finding such a tree with optimal weight. The complexity bounds we prove in this paper are the same as they are known for context-free recognition [3][6].

The parallel algorithm runs on a Concurrent-Read, Exclusive-Write Parallel RAM model (CREW PRAM) in $O(log^2 n)$ time using $O(n^6)$ processors. The sequential algorithm runs in $O(n^3)$ time. In this paper, section 2 gives an introduction in notational conventions and describes the sequential algorithm. Section 3 presents the parallel algorithm in detail. For easier understanding all algorithms and definitions are given for convex polygons. Some additional difficulties occuring in the case of arbitrary polygons can be overcome by preprocessing, which will not increase the complexity bounds of the algorithms.

2 Definitions and Sequential Algorithm

In the following, we present some notational conventions and rules about partitioning polygons.

Definition 1: A polygon is a sequence $v_0, v_1, \ldots, v_{n-1}$ of points in the plane. The v_i's are the vertices of the polygon. The line segment (v_i, v_{i+1}) is the set of points on the line passing through v_i and v_{i+1} and lying between v_i and v_{i+1}. A polygon is simple if its vertices are all different and line segments (v_i, v_{i+1}) intersect at most in a common endpoint. A simple polygon is called convex, if its interior is a convex region.

In the following, we consider only convex simple polygons. As we mentioned already in the introduction, our results also hold for arbitrary simple polygons.

Definition 2: Let P be a simple polygon with vertex set $\{v_0, v_1, \ldots, v_{n-1}\}$ in clockwise order. A triangulation of vertex set $\{v_0, v_1, \ldots, v_{n-1}\}$ is a maximal set of non-intersecting straightline segments between points in this set. A triangulation of polygon P is a triangulation of its vertex set such that all edges of the polygon are edges of the triangulation. An inner triangulation of P consists of all triangles of a triangulation, which are inside P.

For a simple polygon P let us call $\mathcal{T}^P$ the set of all inner triangulations of P and let $P^t \in \mathcal{T}^P$ be an arbitrary inner triangulation of P. The set of all triangulated polygons is denoted by $\wp^t$.

Let P^t be an inner triangulated simple polygon. Every interior line segment belonging to the triangulation defines a partition into P_1 and P_2. Let P_1^t, P_2^t be the triangulations of P_1 and P_2 respectively, induced by P^t. We call (P_1^t, P_2^t) an admissible partition of P^t.

Definition 3: Let P be a simple polygon and P^t its inner triangulation. A function f is called a weight-function of P^t, if it fulfills the following conditions:

a) $f : \wp^t \to I\!R$.

b) There exists $g : I\!R^2 \to I\!R$, g associative and commutative such that
$f(P^t) = g(f(P_1^t), f(P_2^t))$, where (P_1^t, P_2^t) is any admissible partition of P^t.

Then the optimal weighted inner triangulation OPT can be found by computing the minimum of f over all inner triangulations P^t of P.

$$OPT := \min_{P^t \in T^P} f(P^t)$$

Observation: Note, that f is unique for all $P^t \in \wp^t$. $g(x, y)$ can be written as $(x \circ y)$ for all $x, y \in I\!R$. With this notation, we can write $f(P^t)$ in the form:

$$f(P^t) := f(\triangle_1) \circ f(\triangle_2) \circ \ldots \circ f(\triangle_{n-2}).$$

Because g is associative, we can omit the brackets. From g associative and commutative follows, that it is possible to rearrange the elements $(f(\triangle_1), \ldots, f(\triangle_{n-2}))$ following an arbitrary permutation $\pi : \{1..n-2\} \to \{1..n-2\}$:

$$f(P^t) := f(\triangle_{\pi(1)}) \circ f(\triangle_{\pi(2)}) \circ \ldots \circ f(\triangle_{\pi(n-2)})$$

With such a permutation π it is possible to represent every partition Q into Q_1 and Q_2. Hence, $f(P^t) = g(f(P_1^t), f(P_2^t))$ is unique.

Examples:

a) Triangulation with all triangles nearly equiangular

$f(P^t) := \sum_{\triangle \in P^t} (\sum_{i=1}^{3} (\alpha_i - 60)^2)$, with

α_i = angles in $\triangle$, $i = 1, 2, 3$, and

$g(x, y) := x + y$

b) Triangulation with one nearliest equiangular triangle:

$f(P^t) := \min_{\triangle \in P^t} (\sum_{i=1}^{3} (\alpha_i - 60)^2)$, with

α_i = angles in $\triangle$, $i = 1, 2, 3$, and

$g(x, y) := min(x, y)$

Definition 4: A Polygon with n points is called a n-gon. Let P be a simple n-gon and Q be a simple m-gon. Q is called an object of P, if the set of vertices of Q is a subset of the set of vertices of P.

Definition 5: Let P, Q be simple polygons and let Q be an object of P. A chain is a maximal sequence of consecutive vertices of P belonging also to Q. A gap is a maximal sequence of consecutive vertices of P not belonging to Q.

Gaps will be very important in regard to the complexity bounds of the parallel algorithms, because the number of the calculating processors will be dependent on the number of gaps of the objects we get by partitioning.

Definition 6: Let P, Q be simple polygons and let Q be an object of P. If Q has one gap, then Q is denoted by $Q(v_i, v_j)$, that means Q consists of the sequence of consecutive vertices $v_i, v_{i+1}, \ldots, v_j$ in circular order. If Q has two gaps, then Q is denoted by $Q(v_i, v_j, v_k, v_l)$, that means Q consists of the sequence of consecutive vertices $v_i, v_{i+1}, \ldots, v_j, v_k, v_{k+1}, \ldots, v_l$ in circular order (figure 1).

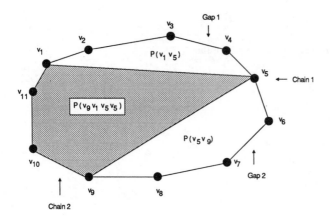

Figur 1:

The gaps and chains are relative to polygon $P(v_9, v_1, v_5, v_5)$, which is an object of $P(v_1, v_{11})$.

We conclude this section with a description of the sequential algorithm, which is designed by using the *dynamic programming method*. That means, weights for smaller objects are stored away to be used later for calculating weights for larger objects.

Theorem 1: Let f be some weight function. The optimal inner triangulation of an arbitrary simple n-gon P with respect to f can be found sequentially in time $O(n^3)$.

Proof: Before we describe the algorithm, which computes the optimal weighted inner triangulation of an arbitrary simple n-gon P, we introduce the method of partitioning P into smaller objects to get a given inner triangulation P^t of P.

Let P be a simple n-gon with an arbitrary inner triangulation P^t and let $v_0, v_1, \ldots, v_{n-1}$ be its vertices in clockwise order. The inner triangulation will be constructed by inserting a triangle into P such that P will be partitioned into P_1 and P_2 with not more than one gap each. Recursively P_1 and P_2 will be partitioned in exactly the same way into smaller polygons with not more than one gap each. The following algorithm will illustrate this proceeding. It is initially started with parameter $P^t(v_0, v_{n-1})$:

Look_Up_Triangulation($P^t(v_i, v_j)$)

1. Let $\triangle(v_i, v_k, v_j)$:=triangle incident to base (v_i, v_j)
2. /* v_k between v_i and v_j in circular order */
3. Print($\triangle(v_i, v_k, v_j)$)
4. $P_1 := P(v_i, v_k)$
5. $P_2 := P(v_k, v_j)$
6. Look_Up_Triangulation(P_1)
7. Look_Up_Triangulation(P_2)

This algorithm triangulates the interior of a polygon P with only $O(n^3)$ triangles and $O(n^2)$ objects with at most one gap.

Let g be a function as defined in Definition 3. The following algorithm shows a way to find the optimal weighted inner triangulation:

Get_Sequential_Triangulation($P(v_0, v_{n-1})$)

1. VAR wg[]: INTEGER; /* actual weights of Objects */
2.   best[]: INTEGER; /* best partition of Objects */
3. Initially set all weights in wg[] to MAXINT
4. Calculate the weights of all $O(n^3)$ triangles in $P(v_0, v_{n-1})$
5. FOR i:=3 TO n DO /* Calculate all objects with one gap and i vertices */
6.   FOR j:=0 TO n-1 DO /* At all positions in polygon */
7.   FOR k:=j+1 TO j+i-2 DO /* Calculate all subpartitions */
8.   BEGIN
9. $t := g(wg[P(v_j, v_k)], g(wg[\triangle(v_j, v_k, v_{j+i-1})], wg[P(v_k, v_{j+i-1})]))$
10.   /* Calculating the minimum */
11.   IF ($t < wg[P(v_j, v_{j+i-1})]$) THEN
12.   BEGIN
13. $wg[j, j + i - 1] := t$
14. $best[j, j + i - 1] := k$
15.   END
16.   END

The calculated inner triangulation can be found with Look_Up_Triangulation. Note, that $best[i, j]$ always gives the best partition of the polygon $P(v_i, v_j)$. Look_Up_Triangulation is initially started with parameter $P(v_0, v_{n-1})$. Get_Sequential_Triangulation runs in $O(n^3)$ time, Look_Up runs in $O(n)$ time. □

3   The Parallel Algorithm

Before the parallel algorithm can be described, some additional information about partitioning polygons have to be mentioned.

Lemma 1: Let P be a simple n-gon with an arbitrary inner triangulation P^t. Then there exists a triangulation line L in P^t, which divides P into a q-gon P_1 and a p-gon P_2 such that $\frac{n}{3} \leq p, q \leq \frac{2n}{3}$ and $p + q = n + 2$.

Proof: Let P be a simple n-gon with an arbitrary inner triangulation P^t and let $v_0, v_1, \ldots, v_{n-1}$ be its vertices in clockwise order. The following algorithm shows a simple way to get the claimed triangulation line L.

Get_Triangulation_Line(P^t)

1. /* Initialize trianluation line L */
2. $(v_i, v_j) := (v_{n-1}, v_0)$
3. /* Search for claimed triangulation line L */
4. REPEAT
5.   IF $\Delta(v_i, v_j, v_{j+1}) \in P^t$ THEN $(v_i, v_j) := (v_i, v_{j+1})$
6.   ELSEIF $\Delta(v_{i-1}, v_i, v_j) \in P^t$ THEN $(v_i, v_j) := (v_{i-1}, v_j)$
7.   ELSE
8.   /* There exist $\Delta(v_j, v_k, v_i) \in P^t$ with $v_j < v_k < v_i$ in cyclic order */
9.   BEGIN
10.   IF $distance(v_k - v_j) \leq distance(v_i - v_k)$
11.   THEN $(v_i, v_j) := (v_i, v_k)$
12.   ELSE $(v_i, v_j) := (v_k, v_j)$
13.   END
14. UNTIL (Number of vertices in $P(v_i, v_j) \geq \frac{n}{3}$
15. END.

Line 5 and 6 add one point after another to polygon $P(v_i, v_j)$. So the algorithm will stop the WHILE-loop to the right time. Line 9 to 13 add at most $\frac{n-l}{2}$ vertices to $P(v_i, v_j)$, where l is the number of vertices of $P(v_i, v_j)$. So $P(v_i, v_j)$ has at most $l + \frac{n-l}{2} = \frac{n+l}{2} \leq \frac{2n}{3}$ vertices. When the algorithm has finished $P_1 := P(v_j, v_i)$, $P_2 := P(v_i, v_j)$ (figure 2). $\square$

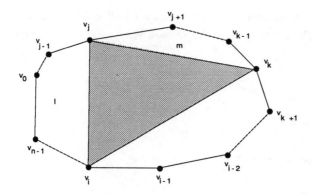

Figur 2:

$m \leq \frac{n-l}{2}, l \leq \frac{n}{3}$, so $P(v_i, v_k)$ has not more than $l + m \leq \frac{2n}{3}$ vertices.

Lemma 2: Let P be a simple n-gon with an arbitrary inner triangulation P^t and three gaps. Then there exists a triangle $\triangle$ in P^t, which divides P into at most three polygons P_1, P_2, P_3 with at most 2 gaps each.

Proof: Let P be a simple n-gon with an arbitrary inner triangulation P^t and let $v_0, v_1, \ldots, v_{n-1}$ be its vertices in cyclic order. Let (v_r, v_s) be one of the gaps and denote the three chains by C_1, C_2, C_3. Note that if (v_i, v_j) is a line in the triangulation P^t, then there exists exactly one vertex v_k such that (v_i, v_j, v_k) form a triangle in P^t and $v_i < v_j < v_k$ in cyclic order. The following algorithm shows a simple way how to get the claimed triangle $\triangle$.

Get_Triangle(P^t)

1. /* Initialize $\triangle$ */
2. $\triangle(v_i, v_j, v_k) :=$ Triangle in P^t with base (v_r, v_s)
3. WHILE (v_i, v_j, v_k not all in different chains)
4.   BEGIN
5.   IF v_i and v_k are in the same chain
6.   THEN $\triangle(v_i, v_j, v_k) := \triangle(v_j, v_l, v_k)$ with $v_j < v_l < v_k$
7.   ELSE $\triangle(v_i, v_j, v_k) := \triangle(v_k, v_l, v_i)$ with $v_k < v_l < v_i$
8.   END

W. l. o. g. let v_r be in C_1 and v_s be in C_2. The initial triangle $\triangle(v_r, v_s, v_k)$ has the third point v_k in one of the three chains C_1, C_2 or C_3. If v_k is in C_3 then we have finished. In the case of v_k in C_1, there exists the triangulation line (v_k, v_s), because it is one side of $\triangle$. (v_k, v_s) has one endpoint in C_1, the other in C_2. This line will be the base for the new triangle $\triangle$ in the WHILE-Loop. The same reason holds for v_k in C_2. Hence, if v_k is not in C_3, all bases for $\triangle$ have one point in C_1 and the second in C_2, in direction to the end of the two chains. The last possible choice for the base of $\triangle$ consists of the last point in C_1 and C_2. In the next step, v_k must be in C_3 (figure 3). $\square$

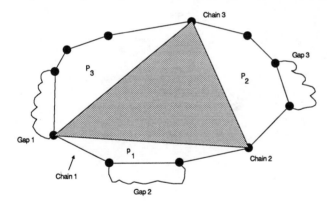

Figur 3:

$\triangle$ divides P into three polygons P_1, P_2, P_3 with at most two gaps each.

Lemma 3: Let P be a simple n-gon with an arbitrary inner triangulation P^t. P^t can be represented by a partition tree with height $O(log\,n)$ and with nodes taken from the set of objects of P with at most two gaps.

Proof: Let P be a simple n-gon, P^t its inner triangulation and let $v_0, v_1, \ldots, v_{n-1}$ be the vertices of P in clockwise order. Lemma 1 shows, that P can be partitioned into P_1 and P_2 with the number n_1, n_2 of vertices, which fulfill the condition $\frac{n}{3} \leq n_1, n_2 \leq \frac{2n}{3}$. Dividing P in this manner defines a partition tree with height less or equal than $log_{\frac{3}{2}} n$. But there are possibly objects with more than two gaps.

Let L be a dividing line constructed by Lemma 1, which produces a third gap in P. Let C_1, C_2, C_3 be the chains of P, C_1 between the first two gaps. Lemma 2 says, that there exists a triangle $\triangle$, with one vertex in C_1, C_2, C_3 each. Let L_1, L_2, L_3 the lines building $\triangle$, L_1, L_2 with one endpoint in C_1.

Now, increase the partition tree in the following way: First insert L_1, L_2. Then insert L_3 and at least insert L. Following this order, P is never divided into smaller objects with more than two gaps. The height of the partition tree is bounded by $4 * log_{\frac{3}{2}} n = O(log\,n)$ (figure 4). $\square$

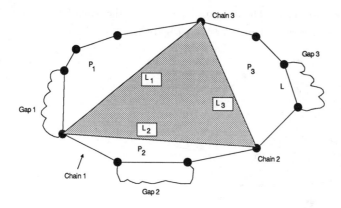

Figur 4:

First insert L_1, L_2, then L_3 and at last L. P_1, P_2, P_3 are separated from the polygon one after another, with at most two gaps each.

Theorem 2: Let f be some weight function. The optimal inner triangulation of an arbitrary simple n-gon P with respect to f can be found in parallel on a CREW PRAM with $O(n^6)$ processors in $O(log^2 n)$ time.

Proof: Let P be a simple n-gon, Q be an object of P. Define $\mathcal{M}_Q := \{L : L$ divides Q into polygons R, S with at most 2 gaps each$\}$. The algorithm calculates in parallel the optimal weight for every object Q by calculating the optimal weight of every partition of Q.

Get_Parallel_Triangulation($P(v_0, v_{n-1})$)

1. VAR wg[]: INTEGER; /* actual weights of objects */
2.   best[]: LINE; /* best partition line of objects */
3   Initially set all weights in wg[] to MAXINT
4. Calculate the weights of all $O(n^3)$ triangles in $P(v_0, v_{n-1})$
5. FOR $(4 * log_{\frac{3}{2}} n)$ TIMES DO
6.   BEGIN
7.   FOR all objects Q of P with at most 2 gaps DO IN PARALLEL
8.   BEGIN
9. $t := \min_{L \in \mathcal{M}_Q} \{g(wg[Q_1], wg[Q_2]) : L$ divides Q into Q_1 and $Q_2\}$
10.   IF $t < wg[Q]$ THEN
11.   BEGIN
12. $wq[Q] := t$
13. $best[Q] :=$ Dividing line L, which fulfills Minimum in Line 9
14.   END
15.   END
16.   END

There are at most $O(n^2)$ choices of dividing a polygon into two parts. Hence, $\|\mathcal{M}_Q\| = O(n^2)$. The weight of each division could be calculated in constant time with one processor. If every object Q of P gets $O(n^2)$ processors then the optimal weight of one object can be calculated in $O(log n)$ time by calculating all $O(n^2)$ partitions in parallel. There exist at most $O(n^4)$ objects with at most two gaps. So line 7 to 15 take $O(log n)$ time with $O(n^6)$ processors. Following Lemma 3, there exist a partition tree with height less or equal than $4 * log_{\frac{3}{2}} n$ and therefore we need maximal $4 * log_{\frac{3}{2}} n$ loops to get the optimal weight of P^t. So the algorithm takes $O(log^2 n)$ time with $O(n^6)$ processors.

The inner triangulation can be looked up in the table best[ ]. Initially, the recursive function Look_Up-_Triangulation is started with parameter $P(v_0, v_{n-1})$. Look_Up_Triangulation() runs in $O(log\ n)$ time. □

Look_Up_Triangulation($P(v_i, v_j)$)

1. P_1 :=first polygon that results of partition with line $best[P(v_i, v_j)]$
2. P_2 :=second polygon that results of partition with line $best[P(v_i, v_j)]$
3. IF P_1 is triangle
4.   THEN print(P_1)
5.   ELSE Look_Up_Triangulation(P_1)
6. IF P_2 is triangle
7.   THEN print(P_2)
8.   ELSE Look_Up_Triangulation(P_2)

Corollar: Let f be some weight function. The optimal inner triangulation of an arbitrary simple n-gon P with respect to f can be found in parallel on a CRCW PRAM with $O(n^8)$ processors in $O(log\ n)$ time.

Proof: Because the minimum of r numbers can be calculated with r^2 processors on a CRCW PRAM in constant time, line 9 in Get_Parallel_Triangulation computes the minimum in constant time with $O(n^4)$ processors and Get_Parallel_Triangulation runs in $O(log\ n)$ time with $O(n^8)$ processors. □

Acknowledgement: We wish to thank C. Hornung (Nixdorf Computer Company), who posed this problem to us. In particular thanks to W. Preilowski for his valuable comments and suggestions and his help in improving the manuscript.

References:

1. H. Edelsbrunner, Algorithms in Combinatorial Geometry, Springer 1987, p. 302.

2. H. ElGindy, An Optimal Speed-Up Parallel Algorithm for Triangulating Simplicial Point Sets in Space, International Journal of Parallel Programming, Vol. 15, No. 5, 1986, 389-398.

3. A. Gibbons, W. Rytter, Efficient Parallel Algorithms, Cambridge University Press, 1988.

4. K. Mehlhorn, Data Structures and Algorithms 3: Multi-dimensional Searching and Computational Geometry, Springer1977.

5. E. Merks, An Optimal Parallel Algorithm for Triangulating a Set of Points in the Plane, International Journal of Parallel Programming, Vol. 15, No. 5, 1986, p. 399-410.

6. W. Ruzzo, On the complexity of general context-free language parsing and recognition. Automata, languages and programming, Lecture Notes in Computer Science, (1979), p. 489-499.

IMPLEMENTING DATA STRUCTURES ON A HYPERCUBE MULTIPROCESSOR, AND APPLICATIONS IN PARALLEL COMPUTATIONAL GEOMETRY

FRANK DEHNE[*] AND ANDREW RAU-CHAPLIN

Center for Parallel and Distributed Computing
School of Computer Science, Carleton University, Ottawa, Canada K1S 5B6

Abstract. In this paper, we study the problem of implementing standard data structures on a hypercube multiprocessor. We present a technique for efficiently executing multiple independant search processes on a class of graphs called ordered h-level graphs. We show how this technique can be utilized to implement a segment tree on a hypercube, thereby obtainig $O(\log^2 n)$ time algorithms for solving the next element search problem, the trapezoidal decomposition problem, the triangulation problem, and the (multiple) planar point location problem.

1 INTRODUCTION

One of the main differences, besides the difference in communication delay, between the parallel random access machine (CREW-PRAM) and the hypercube processor network is that the PRAM has one large (shared) memory similar to a standard sequential computer, whereas the hypercube has its memory divided into pieces of constant size and distributed over the network.

The fact that the PRAM memory resembles the structure of the standard sequential machine memory has been extensively used for the design of efficient PRAM algorithms. It allows the implemention, on a PRAM, of well established data structures like, e.g., segment trees [ACG], [ACGD], [G] or subdivision hierarchies [DK]. Once such a data structure has been built, each processor can search in it, independently of the others, in the standard manner.

In this paper, we show that a similar paradigm can also be used for a hypercube multiprocessor of size N; i.e., a set of N processors PE(i), $0 \leq i \leq N-1$, where two processors PE(i) and PE(j) are connected by a communication link if the binary representations of i and j differ in exactly one bit.

We define a class of graphs called *ordered h-level graphs* which includes most of the standard data structures (in particular, all k-nary search trees for k=O(1)) and show that for such a graph with n nodes stored on a hypercube multiprocessor, O(n) search processes can be efficiently executed independantly and in parallel.

We apply this method to implement a segment tree [BW], [M], [PS] for next element search on a hypercube. Since the total length of all lists attached to the nodes of a segment tree (for n segments) is O(n log n), we can not construct the entire segment tree before starting the search processes (as in [ACG], [ACGD], [G]), since this would require a hypercube of size n log n. Instead, we first build the segment tree without the node lists, and then show how during the execution of the search processes these lists can be dynamically created only for those nodes currently visited by the search queries (thus, not exceeding a total length of O(n)).

Our approach provides $O(\log^2 n)$ time hypercube algorithms for the next element search problem, the trapezoidal map construction problem, the triangulation problem, and the planar (multiple) point location problem.

The paper is organized as follows: In Section 2, we define ordered h-level graphs and the associated m-way search problem, and also review some standard data movement operations. In Section 3, we present an efficient hypercube algorithm for m-way search on ordered h-level graphs and, in Section

---

[*] Research partially supported by the Natural Sciences and Engineering Research Council of Canada under Grant A9173.

4, we show how to use this method to implement a segment tree for next element search on a hypercube. Finally, in Section 5, we describe how this algorithm can be used to obtain efficient hypercube algorithms for the other geometric problems listed above.

2 DEFINITIONS AND BASIC HYPERCUBE OPERATIONS

In this section, we will first define ordered h-level graphs and the associated m-way search problem. Then, some basic standard hypercube data movement operations are reviewed, as these operations will be used in the remainder of this paper.

2.1 ORDERED H-LEVEL GRAPHS

Assume we are given a directed graph G=(V,E) with vertex set V and edge set E. An *ordered h-partitioning* of G is a partitioning of V into an ordered sequence of h disjoint sets $L_1,...,L_h$ together with an ordering of the elements in each subset L_i ($1 \leq i \leq h$).

For every ordered h-partitioning of G we define, for every $v \in V$, three numbers *Level(v)*, *Levelindex(v)* and *Index(v)* as follows:

- Level(v) = i if and only if $v \in L_i$,
- Levelindex(v) is the rank of v with respect to the ordering of the vertices in $L_{Level(v)}$, and
- Index(v) = ($\sum_{1 \leq i \leq Level(v)-1} |L_i|$) + Levelindex(v)

A directed acyclic graph G = (V,E) is called an *ordered h-level graph* if it has the following properties (see Figure 1 for an illustration):
(1) There exists a constant k = O(1) such that every node of G has an out-degree of at most k.
(2) There exists an ordered h-partitioning $L_1,...,L_h$ of G such that
    (a) every source of G is contained in L_1,
    (b) if (v,w) is an edge of G then Level(w) = Level(v) + 1, and
    (c) if (v,w), (v',w') are two edges of G with Index(v) < Index(v'), then Index(w) $\leq$ Index(w').

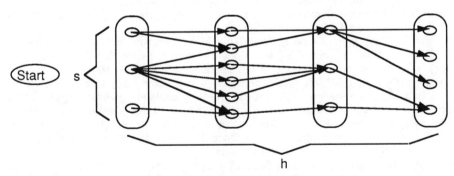

Figure 1. An Ordered h-Level Graph.

We observe that ordered h-level graphs are acyclic and planar, and that any k-nary tree (k=O(1)) is an ordered h-level graph.

2.2 THE M-WAY SEARCH PROBLEM FOR ORDERED H-LEVEL GRAPHS

Let G = ($V=L_1\cup...\cup L_h$, E) be an ordered h-level graph (with maximum out-degree k), and let U is a universe of possible search queries on G.

A *search path* for a query q $\in$ U is a sequence path(q)=(v_1, ..., v_h) of h vertices of G defined by a *successor* function f: (V $\cup$ {start}) x U $\Rightarrow$ N (i.e., a function with the property that f(start,q) $\in L_1$ and for every vertex v$\in$ V, (v,f(v,q)) $\in$ E) as follows:

- $\text{Index}(v_1) = f(\text{start}, q)$
- $\text{Index}(v_{i+1}) = f(v_i, q)$, $1 \leq i < h$.

We also define an associated *successor rank* function g: $(V \cup \{\text{start}\}) \times U \Rightarrow \{1,...,k\}$ as follows:
- $g(\text{start}, q)$ is the rank of $f(\text{start}, q)$ in the set $\{\text{Index}(v) | v \in L_1\}$, and
- $g(v_i, q)$ is the rank of $f(v_i, q)$ in the set $\{\text{Index}(w) | (v_i,w) \in E\}$, $1 \leq i < k$.

For example, if G is a binary search tree then, for every query q, $f(\text{start},q)$ is the root of the tree; for every node v, $g(v,q) \in \{1,2\}$ indicates whether the left or right child is to be visited next and $f(v,q)$ is the index of (or pointer to) that child.

Given an ordered h-level graph G with n nodes stored in a hypercube multiprocessor such that the node v with $\text{Index}(v)=i$ is stored in processor PE(i), then a *search process* for a query q with search path $(v_1, ..., v_h)$ is a process divided into h time steps $t_1 < t_2 < ... < t_h$ such that at time t_i, $1 \leq i \leq h$, there exists a processor which contains a description of both, the query q and the node v_i. Note, however, that we do not assume that the search path is given in advance; we assume that it is constructed during the search by successive applications of the functions f and g.

Given an ordered h-level graph G with n nodes and a set $Q = \{q_1,...,q_m\} \subseteq U$ of m queries, m=O(n), then the *m-way search problem* consists of executing (in parallel) all m search processes induced by the m queries.

2.3 BASIC HYPERCUBE OPERATIONS

The m-way search algorithm described in the next section uses slightly generalized versions of eight well-defined hypercube data movement operations; in addition to those registers listed below, their implemenation requires a constant number of auxiliary registers. In the following, for every register A available at every processor, *A(i)* refers to register A at processor PE(i).

Rank(Reg(i),Cond(i)): Compute, in time O(log N), in register Reg(i) of every processor PE(i) the number of processors PE(j) such that j< i and Cond(j) is true [NS].

Number(Reg(i),Cond(i)): Compute, in time O(log N), in register Reg(i) of every processor PE(i) the number of processors PE(j) such that Cond(j) is true.

Concentrate([Reg$_1$(i),...,Reg$_z$(i)],Cond(i)): This operation includes an initial Rank(R(i), Cond(i)) operation. Then for each PE(i) with Cond(i) = true, registers Reg$_1$(i),...,Reg$_z$(i) are copied to PE(R(i)), z=O(1). The time complexity of this operation is also O(log N) [NS].

Route([Reg$_1$(i),...,Reg$_z$(i)],Dest(i),Cond(i)): Every processor PE(i) has z=O(1) data registers Reg$_1$(i),...,Reg$_z$(i), a destination register Dest(i), and a boolean condition register Cond(i). It is assumed that the destinations Dest(i) are monotonic; i.e., if i<j then Dest(i)<Dest(j). This operation routes, for every processor PE(i) with Cond(i) = true, all registers Reg$_1$(i),...,Reg$_z$(i) to processor PE(Dest(i)); it can be implemented with an O(log N) time complexity by using a Concentrate operation followed by a Distribute operation described in [NS].

RouteAndCopy([Reg$_1$(i),...,Reg$_z$(i)],Dest(i),Cond(i)): Under the same assumptions as for the Route operation, this operation routes, for every processor PE(i) with Cond(i) = true, a copy of registers Reg$_1$(i),...,Reg$_t$(i) to processors PE(Dest(i - 1) + 1), ..., PE(Dest(i)), each; it can be implemented with an O(log(N)) time complexity by using a Concentrate followed by a Generalize operation described in [NS].

Reverse([Reg$_1$(i),...,Reg$_z$(i)],Start,End): This operation routes for every PE(i) with Start $\leq$ i $\leq$ End, its registers Reg$_1$(i), ..., Reg$_z$(i), z=O(1), to PE(Start + End - i); i.e., it reverses the contents of those registers for the sequence of processors between PE(Start) and PE(End). Reversing, in the entire hypercube, a sequence of n values (each stored in one processor) corresponds to routing each value stored at processor PE(i) to processor PE(i'), where i' is obtained from i by inverting all bits in its binary representation. Hence, this operation can be implemented in time log(n) similarly to the Concentrate/Distribute operation described in [NS].

BitonicMerge([Reg₁(i),...,Regz(i)],Key(i),Left,Peak,Right): This operation is the well known bitonic merge [B]. It converts in time O(log N) a bitonic sequence (with respect to register Key(i)) into a sorted sequence; it simultaneously permutes the registers $Reg_1(i),...,Reg_t(i)$ ($z=O(1)$). Here, we apply it to a particular bitonic sequence consisting of an increasing sequence starting at PE(Left) and ending at PE(Peak) followed by a decreasing sequence starting at PE(Peak+1) and ending at PE(Right).

Sort([Reg₁(i),...,Regz(i)],Key(i)): This operation refers to $O(\log^2 n)$ time bitonic sort [B] with respect to Key(i); it simultaneously permutes the registers $Reg_1(i),...,Reg_z(i)$ ($z=O(1)$).

3 AN $O(\min\{s \log N, \log^2 N\} + h \log N)$ TIME HYPERCUBE ALGORITHM FOR M-WAY SEARCH ON ORDERED H-LEVEL GRAPHS

Let $G = (V = L_1,...,L_h, E)$ be an ordered h-level graph (with h-partitioning $V = L_1,...,L_h$), where $|V| = n$, and such that every node has an out-degree of at most $k=O(1)$ and can be stored using O(1) space. For the remainder, s denotes the number of sources of G (i.e., the number of $v \in V$ with Level(v)=1).

Furthermore, let U be a universe of search queries, each of which can also be stored using O(1) space, and let f: $(V \cup \{start\}) \times U \Rightarrow N$ and g: $(V \cup \{start\}) \times U \Rightarrow \{1,...,k\}$ be the successor function and successor rank function, respectively, describing the search path in G associated with every search query. We assume that f(x,q) and g(x,q) can be computed in constant time for any $x \in V \cup \{start\}$, $q \in U$.

In this section, we consider the problem of solving, on a hypercube multiprocessor, the m-way search problem for a set $Q=\{q_1,...,q_m\} \subseteq U$ of m queries. We present an $O(\min\{s \log N, \log^2 N\} + h \log N)$ time algorithm for solving the m-way search problem on a hypercube of size N, where $N=\max\{n,m\}$.

For the remainder we assume, w.l.o.g., that $n=m=N=2^d$; all results obtained can be easily generalized. In the following Section 3.1 we give an overview of the algorithm, including the assumed initial configuration of the hypercube, and how the result, i.e. the m-way search, is reported. In Sections 3.2 and 3.3 we will then present the details of the algorithm. Section 3.4 summarizes the results.

3.1 ALGORITHM OVERVIEW

The graph G is assumed to be stored in the hypercube such that each vertex v with Index(v)=i is stored in register v(i) of processor PE(i); register v(i) contains fields v.data(i), v.Level(i), v.Levelindex(i), and v.Index(i), storing a constant amount of data associated with vertex v, its level, levelindex and index, respectively. The edges of G are stored as adjacency lists. That is, for every vertex v stored in register v(i), the indices of the at most k successors of v (i.e.; the indices of the vertices w such that $(v,w) \in E$) are stored in the fields v.successor1(i), ..., v.successork(i), respectively; see Figure 2.

The set $Q=\{q_1,...,q_m\}$ of m queries is stored in arbitrary order, and such that every processor PE(i) stores one query in its register q(i).

Figure 2 shows the set of registers necessary at every processor PE(i). In addition to the registers v(i) and q(i) mentioned above, the algorithm assumes that every processor also has a register v'(i) to store another vertex of G as well as other auxiliary registers which will be described later.

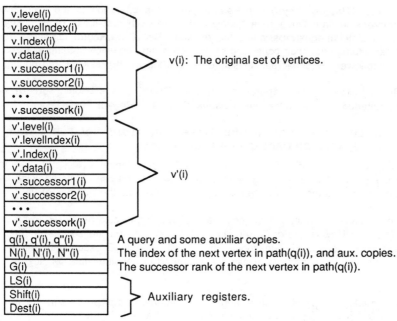

| v.level(i) |
| v.levelIndex(i) |
| v.Index(i) |
| v.data(i) |
| v.successor1(i) |
| v.successor2(i) |
| ••• |
| v.successork(i) |

v(i): The original set of vertices.

| v'.level(i) |
| v'.levelIndex(i) |
| v'.Index(i) |
| v'.data(i) |
| v'.successor1(i) |
| v'.successor2(i) |
| ••• |
| v'.successork(i) |

v'(i)

q(i), q'(i), q''(i)	A query and some auxiliar copies.
N(i), N'(i), N''(i)	The index of the next vertex in path(q(i)), and aux. copies.
G(i)	The successor rank of the next vertex in path(q(i)).
LS(i)	
Shift(i)	Auxiliary registers.
Dest(i)	

Figure 2. The Registers Required at Each Processor PE(i,j).

The global structure of the m-way search algorithm is described in Figure 3. The m search processes for all m queries $q_1, ..., q_m$ are executed in h phases; each phase moves all queries one step ahead in their search paths.

Procedure **M-Way-Search**:
(1)   Phase$_1$ {Match every query with the 1st node in its search path.}
(2)   For x := 2 to h do
(3)   Phase$_x$ {Match every query with the xth node in its search path.}

Figure 3: Global Structure of the M-Way-Search Algorithm

The algorithm permutes the queries (in registers q(i)) and copies some nodes into the registers v'(i) such that at the end of phase x ($1 \leq x \leq h$):
- all queries are sorted with respect to the index of the xth node in their search path, and
- each processor PE(i) containing a query q in its register q(i), contains in its register v'(i) a copy of the xth node in the search path of q (this is called a *match* of q and the xth node in its search path).

A typical situation at the end of a phase is depicted in the following Figure 4; each vertical column represents the registers q(i) and v'(i) of a processor PE(i).

PE(0)   ...   PE(15)

q(i)	q5	q7	q15	q1	q4	q6	q8	q9	q12	q14	q16	q2	q3	q10	q11	q13
v'.Index(i)	1	1	1	2	2	2	2	2	2	2	2	3	3	3	3	3
v'.Data(i)	v1	v1	v1	v2	v2	v2	v2	v2	v2	v2	v2	v3	v3	v3	v3	v3

Figure 4. A Typical Situation at the End of a Phase.

In the following Sections 3.2 and 3.3, we will describe the details of Phase 1 and Phase x ($2 \leq x \leq h$), respectively. The first phase is different from the remaining phases. When ordering the queries with respect to the index of the first node in their search path, the first phase has to start with an arbitrary permutation of the queries, whereas each subsequent phase will utilize the ordering of the previous phase (in order to improve the time complexity of the algorithm).

3.2 PHASE 1 OF THE M-WAY SEARCH ALGORITHM

An outline of Phase 1 is given in Figure 5. The algorithm consists of four basic steps. First, every processor PE(i) calculates the index of the first node in the search path of its query q(i), and stores this value in an auxiliary register N(i). Then, in Step 2, the number of sources is calculated (here represented by a variable s). In Step 3, the queries are sorted by the index of the first node in the search path, i.e. N(i); this ordering is performed in one of two possible ways depending on the number of sources. If $s \geq \log(N)$ then bitonic sorting [B] is used; if $s < \log(N)$, a procedure called SortBySourceIndex is used which sorts the queries in $O(s \log N)$ time and will be described below. Note that in many application s is a constant (e.g. for search trees) and, thus, the sorting step is performed in time $O(\log N)$.

Finally, in Step 4, the source nodes are copied to the queries for which they are the first node in their search path. Because of the ordering of the queries, this step can be performed in $O(\log(N))$ time using a procedure MoveVerticesToQueries which will also be described below.

```
Procedure Phase₁:
(1)    Every  PE(i):  N(i):=f(Start,q(i))
(2)    Number(LS(i),  v.Level(i)=1)
       s:=LS(0)                          {Note:  LS(i)=LS(i') for  all  i,i'}
(3)    IF s≥log(N) THEN
               Sort([q(i),  N(i)],  N(i))
       ELSE
               SortBySourceIndex(s)
( 4)   MoveVerticesToQueries(1)
```
Figure 5. Outline of Phase 1.

We first discuss the details of Procedure SortBySourceIndex; see Figure 6a. The procedure uses a register Shift(i) at each processor which stores the number of queries that have already been sorted. In Step 1, all registers Shift(i) are initialized to 0. Then, Steps 3 to 8 are executed for each source of the graph. In each iteration, the queries q(i) that need to be matched with that source [as well as the associated source indices N(i)] are copied into registers q'(i) [N'(i)] of the same processors, concentrated (Steps 3 and 4), and then appended at the end of the sequence of queries ordered so far (Steps 5 and 6); finally the registers Shift(i) are updated (Steps 7 and 8). These steps produce, in the registers q"(i), the correct permutation of the queries, which are then copied back into the registers q(i) [N(i)]; see Step 9. Obviously, each iteration takes $O(\log N)$ time and, hence, the time complexity of procedure SortBySourceIndex(s) is $O(s \log N)$.

```
Procedure  SortBySourceIndex(s):
(1)    Every PE(i):  Shift(i):=0
(2)    FOR r:=1 TO s DO
(3)            Every PE(i) with  N(i)=r:  q'(i):=q(i),  N'(i):=N(i)
(4)            Concentrate([q'(i),  N'(i)],  N'(i)=r)
(5)            Route([q'(i),  N'(i)],  i+Shift(i),  N'(i)=r)
(6)            Every PE(i)  with  N'(i)=r:  q"(i):=q'(i),  N"(i):=N'(i)
(7)            Number(H(i),  N(i)=r)
(8)            Every  PE(i):  Shift(i):=Shift(i)+H(i)
(9)    Every PE(i):  q(i):=q"(i),  N(i):=N"(i)
```
Figure 6a. Detailed Description of Procedure SortBySourceSelected.

```
Procedure  MoveVerticesToQueries(CurrentLevel):
(1)    Every PE(i): N'(i):=N(i)
(2)    Route([N'(i)],  i-1,  i>0)
(3)    PE(1): N'(N):=N(N) + 1
(4)    Every PE(i) with N'(i)≠N(i): Dest(i):=i
(5)    Route([Dest(i)],  N(i),  N'(i)≠N(i))
(6)    Every PE(i): v'(i):=v(i)
(7)    RouteAndCopy([v'(i)],  Dest(i),  v.Level(i)=CurrentLevel)
```

Figure 6b. Detailed Description of Procedure MoveVerticesToQueries.

Once the queries have been sorted by the index of the first vertex in their search path, the matching process between each query and the first node in its search path can be performed in time O(log N) using the procedure MoveVerticesToQueries described in Figure 6b. The parameter CurrentLevel (which is one for all sources) denotes the level of the nodes to which the queries are to be routed (i.e., to be matched with). The idea is to identify for each node, the largest address of a query to be matched with that node (Steps 1 to 5), and then use the procedure RouteAndCopy to broadcast each node to the block of queries to be matched with (Steps 6 and 7). The time complexity of this process is O(log N).

3.3 PHASE x (2≤x≤h) OF THE M-WAY SEARCH ALGORITHM

As indicated in Section 3.1, the purpose of each subsequent phase is to advance, in time O(log N) all queries by one step in their search path. After Phase x-1 has been completed, all queries are sorted with respect to the index of the $(x-1)^{th}$ node in their search path, and each processor PE(i) contains a query q in its register q(i) together with a copy of the $(x-1)^{th}$ node in the search path of q in its register v'(i). The desired effect of Phase x is to have all queries sorted with respect to the index of the xth node in their search path, and have each processor PE(i) contain, in its register v'(i), a copy of the xth node in the search path of the query q(i).

```
Procedure  Phase(x),  2≤x≤h:
(1)    Every PE(i): N(i):=f(v'(i),q(i)),  G(i):=g(v'(i),q(i))
(2)    OrderQueriesByNextVertex
(3)    MoveVerticesToQueries(x)
```

Figure 7. Overview of Phase x, 2≤x≤h.

An outline of the algorithm for Phase x is given in Figure 7. First (in Step 1), every PE(i) computes for the query currently stored in its register q(i) the index of the next node in its search path as well as the successor rank of that node (see Section 2.2) and stores these two numbers in the auxiliary registers N(i) and G(i), respectively. In Step 2, all queries are sorted by the index of the next node in their search paths. This sorting operation is performed by a procedure OrderQueriesByNextVertex in time O(log N) by using the properties of the previous permutation of the queries. Once this ordering has been obtained, the nodes can be matched with the queries in time O(log N) in the same way as described in Section 3.2.

What remains to be discussed are the details of procedure OrderQueriesByNextVertex. This procedure, which is described in Figure 8, creates in time O(log N) the new ordering of the queries with respect to the indices of the next nodes in the search paths.

Consider all edges (v,w) and (v',w') where Level(v)=Level(v')=x-1 and w and w' have the same successor rank. If Index(v) < Index(v') then Index(w) ≤ Index(w'). Therefore, the subsequence of queries for which the successor rank of the next vertex in their search path has the same value r is already sorted with respect to the index of the next vertex. Furthermore notice that, since each node has an outdegree of at most k, there are at most k=O(1) such subsequences. The idea for creating, in time O(k log N)=O(log N), the new ordering of the queries is therefore to extract these k ordered subsequences and merge them in k bitonic merge steps. The details are shown in Figure 8: for each of the k possible successor ranks, the respective subsequence of queries is extracted (Steps 4 and 5), inverted (Step 9), appended to the sequence of queries already ordered (Steps 10 and 11), and finally the so created bitonic sequence is converted into a sorted sequence (Step 12).

```
Procedure OrderQueriesByNextVertex:
(1)    Initialize all shift registers.
(2)    Every PE(i): Shift(i):=0
(3)    FOR r:=1 TO k DO
(4)        Every PE(i) with G(i)=r:  q'(i):=q(i),  N'(i):=N(i)
(5)        Concentrate([q'(i),  N'(i)],  N'(i)=r)
(6)        Number(LS(i),  N(i)=r)
(7)        ls := LS(0)
(8)        shift := Shift(0)
(9)        Reverse([q'(i),  N'(i)],0,ls)
(10)       Route([q'(i),  N'(i)],  i+Shift(i),  N'(i)=r)
(11)       Every PE(i) with N'(i)=r:  q"(i):=q'(i),  N"(i):=N'(i)
(12)       BitonicMerge([q"(i),  N"(i)],N"(i),0,shift,shift+ls)
(13)       Every PE(i): Shift(i):=Shift(i)+LS(i)
(14)   Every PE(i):  q(i):=q"(i),  N(i):=N"(i)
```

Figure 8. Details of Procedure OrderQueriesByNextVertex.

3.4 SUMMARY

We obtain the following

Theorem 1. *The m-way search problem for an ordered h-level graph with n nodes and s sources can be solved on a hypercube multiprocessor of size N, N=max{n,m}, in time O(min{s log N, log^{2}N} + h log N).*

The algorithm presented in Sections 3.1 to 3.3 has the additional property that it consists of h phases such that at the end of phase x (1≤x≤h) all queries are sorted with respect to the index of the xth node in their search path, and each processor PE(i) contains in its register v'(i) a copy of the xth node in the search path of the query currently stored in its register q(i).

4 A HYPERCUBE IMPLEMENTATION OF A SEGMENT TREE FOR NEXT ELEMENT SEARCH

We will now apply the results obtained in Section 3 and present an efficient parallel implementation, for the hypercube multiprocessor, of a well known data structure: the segment tree [BW]. The segment tree is a widely used structure which has for example been used to obtain efficient implementations of plane sweep algorithms in computational geometry [BW], [M], [PS]. Here, we consider an application of the segment tree to the next element search problem.

In the following, we will first review the definition of the next element search problem as well as the definition and some basic properties of segment trees. We will then show how to implement a segement tree on a hypercube multiprocessor, using the parallel m-way search algorithm for ordered h-level graphs, and obtain an efficient solution for the next element search problem.

The *next element search* problem is a well known problem in computational geometry. Given a set S of n non intersecting line segments $l_1,...,l_n$ and a direction D_{next} (without loss of generality we will assume that D_{next} is the direction of the positive Y-axis), the next element search problem consists of finding for each point p_i of a set of m query points $p_1,...,p_m$ the line segment l_j first intersected by the ray starting at p_i in direction D_{next} (m=O(n)), as illustrated in Figure 9.

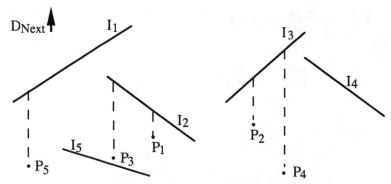

Figure 9. The Next Element Search Problem.

An obvious method for solving the next element search problem is to apply a plane sweep in direction D_{next} using a *segment tree* [BW], [M], [PS].

Let $l_i^{(x)}$ [$p_i^{(x)}$] be the projection of line segment l_i [point p_i, respectively] onto the x-axis, and let $(x_1,x_2,...,x_{2n})$ be the sorted sequence of the projections of the 2n endpoints of $l_1,...,l_n$ onto the x-axis. The segment tree $T(S) = (V_S,E_S)$ for S is the complete binary tree with leaves $x_1,...,x_{2n}$. For every node v of $T(S)$, an interval xrange(v) is defined as follows:

- if v is a leaf x_i, then xrange(v) = $[x_i,x_{i+1})$. ($[x_{2n},x_{2n+1}]=[x_{2n},x_{2n}]$)
- if v is an internal node, then xrange(v) is the union of all intervals xrange(v') such that v' is a leaf of the subtree of $T(S)$ rooted at v.

With every node v of a segment tree $T(S)$ there is associated a node list NL(v) $\subseteq$ S which is defined as follows:

NL(v) = { $l \in$ S | xrange(v) $\subseteq l^{(x)}$ and not (xrange(father of v) $\subseteq l^{(x)}$)}.

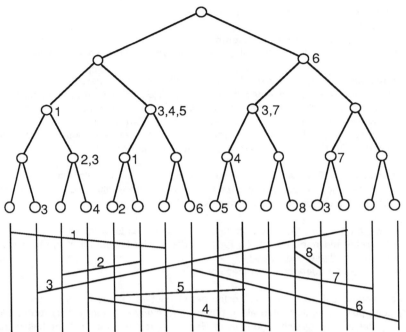

Figure 10. A Segment Tree (The Numbers Associated With The Nodes Represent the Node Lists).

A segement tree T(S) is an ordered h-level graph where h is the height of T(S). For any node v, Level(v) is the height of v in T(S), Levelindex(v) is the rank of v in {v'|Level(v')=Level(v)} with respect to the ordering of these nodes by increasing x-coordinate of xrange(v'), and Index(v) is defined by the above as described in Section 2.1.

For every query point p, we define path(p) to be the path in T(S) from the root to the leaf v such that $p^{(x)} \in$ xrange(v). In order to solve the next element search problem, each query point p is routed along path(p). At every node v on the path, the next element of p in NL(v) is determined (this process will be referred to as *locating* p in NL(v)). For each query point, the final result to be reported is obtained by maintaining, throughout the process, the closest next element found so far. The routing of all query points along their paths can be implemented in time $O(\log^2 N)$ using the m-way search algorithm of Section 3. What remains to be described is how to build the node lists NL(v) and how to locate the query points in the node lists on their paths.

Note that each line segment can occur in O(log n) node lists and, thus, the sum of the lengths of all node lists is O(n log n) [M]. Hence, storing the segment tree with all its node lists in a hypercube multiprocessor would require O(n log n) processors.

Fortunately, the sum of the lengths of the node lists of all nodes with the same level (height) is O(n). Hence, the strategy for solving the next element search problem will be to first construct the segment tree without its node lists (which is trivial) and, while the query points are routed through the tree only the node lists at the level currently reached by the query points are constructed.

For a segment l ∈ S with $l^{(x)}$ = [a,b] we define *l-path(l)* to be the path from the root of T(S) to the leaf v of T(S) with a∈ xrange(v). Likewise we define *r-path(l)* to be the path from the root of T(S) to the leaf v of T(S) with b∈ xrange(v). We observe that, if a line segement l is contained in a node list NL(v), then exactly one of the following four cases applies:

(1) v ∈ l-path(l)

(2) v is the right child of a node v' $\in$ l-path(l)
(3) v $\in$ r-path(l)
(4) v is the left child of a node v' $\in$ r-path(l)
We define $NL_r(v)$, $r \in \{1,2,3,4\}$, be the set of all l $\in$ NL(v) for which case r applies.

The algorithm for solving the next element search problem consists of four parts. In Part r, $1 \leq r \leq 4$, all query points are routed along their paths in T(S). When they arrive at a level i, $1 \leq i \leq h$, the node list $NL_r(v)$ of all nodes v at level i are created and each query point is located in the node list $NL_r(v)$ of the node v to which it was routed. At the end of Part 4, for every query point four partial results (i.e., next elements with respect to partial node lists) have been obtained; the final result can then be computed in O(1) time by comparing these four partial results.

In the remainder, we will show how to execute Parts 1 and 2 of the algorithm; Parts 3 and 4 follow by symmetry. We assume a hypercube of size N=max{n,m} where initially every processor stores one line segment and one query point; w.l.o.g., $m=n=N=2^d$.

We first show <u>Part 1</u> of the algorithm. There are two problems to be solved:
(1) how to create the node list $NL_1(v)$ for all nodes v with Level(v)=i at the time when the query points arrive at these nodes, and
(2) how to locate the query points in these node lists.

The first problem is solved by increasing the set of queries for the m-way search to $Q \cup S$, where every query point $p \in Q$ is again routed along path(p) and every segment $l \in S$ is routed along l-path(l). When applying the m-way search algorithm of Section 3 to the tree T(S) for this increased set of queries, every query point $p \in Q$ reaches a node v on path(p) in the same phase as all line segments $l \in S$ with $l \in NL_1(v)$. Thus, at the end of Phase i, $1 \leq i \leq h$, for every node v with Level(v)=i there exists a block of consecutively numbered processors containing all $p \in Q$ for which v is the ith node in path(p) and all line segments $s \in NL_1(v)$.

What remains to be shown is how to solve the second problem; i.e., how to locate for each level of nodes all queries in the respective node lists. In order to obtain a $O(\log^2 N)$ time complexity for the entire algorithm, this operation must be executed in time O(log N) for each level. We observe that the m-way search algorithm of Section 3 applied to a segment tree T(S) is stable in the following sense: if two queries q_1 and q_2 (either points or line segments) are initially stored in processors $PE(j_1)$ and $PE(j_2)$ with $j_1 < j_2$, and the i_{th} node in path(q_1) is the same as the ith node in path(q_2), then at the end of Phase i the queries q_1 and q_2 are stored in two processors $PE(j_1')$ and $PE(j_2')$ with $j_1' < j_2'$. Therefore, we initially sort all query points by their y-coordinates and all line segments by the y-coordinates of their left endpoints. Then, at the end of each Phase i all query points and line segments, respectively, which were routed to a node v are ordered by y-coordinate. Note that for each such point p and line segment l, $p^{(x)} \subseteq$ xrange(v) and $l^{(x)} \subseteq$ xrange(v), and that no two line segments intersect. Consider the total ordering of all line segments and points routed to v obtained by comparing line segments by the y-coordinate of their left endpoint, points by their y-coordinate, and lines vs. points by the obvious above/below relation. For every query point, its next element in $NL_1(v)$ is the next line segment in this ordering. Furthermore, this total ordering can be obtained by simply merging the already available sorted sequence of points and line segments, respectively, that were routed to node v.

For all nodes v with Level(v)=i, these merges can be executed in parallel by applying one bitonic merge procedure with the levelindex of the respective node as the major and the rank in the respective sequence as the minor key of each point or line segment.

Summarizing, we obtain

Lemma 1. *Part 1 of the next element search algorithm can be executed in time $O(\log^2 N)$ on a hypercube of size N.*

We now turn to <u>Part 2</u> of the algorithm; i.e., locating each query point p in the node lists $NL_2(v)$ of all nodes in path(p). As in Part 1, each line segment I in a node list $NL_2(v)$ still has the property that $xrange(v) \subseteq I^{(x)}$, but in contrast to the former case a total ordering of the line segments in $NL_2(v)$ and points routed to v can not be obtained by using the sorted order of the left (or right) endpoints of the segments.

Let $v[i,j]$ be the vertex v of T(S) with Level(v)=i and Levelindex(v)=j. We introduce h-1 new vertice $v[i,2^{i-1}]$, $1 \leq i \leq h-1$, and define a new tree T'(S) as follows (see Figure 11);
$T'(S)=(V_s', E_s')$ where

- $V_s' = (V_s - \{v[i,0]|1 \leq i \leq h-1\}) \cup \{v[i,2^{i-1}]|1 \leq i \leq h-1\}$, and
- $(v[i,j], v[i',j']) \in E'$ if and only if
$$((i = h, j \bmod 2 = 1, i' = i-1, \text{ and } j' = \frac{j+1}{2})$$
$$\text{or } (1 < i < h, i' = i-1, \text{ and } j' = \lfloor \frac{\lfloor \frac{j+1}{2} \rfloor + 1}{2} \rfloor)).$$

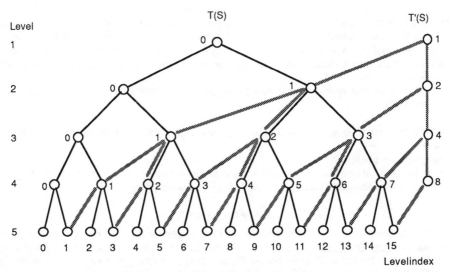

Figure 11. The Trees T(S) and T'(S).

Consider the graph G=(V,E) where $V = V_s \cup \{v[i,2^{i-1}]|1 \leq i \leq h-1\}$ and E is the set of edges of T(S) with their direction reversed together with all edges of T'(S). For a query point $p \in Q$ let path'(p) be the reverse of path(p); for each line segment $I \in S$ let I-path'(I) be the path in T'(S) either from the last node v of I-path(I), if v is a right child in T(S), or otherwise from the right sibling of v in T(S), to the root of T'(S).

From the above definitions it follows that if a segment $I \in S$ is in a node list $NL_2(v)$ then v is a node in I-path'(I).

We also observe that if I-path'(I)=$(w_1, ..., w_h)$ and $I \notin NL_2(w_i)$, then $I \notin NL_2(w_j)$ for all $j \geq i$. Thus, for three nodes w_0, w_1, and w_2 in T'(S) such that $(w_1,w_0) \in E_s'$ and $(w_2,w_0) \in E_s'$ it follows that $NL_2(w_0) \subseteq NL_2(w_1) \cup NL_2(w_2)$. Let $xrange(w_0)=[a,b]$ and consider the ordering of $NL_2(w_0)$ obtained by sorting the line segments by the y-coordinate of their intersection with the line x=a. We observe that this ordering can be constructed from the analogous orderings of $NL_2(w_1)$ and $NL_2(w_2)$ by eliminating from these sequences the elements not contained in $NL_2(w_0)$ and merging the so obtained subsequences.

The idea for Part 2 of the algorithm is to route all query points p along path'(p) and all line segments along I-path'(I). Since G is an ordered h-level graph with O(n) sources, this can be implemented in time $O(\log^2 N)$ using the m-way search procedure of Section 3. It is easy to see that during this search, it is possible to delete some line segments (i.e., eliminate them from further consideration) in any phase of the m-way search algorithm without changing the time complexity. In this particular case, we delete a line segment $l \in S$ if it has been routed to some node v with $l \notin NL_2(v)$.

At the end of Phase i, $1 \leq i \leq h$, for each node w_0 in G with Level(w_0)=h-i+1 there exists a consecutive sequence of processors containing all query points p such that w_0 is the ith node in path'(p) and all line segments $l \in NL_2(w_0)$. In Phase i-1, these line segments have been routed to at most two different nodes w_1 and w_2. If $NL_2(w_1)$ and $NL_2(w_2)$ where previously ordered as described above, then the same ordering for $NL_2(w_0)$ can be obtained by extracting the two subsequences of segments previously routed to $NL_2(w_1)$ and $NL_2(w_2)$, respectively, and merging these subsequences using a bitonic merge. Since only two line segments (could be ordered in constant time) where initially routed to every source of T'(S), the orderings of all lists $NL_2(v)$ can be maintained through all phases with an overhead of O(log N) steps per phase.

Similarly, every query point p is routed along its path'(p), and such that at every level of T(S), the queries routed to each node of that level are ordered with respect to their y-coordinate.

Hence, at the end of each phase i, $1 \leq i \leq h$, we are now in the same situation as in Part 1 of the algorithm; i.e., the line segments and queries which were routed to each node v at level i are ordered such that the next element search problem for each query (with respect to these line segments) can be solved by merging the two ordered sequences (of line segements and query points, respectively) using a bitonic merge procedure.

Lemma 2. *Part 2 of the next element search algorithm can be executed in time $O(log^2 N)$ on a hypercube of size N.*

As already indicated above, Parts 3 and 4 of the algorithm are symmetric to Parts 1 and 2, respectively.

Summarizing, we obtain

Theorem 2. *The next element search problem for a set of n disjoint line segments and m query points can be solved on a hypercube of size N in time $O(log^2 N)$; $N=max\{m,n\}$.*

5 APPLICATIONS

Consider a subdivision of the plane consisting of n edges, and a set of O(n) query points. The (multiple) *planar subdivision search* problem consists of identifying for each query point p the face of the subdivision containing p [DK]. Goodrich [G] presented an O(log n) time solution of this problem for a PRAM with n processors (and O(n log n) memory space); an $O(\sqrt{n})$ time algorithm for the $\sqrt{n} \times \sqrt{n}$ mesh-of-processors is described in [JL].

The (multiple) planar subdivision search problem can obviously be reduced to the next element search problem. Hence, as a consequence of Theorem 2, we obtain

Corollary 1. *The (multiple) planar subdivision search problem for a subdivision with n edges, and O(n) query points, can be solved on a hypercube of size n in time $O(log^2 n)$.*

Theorem 2 also implies an efficient hypercube solution for another fundamental geometric problem: the construction of the *trapezoidal map* [TW].

Given a set S of n disjoint line segments in the plane; for any endpoint p of a segment in S, the trapezoidal segments for p are the (at most two) line segments first intersected by the rays emanating from p in direction of the positive and negative y-axis, respectively. The construction of the trapezoidal map consists of finding for each endpoint of the segments in S its trapezoidal segments.

This problem is fundamental in computational geometry and is frequently used to solve other geometric problems; see e.g. [G], [TW], and [Y]. Atallah, Cole, and Goodrich [ACG], [G] presented an O(log n) time algorithm for computing the trapezoidal decomposition on a PRAM with O(n) processors (and O(n log n) space). As a consequence of Theorem 2, we obtain

Corollary 2. *For a set of n disjoint line segments, the trapezoidal map can be computed on a hypercube with n processors in time $O(log^2 n)$.*

Yap [Y] has shown that on a PRAM with $O(n)$ processors (and $O(n \log n)$ space), the _triangulation_ of a simple polygon (see [TW]) can be computed in time $O(\log n)$ by essentially applying two calls of the trapezoidal map algorithm (of [ACG], [G]). By combining the result in [Y] with Corollary 2, we obtain

Corollary 3. _An n-vertex simple polygon can be triangulated on a hypercube multiprocessor of size n in time $O(\log^2 n)$._

6 CONCLUSION

In this paper, we have presented a general technique for implementing standard data structures on a hypercube multiprocessor.

A paradigm frequently used for the design of efficient PRAM algorithms is to use well established standard data structures in a parallel environment by executing (in parallel) a linear number of independant search processes on these structures. We have shown that this paradigm can also be efficiently applied to hypercube multiprocessors for the class of data structures that can be represented by ordered h-level graphs. This follows from an algorithm presented here that solves the m-way search problem for ordered h-level graphs with n nodes and s sources on a hypercube multiprocessor of size N, $N=\max\{n,m\}$, in time $O(\min\{s \log N, \log^2 N\} + h \log N)$.

We applied this method to the implemention of a segment tree for next element search on a hypercube, and showed that our approach provides an $O(\log^2 n)$ time hypercube algorithm for the next element search problem. As a consequence of this result, we also obtained $O(\log^2 n)$ time solutions to the planar (multiple) point location problem, the trapezoidal map construction problem, and the triangulation problem.

REFERENCES

[ACG]   M.J. Atallah, R. Cole, and M.T. Goodrich, "Cascading divide-and-conquer: a technique for designing parallel algorithms", Technical Report CSD-TR-665, Deparetment of Computer Science, Purdue University, 1987.

[ACGD]   A. Aggarwal, B. Chazelle, L. Guibas, C. O'Dunlaing, and C. Yap, "Parallel computational geometry", Algorithmica 3:3, 1988, pp. 293-327.

[B]   K.E. Batcher, "Sorting networks and their applications", in Proc. AFIPS Spring Joint Computer Conference, 1968, pp. 307-314.

[BW]   J.L. Bentley and D. Wood, "An optimal worst case algorithm for reporting intersections of rectangles", IEEE Transactions on Computers 29:7, 1980, pp. 571-576.

[DK]   N. Dadoun and D.G. Kirkpatrick, "Parallel processing for efficient subdivision search", in Proc. ACM Symp. on Computational Geometry, 1987, pp. 205-214.

[G]   M.T. Goodrich, "Efficient parallel techniques for computational geometry", Ph.D. thesis, Department of Computer Science, Purdue University, 1987.

[JL]   C-S. Jeong and D.T. Lee, "Parallel geometric algorithms on mesh-connected computers", in Proc. Fall Joint Computer Conf., 1987.

[M]   K. Mehlhorn, "Data structures and algorithms 3: multi-dimensional searching and computational geometry", Springer Verlag, 1984.

[NS]   D. Nassimi, S. Sahni, "Data broadcasting in SIMD computers", IEEE Trans. on Computers 30:2, 1981, pp. 101-106.

[PS]   F.P. Preparata and M.I. Shamos, "Computational geometry - an introduction", Springer Verlag, 1985.

[TW]   R.E. Tarjan and C.J. Van Wyk, "An $O(n \log \log n)$ time algorithm for triangulating a simple polygon", SIAM Journal of Computing 17, 1988, 143-178.

[Y]   C.-K. Yap, "Parallel triangulation of a polygon in two calls to the trapezoidal map", Algorithmica 3:2, 1988, pp. 279-288.

k – Nearest – Neighbor Voronoi Diagrams for Sets of Convex Polygons, Line Segments and Points

Thomas Roos *

Abstract

The notion k - th order Voronoi diagram of a finite set of points in the Euclidean plane $\mathbb{E}^2$ is generalized to the k - nearest - neighbor Voronoi diagram of a finite set of convex polygons, line segments and points and they are characterized by some interesting theorems.

Furthermore given n convex polygons with a maximum number of m vertices and a total number of M vertices, we present an algorithm for constructing simultaneously all k - nearest - neighbor Voronoi diagrams, $k \in \{1, \ldots, n-1\}$, that takes $O(n^2(n + m)M)$ time and $O(n^2(n^2 + M))$ space.

We can also apply that algorithm efficiently to a CREW-PRAM with $2 \binom{n}{3}$ processors, where it runs in $O(m(n + m))$ time and $O(n^2(n^2 + M))$ space.

The algorithm is also shown to be applicable under more general convex objects if certain conditions are satisfied.

There are several applications in motion planning, pattern recognition, clustering algorithms, etc.

1 Introduction

One of the most fundamental problems in *computational geometry* is the k - *nearest - neighbor problem*, a variant of the classical nearest - neighbor problem.

The problem is to find, among a set S of n objects in a space E, the k nearest objects to a given test point $q \in E$ with regard to a general distance measure d.

To solve the k - *nearest - neighbor problem* for a finite set of points in the Euclidean plane $\mathbb{E}^2$, Shamos and Hoey [ShHo 75] proposed an approach using Voronoi diagrams. They introduced the k - *nearest - neighbor Voronoi diagram* $V_k(S)$, that subdivides $\mathbb{E}^2$ into maximal regions, so that all points within a given region have the same k nearest neighbors.

---

*Lehrstuhl für Informatik I, Universität Würzburg, Am Hubland, D - 8700 Würzburg, West Germany. This work was supported by the Deutsche Forschungsgemeinschaft (DFG) under contracts (No 88/6 - 1) and (No 88/6 - 2).

The first algorithm for computing $V_k(S)$ was presented by Lee [Le 82]; this method required $O(k^2 n \log n)$ time and $O(k^2 (n-k))$ space. Later Edelsbrunner [Ed 86] reported an other technique to construct the diagram in $O(k (n-k) \sqrt{n} \log n)$ time and optimal $O(k (n-k))$ storage. Shortly afterwards Chazelle and Edelsbrunner [ChEd 87] presented two versions of an algorithm for constructing $V_k(S)$: the first one requires $O(n^2 \log n + k (n-k) \log^2 n)$ time and optimal $O(k (n-k))$ storage, while with additional $O(n^2)$ preprocessing and storage, the other version speeds up the computation to $O(n^2 + k (n-k) \log^2 n)$.[1]

On the other hand Dehne [De 83] evolved an algorithm for constructing all Voronoi diagrams $V_1(S), \ldots, V_{n-1}(S)$ in $O(n^4)$ time and space, which was later improved by an algorithm of Edelsbrunner and Seidel [EdSe 86] that takes only optimal $O(n^3)$ time and storage.

The main result of the present work consists in the generalization of the underlying objects. We present k - nearest - neighbor Voronoi diagrams of convex polygons, line segments and points in the Euclidean plane $\mathbb{E}^2$. After a detailed examination of the bisector of two convex polygons in the second section and the Voronoi diagrams in the third section we develop an algorithm for constructing simultaneously all Voronoi diagrams $V_1(S), \ldots, V_{n-1}(S)$ in $O(n^2 (n+m) M)$ time and $O(n^2 (n^2 + M))$ storage in section four. At last we apply that algorithm efficiently to a CREW-PRAM with $2 \binom{n}{3}$ processors where it takes $O(m(n+m))$ time and $O(n^2(n^2 + M))$ space.

These new Voronoi diagrams are also a generalization of first order Voronoi diagrams of a set of line segments which were introduced by Drysdale and Lee [DrLe 78] and extensively studied in the last years by Fortune [Fo 86] and Yap [Ya 87].

2 The Bisector of two Convex Polygons

Given a finite set of convex polygons

$$S := \{Pol_1, \ldots, Pol_n\}$$

in the Euclidean plane $\mathbb{E}^2$, $|S| = n \geq 3$, under the two following *assumptions* :

A   *The polygons in S are disjoint.*

B   *There is no point in the Euclidean plane with the same distance to four different polygons in S.*

Let every polygon $Pol_i \in S$ be represented by a cyclically ordered sequence $P_{i1}, \ldots, P_{il(i)}$ of its vertices.[2]

---

[1] A comparison between these four algorithms is drawn in [ChEd 87].

[2] $l(i)$ denotes the number of vertices of $Pol_i \in S$. If Pol_i degenerates into a line segment or a point, then $l(i) = 2$ and $l(i) = 1$ respectively.

First of all let's have a look at the *bisector*

$$B(Pol_i, Pol_j) := \{x \in \mathbb{E}^2 \mid d(x, Pol_i) = d(x, Pol_j)\}$$

of two convex polygons $Pol_i, Pol_j \in S$.[3]

Before examining the bisector $B(Pol_i, Pol_j)$, we need some further definitions:

The definite point $p \in Pol_i$ which is closest to a given point $q \in \mathbb{E}^2$ is called the *image* $I(q, Pol_i)$ of q on Pol_i. Furthermore, for any $q \in \mathbb{E}^2 \setminus Pol_i$, $hg(q, Pol_i)$ denotes the *half line* with starting point $I(q, Pol_i)$ and direction $\overrightarrow{I(q, Pol_i)\,q}$.

Now the following lemma shows that the bisector of two convex polygons has no interior points according to the Euclidean topology.[4]

Lemma 2.1 *Let $Pol_i, Pol_j \in S$ be two convex polygons.*
Then for any $q \in B(Pol_i, Pol_j)$ we have

$$B(Pol_i, Pol_j) \cap hg(q, Pol_i) = \{q\}$$

Therefore there is no further point of $B(Pol_i, Pol_j)$ on the half line $hg(q, Pol_i)$.

Now we define for any polygon $Pol_i \in S$ where $l(i) > 1$ the complete disjoint *partition*[5] $Z(Pol_i)$ of the plane $\mathbb{E}^2 \setminus Pol_i$ in $2\,l(i)$ areas. For any $k \in \{1, \ldots, l(i)\}$ let

$$X(r_{ik}^\circ) := \{q \in \mathbb{E}^2 \setminus Pol_i \mid I(q, Pol_i) \in r_{ik}^\circ\}^6$$
$$X(P_{ik}) := \{q \in \mathbb{E}^2 \setminus Pol_i \mid I(q, Pol_i) = P_{ik}\}$$

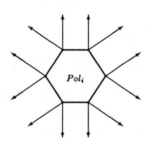

With the help of the two partitions $Z(Pol_i)$ and $Z(Pol_j)$ we construct the *intersectional partition* $SZ(Pol_i, Pol_j)$, which cuts up $\mathbb{E}^2 \setminus (Pol_i \cup Pol_j)$ into disjoint sections. Now we investigate the bisector $B(Pol_i, Pol_j)$ in any nonempty section L of the intersectional partition $SZ(Pol_i, Pol_j)$.

---

[3]d denotes the Euclidean metric; the calculation of $d(q, Pol_i)$ requires $O(l(i))$ time.

[4]This property is lost under the L_1 - metric, as well as with renunciation to assumption (A).

[5]The partition isn't disjoint under starshaped polygons.

[6]$r_{ik} := \overline{P_{ik}P_{ik+1}}$ and $r_{ik}^\circ := r_{ik} \setminus \{P_{ik}, P_{ik+1}\}$.

According to the construction there exists a representation

$$L = X(o_{ik}) \cap X(o_{jl})$$

where $o_{ik} \in \{P_{ik}, r_{ik}^o\}$ and $o_{jl} \in \{P_{jl}, r_{jl}^o\}$ with $k \in \{1, \ldots, l(i)\}$ and $l \in \{1, \ldots, l(j)\}$. Now we have

$$B(Pol_i, Pol_j) \cap L = B(o_{ik}, o_{jl}) \cap L$$

Basically we have to distinguish three cases:

1st Case : $o_{ik} = P_{ik} \wedge o_{jl} = P_{jl}$

$\Rightarrow$ $B(Pol_i, Pol_j) \cap L$ is part of a straight line.

2nd Case : $o_{ik} = r_{ik}^o \wedge o_{jl} = r_{jl}^o$

$\Rightarrow$ $B(Pol_i, Pol_j) \cap L$ is part of a bisector of an angle.

3rd Case : $o_{ik} = r_{ik}^o \wedge o_{jl} = P_{jl}$

$\Rightarrow$ $B(Pol_i, Pol_j) \cap L$ is part of a parabola.[7]

Therefore we have shown that the bisector $B(Pol_i, Pol_j)$ consists of differentiable curve segments. With the help of Lemma 2.1 the proof of the *continuity* of $B(Pol_i, Pol_j)$ can be given. With $T(x)$ denoting the tangent to the bisector $B(Pol_i, Pol_j)$ in $x \in B(Pol_i, Pol_j) \cap L^o$, we are able to show the following *tangent formula* :

$$T(x) \equiv B(I(x, Pol_i), I(x, Pol_j))$$

With the help of this tangent formula finally results the proof of the differentiable transition of the bisector $B(Pol_i, Pol_j)$.[8]

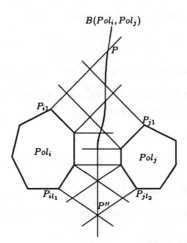

The bisector contains two half rays for the reason of neighboring vertices of different polygons on the convex hull $CH(Pol_i \cup Pol_j)$. Now these half rays serve as a starting-point for the *construction of the bisector* (see fig.).

---

[7]This is easy to show, for the Euclidean metric is invariant with respect to translations and rotations.

[8]The bisector $B(Pol_i, Pol_j)$ isn't differentiable under starshaped polygons.

1. First of all calculate the upper and lower tangent[9] $\overline{P_{i1} P_{j1}}$ and $\overline{P_{il_1} P_{jl_2}}$. Consequently we obtain the *relevant objects*[10] $(P_{i1}, r_{i1}^o, P_{i2}, \ldots, P_{il_1})$ of Pol_i and *relevant half lines* of the partition $Z(Pol_i)$. o_i denotes the *current object* and g_i the *current half line* of $Z(Pol_i)$.[11]

2. Now compute the starting - point P of the upper starting - ray and let $P' := P$.

3. Then calculate gradually the next current point as the "middle"[12] point of the three points P' (the previous point), $B(o_i, o_j) \cap g_i$ and $B(o_i, o_j) \cap g_j$ (on the bisector $B(o_i, o_j)$) with a following update of the current objects.

4. Repeat step (3) until the starting point P'' of the lower half - ray is reached.

Altogether we can prove the following theorem.

Theorem 2.2 *Let $Pol_i, Pol_j \in S$ be two disjoint convex polygons with $l(i)$ and $l(j)$ vertices respectively. Then the bisector $B(Pol_i, Pol_j)$ is a simple differentiable curve, consisting of maximal $l(i) + l(j)$ parts of straight lines and $l(i) + l(j) - 1$ parts of parabolas.[13] $B(Pol_i, Pol_j)$ can be constructed in optimal $O(l(i) + l(j))$ time and storage.*

3 The Generalized Voronoi Diagram

Now let's turn our attention to Voronoi diagrams. For any nonempty subset $A \subset S$

$$v(A) := \{x \in \mathbb{E}^2 \mid \forall_{Pol_i \in A} \forall_{Pol_j \in S \setminus A} \ d(x, Pol_i) \le d(x, Pol_j)\}$$

defines the *set of Voronoi polygons of A*.[14] A is called the *label* of $v(A)$.

The vertices of the Voronoi polygons are called *Voronoi points* and the bisector parts on the boundary are called *Voronoi edges*.

Finally for any $k \in \{1, \ldots, n-1\}$

$$V_k(S) := \{v(A) \neq \emptyset \mid A \subset S \wedge |A| = k\}$$

defines the *Voronoi diagram of order k*.

First of all let's examine the structure and the properties of Voronoi polygons.

---

[9]E.g. with the help of an algorithm presented in [PrHo 77].

[10]I.e. the objects which are relevant for the construction of the bisector.

[11]Analogue expressions for Pol_j.

[12]We can prove that the calculation of the "middle" point requires only calculations of Euclidean distances.

[13]This is a generalization of a result presented by [Fo 86].

[14]In contrast to Euclidean Voronoi diagrams for points, $v(A)$ may here decay into several Voronoi polygons which are generally not convex.

Theorem 3.1 *Let $Pol_i \in S$. Then $v(Pol_i)$ is generalized-starshaped with nucleus Pol_i, i.e. for any point $q \in v(Pol_i)$ there exists a point $p \in Pol_i$[15] so that the line segment $\overline{pq}$ lies completely within $v(Pol_i)$. Therefore $v(Pol_i)$ is connected.*

Voronoi polygons in Voronoi diagrams of order $k \geq 2$ have a description as a finite intersection of generalized-starshaped regions with different nuclei.

Using assumption (B) we can prove the following characterization of Voronoi points.

Theorem 3.2 *Any Voronoi point $v \in V_k(S)$, $k \in \{1, \ldots, n-1\}$, is a point of intersection of bisectors belonging to three different polygons in S.*
If on the other hand $v \in B(Pol_i, Pol_j) \cap B(Pol_j, Pol_l)$, where $Pol_i, Pol_j, Pol_l \in S$ and $H := \{Pol \in S \mid d(v, Pol) < d(v, Pol_i)\}$, as well as $k := |H| \leq n-3$, then v is Voronoi point in $V_{k+1}(S)$ and $V_{k+2}(S)$.

By the way, any Voronoi point has exactly the degree three. Also the tangent formula implies that the participant bisectors have a proper intersection in a Voronoi point.

The following figure shows the Voronoi point v with its three adjacent labels[16] in the Voronoi diagram

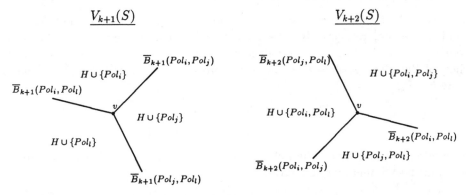

The next theorem provides a quantitative investigation of the Voronoi points.

Theorem 3.3 *Given three convex polygons $Pol_i, Pol_j, Pol_l \in S$, the participant bisectors have at most two points of intersection.*

We can prove this theorem by a suitable partition of the Euclidean plane.

---

[15]Select $p := I(q, Pol_i)$.

[16]For simplification we draw the tangents as a local approximation to the bisectors.

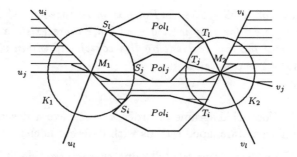

Let $S_x := I(M_1, Pol_x)$ and $T_x := I(M_2, Pol_x)$ be the images of two Voronoi points M_1 and M_2 on the three convex polygons and $u_x := hg(M_1, Pol_x)$ and $v_x := hg(M_2, Pol_x)$ the corresponding half lines, $x \in \{i, j, l\}$. With the help of the tangent formula we can prove that in a neighborhood of M_1 and M_2 the bisector $B(Pol_i, Pol_j)$ lies within the shaded region.[17] Using Lemma 2.1 we have

$$u_i \cap B(Pol_i, Pol_j) = \{M_1\} \quad \wedge \quad u_j \cap B(Pol_i, Pol_j) = \{M_1\}$$

$$v_i \cap B(Pol_i, Pol_j) = \{M_2\} \quad \wedge \quad v_j \cap B(Pol_i, Pol_j) = \{M_2\}$$

Finally assumption (A) and the continuity of the bisector $B(Pol_i, Pol_j)$ imply that $B(Pol_i, Pol_j)$ cannot leave the shaded region. Applying this result to the bisector $B(Pol_j, Pol_l)$ we obtain that there cannot exist further points of intersection.

Now Theorem 3.2 implies that there are at most $2\binom{n}{3}$ Voronoi points which appear in two Voronoi diagrams of successive orders.

Finally we look at some quantitative properties of Voronoi diagrams. Let N_k, I_k, E_k, O_k and Z_k denote, respectively, the number of Voronoi polygons, - points, - edges, unbounded Voronoi polygons and connected components of the Voronoi edges. With the help of *Euler's polyhedron formula* we can prove at first the following two equations. For any $k \in \{1, \dots, n-1\}$

$$E_k = 3N_k - O_k - Z_k - 2$$
$$I_k = 2N_k - O_k - Z_k - 1$$

Furthermore, we can present the following *estimation* for the number of Voronoi polygons N_k in the Voronoi diagram $V_k(S)$:

$$N_1 = n$$

and for $k \in \{2, \dots, n-1\}$:

$$N_k \leq (2k-1)n - (k^2 - k) - \sum_{i=1}^{k-1} (O_i + Z_i)$$

so $N_k \in O(kn)$.

Some further interesting estimations result from that inequation.

---

[17]In the figure the tangents to the bisector are displayed.

4 An Algorithm for Constructing Simultaneously all Voronoi Diagrams $V_1(S), \ldots, V_{n-1}(S)$

In this section we present an algorithm for computing simultaneously all Voronoi diagrams $V_k(S)$, $k \in \{1, \ldots, n-1\}$:

1. For any $i, j \in \{1, \ldots, n\}$ with $i < j$: Construct the bisector $B(Pol_i, Pol_j)$.

2. For any $i, j, l \in \{1, \ldots, n\}$ with $i < j < l$:
   Calculate the possible points of intersection $M_1(i, j, l)$ and $M_2(i, j, l)$ of the bisectors $B(Pol_i, Pol_j)$ and $B(Pol_j, Pol_l)$, as well as their cardinality $A(i, j, l) \in \{0, 1, 2\}$ and the six adjacent parts of bisectors (compare Theorem 3.2).

3. For any $i, j, l \in \{1, \ldots, n\}$ with $i < j < l$ and any $x \in \{1, \ldots, A(i, j, l)\}$: Calculate

$$H_x(i, j, l) := \{y \in \{1, \ldots, n\} \mid d(M_x(i, j, l), Pol_y) < d(M_x(i, j, l), Pol_i)\}$$

   and its cardinality $|H_x(i, j, l)|$.

4. For any $i, j \in \{1, \ldots, n\}$ with $i < j$:
   Calculate the flag

$$F(i, j) := \sum_{l \neq i \wedge l \neq j} |A(\{i, j, l\}^{18})|$$

   If $F(i, j) = 0$
   Calculate $k := |\{y \in \{1, \ldots, n\} \mid d(M, Pol_y) < d(M, Pol_i)\}|$ for some point $M \in B(Pol_i, Pol_j)$ and add the entire bisector $B(Pol_i, Pol_j)$ to the Voronoi diagram $V_{k+1}(S)$.
   else
   For any $l \in \{j + 1, \ldots, n\}$, any $x \in \{1, \ldots, A(i, j, l)\}$ and any of the six adjacent parts of bisectors in $M_x(i, j, l)$ perform the following steps :
   Being $B \subset B(Pol_a, Pol_b)$ with $a, b \in \{i, j, l\}$ such a part of a bisector in $V_k(S)$ and $c := \{i, j, l\} \setminus \{a, b\}$.
   Traverse all existing Voronoi points $M_{x'}(\{a, b, c'\})$ with $(x', c') \neq (x, c)$, $c' \neq a$ and $c' \neq b$ and test whether $M_{x'}(\{a, b, c'\})$ is a Voronoi point in $V_k(S)$ and lies on B.
   If there are several Voronoi points of such a kind, select that Voronoi point $M_{x_0}(\{a, b, c_0\})$ with minimum distance (along B) to $M_x(i, j, l)$.
   Now reduce B to a part of a bisector between $M_x(i, j, l)$ and $M_{x_0}(\{a, b, c_0\})$ and the corresponding part in $M_{x_0}(\{a, b, c_0\})$ too.

We perform the cutting of the bisector in step (4) only if the concerning part of the bisector hasn't been cut yet.

---

[18] $\{i, j, l\}$ denotes here the ascending ordered sequence of the indexes i, j and l.

The two *principal problems* of Voronoi diagrams of sets of convex polygons, line segments and points are on the one hand the possibly missing connection of the Voronoi edges and on the other hand the isolated occurrence (that means without Voronoi points) of some entire bisectors in a Voronoi diagram.

The *correctness* of the steps (1) to (3) follows from Theorem 2.2 for bisectors and the characterization Theorem 3.2 for Voronoi points and the resulting local properties. The assignment of a point of intersection to two Voronoi diagrams of successive orders follows here by Theorem 3.2 . In step (4) we distinguish whether there exists a point of intersection of $B(Pol_i, Pol_j)$ with another bisector of the kind $B(Pol_j, Pol_l)$ with $l \neq i$ and $l \neq j$ and the flag $F(i, j)$ is set respectively. Afterwards, in the respective case, we add the entire bisector to the corresponding Voronoi diagram or do a cutting of the bisectors. The correctness of this step can also be shown easily.

Now we turn to the analysis of the *running time* and the *storage requirements*. Let

$$m := \max_{i \in \{1,\dots,n\}} \{l(i)\}$$

denote the *maximum number of vertices* of a single polygon and

$$M := \sum_{i=1}^{n} l(i)$$

the *total number of vertices* of all polygons in S.

The following diagram shows the requirements for running time and storage of the particular steps of the presented algorithm on a RAM and a CREW-PRAM with $2 \binom{n}{3}$ processors:

Step	Time(RAM)	Time(PRAM)	Storage
(1)	$O(nM)$	$O(m)$	$O(nM)$
(2)	$O(n^2 mM)$	$O(m^2)$	$O(n^2 M)$
(3)	$O(n^3 M)$	$O(M)$	$O(n^4)$
(4)	$O(n^3 M)$	$O(nm)$	$O(n^2)$

Summarizing our results we obtain the following theorem.

Theorem 4.1 *For a set S of n convex polygons, line segments and points the presented algorithm constructs simultaneously all Voronoi diagrams $V_1(S), \ldots, V_{n-1}(S)$ in $O(n^2 (n + m) M)$ time and $O(n^2 (n^2 + M))$ storage.*

On a CREW-PRAM with $2 \binom{n}{3}$ processors[19] we obtain the parallel, simultaneous calculation of all Voronoi diagrams $V_1(S), \ldots, V_{n-1}(S)$ in $O(m(n + m))$ time and $O(n^2 (n^2 + M))$ storage.

Finally we obtain the following almost optimal *efficiency*

$$E = \frac{C_1 n^2 (n + m) M}{C_2 \binom{n}{3} m (n + m)} = C \frac{M}{n m}$$

or in other words the *"speed – up"* is nearly as high as the number of processors.[20]

5   Concluding Remarks and Open Problems

We presented in our work an algorithm for computing simultaneously all Voronoi diagrams $V_1(S), \ldots, V_{n-1}(S)$. This algorithm is not only easy to implement but also parallelizable with a nearly optimal efficiency.

Furthermore, we can apply that algorithm to any convex objects in the Euclidean plane under the assumptions that the bisector of two objects as well as the intersection of two bisectors are calculable.

Today the Voronoi diagram belongs to the most powerful data structures in computational geometry. With the help of our algorithm we can present an efficient solution to many problems as the classical Voronoi diagram did before (see e.g. [ShHo 75] or [PrSh 85]).

Voronoi diagrams as they appear today are still a wide field of research with many unsolved problems. So far, there is no algorithm for the separate computation of the order k Voronoi diagram of a set of line segments. An extension of our algorithm to other metrics (e.g. the L_1 – metric) is also desirable.

Acknowledgement

The author wishes to thank Professor Hartmut Noltemeier and Hugo Heusinger for their helpful comments.

---

[19]Here we demand $n \geq 4$.
[20]C_1, C_2 and C are constants depending on the underlying machines.

References

[ChEd 87]   B. Chazelle and H. Edelsbrunner, *An Improved Algorithm for Constructing k th - Order Voronoi Diagrams*, IEEE Transactions on Computers, Nov. 1987, Vol. C-36, No. 11, pp 1349 – 1354

[De 83]   F. Dehne, *An $O(n^4)$ Algorithm to Construct all Voronoi Diagrams for k Nearest Neighbor Searching*, Proc. 10 th Colloquium on Automata, Languages and Programming, 1983

[DrLe 78]   R.L. Drysdale III and D.T. Lee, *Generalized Voronoi Diagrams in the Plane*, Proc. 16 th Annual Allerton Conference on Communications, Control and Computing, Oct. 1978, pp 833 – 842

[Ed 87]   H. Edelsbrunner, *Algorithms in Combinatorical Geometry*, EATCS Monographs in Computer Science, Springer - Verlag, Berlin - Heidelberg, 1987

[Ed 86]   H. Edelsbrunner, *Edge-Skeletons in Arrangements with Applications*, Algorithmica 1986, Vol. 1, pp 93 – 109

[EdSe 86]   H. Edelsbrunner and R. Seidel, *Voronoi Diagrams and Arrangements*, Proc. 2 nd Annual ACM Symp. Computational Geometry, Yorktown Heights, 1986

[Fo 86]   S. Fortune, *A Sweepline Algorithm for Voronoi Diagrams*, Proc. 2 nd Annual ACM Symp. Computational Geometry, Yorktown Heights, 1986, pp 313 – 322

[Le 82]   D.T. Lee, *On k - Nearest Neighbor Voronoi Diagrams in the Plane*, IEEE Transactions on Computers, June 1982, Vol. C–31, No. 6, pp 478 – 487

[LeDr 81]   D.T. Lee and R.L. Drysdale III, *Generalisation of Voronoi Diagrams in the Plane*, SIAM J. Comput., Feb. 1981, Vol. 10, No. 1, pp 73 – 87

[PrHo 77]   F.P. Preparata and S.J. Hong, *Convex Hulls of Finite Sets of Points in Two and Three Dimensions*, Comm. ACM, Feb. 1977, Vol. 20, No. 2, pp 87 – 93

[PrSh 85]   F.P. Preparata and M.I. Shamos, *Computational Geometry – An Introduction*, Springer - Verlag, New York, 1985

[ShHo 75]   M.I. Shamos and D. Hoey, *Closest - Point Problems*, Proc. 16 th Annual Symp. on FOCS, 1975, pp 151 – 162

[Ya 87]   C.K. Yap, *An $O(n \log n)$ Algorithm for the Voronoi Diagram of a Set of Simple Curve Segments*, Discrete & Computational Geometry, 1987, Vol. 2, pp 365 – 393

Finding Squares and Rectangles
in Sets of Points

Marc J. van Kreveld   Mark T. de Berg*

Dept. of Computer Science, University of Utrecht,
P.O.Box 80.089, 3508 TB Utrecht, the Netherlands.

Abstract

The following problem is studied: Given a set S of n points in the plane, does it contain a subset of four points that form the vertices of a square or rectangle. Both the axis-parallel case and the arbitrarily oriented case are studied. We also investigate extensions to the d-dimensional case. Algorithms are obtained that run in $O(n^{1+1/d} \log n)$ time for axis-parallel squares and $O(n^{2-1/d})$ time for axis-parallel rectangles. For arbitrarily oriented squares the time bounds are $O(n^2 \log n)$, $O(n^3)$ and $O(n^{d-1/2}\beta(n))$ for $d = 2$, $d = 3$ and $d \geq 4$, respectively (where $\beta(n)$ is related to the inverse of Ackermann's function), whereas the algorithm for arbitrarily oriented rectangles takes time $O(n^d \log n)$. Furthermore, it is shown that recognizing axis-parallel rectangles is equivalent to recognizing a $K_{2,2}$-subgraph in a bipartite graph, resulting in a $O(|E|\sqrt{|E|})$ time and $O(|V| + |E|)$ space solution to this problem. Also, combinatorial results on the maximal number of squares and rectangles any point set can contain are given.

1   Introduction

In this paper we study the problem of determining whether a set of points contains some special configuration of points. This problem in pattern recognition is useful, e.g., when such a 'degeneracy' in a set of points needs special treatment by an algorithm. In particular we will search for vertices of squares or rectangles. The following problems are considered.

Problem 1 *Given a set S of points in the plane, determine whether S contains a subset of four points that form the vertices of an axis-parallel square.*

Problem 2 *Given a set S of points in the plane, determine whether S contains a subset of four points that form the vertices of an axis-parallel rectangle.*

A trivial method, exploiting the fact that two opposite vertices define the square or rectangle, yields an $O(n^2 \log n)$ time bound for the problems. We present algorithms that solve the problems in time $O(n\sqrt{n} \log n)$ and $O(n\sqrt{n})$, respectively. We will show that the rectangle problem is equivalent to the problem of determining whether a bipartite graph contains a $K_{2,2}$ subgraph, and obtain an $O(|E|\sqrt{|E|})$ time solution.

*The work of this author is supported by the Dutch Organisation for Scientific Research (N.W.O.).

The results extend to d-dimensional space, and we obtain an $O(n^{1+1/d} \log n)$ time algorithm for the square problem, and an $O(n^{2-1/d})$ time algorithm for the rectangle problem. Note that the square problem becomes easier to solve, while the rectangle problem becomes harder to solve for higher dimensions. Both algorithms can be adapted to solve the 'report all squares/rectangles' problem. Then the output size is added as a linear term in the time bound.

The combinatorial side of the problems is studied as well: What is the maximal number of subsets of a set of n points that are the vertices of a square or rectangle. We prove tight bounds of $\Theta(n^{1+1/d})$ and $\Theta(n^2)$, respectively.

We also study the following related problems.

Problem 3 *Given a set S of points in the plane, determine whether S contains a subset of four points that form the vertices of an arbitrarily oriented square.*

Problem 4 *Given a set S of points in the plane, determine whether S contains a subset of four points that form the vertices of an arbitrarily oriented rectangle.*

Note that in these problems the restiction that the object is axis-parallel has been dropped.

In the d-dimensional case we find rectangles in $O(n^d \log n)$ time, whereas the square problem can be solved in time $O(n^2 \log n)$ $(d = 2)$, $O(n^3)$ $(d = 3)$, or $O(n^{d-1/2} \beta(n))$ $(d \geq 4)$, where $\beta(n)$ is related to the extremely slowly growing inverse of Ackermann's function. Again, our methods can easily be adapted to report all squares and rectangles. Also, we show that the maximal number of squares in the plane is $\Theta(n^2)$. Tight bounds on the maximal number of rectangles in the plane are not known: we give an $\Omega(n^2 \log n)$ lower bound, whereas we prove an upper bound of $O(n^2 \sqrt{n})$.

This paper is organized as follows.

In section 2, the axis-parallel square problem is considered. We obtain combinatorial results on the maximal number of squares any planar set of points can contain, and we give an algorithm for determining whether a planar set of points contains the vertices of a square. Both results are extended to higher dimensions.

In section 3, the axis-parallel rectangle problem is considered, and all problems of section 2 are now studied for rectangles instead of squares. Additionally, we show equivalence of the rectangle determination problem to finding a $K_{2,2}$ subgraph in a bipartite graph.

Section 4 deals with arbitrarily oriented squares and rectangles in sets of points. We give combinatorial results in the planar case, and algorithms in the arbitrarily dimensional case.

In section 5 we conclude the paper by mentioning some directions for further research.

2 Axis-parallel squares

When we restrict ourselves to axis-parallel shapes, we can make use of the fact that these shapes contain points with a number of coordinates equal-valued. We use this fact in this section and the next for recognizing these shapes. In this section we will study the problem of recognizing squares in the plane and more-dimensional space. We begin with a combinatorial result.

Theorem 1 *For a set S of n points in d-dimensional space, the maximal number of subsets of 2^d points that form the vertices of an axis-parallel hypercube is $\Theta(n^{1+1/d})$.*

Proof: We consider the planar case. To prove the lower bound, one can place the points of S on a square grid G.

To prove the upper bound, partition S in a number of subsets $S_1, \ldots, S_k$, such that two points p and q are in the same subset if and only if they have equal first coordinate. If some subset S_i contains no more than $\sqrt{n}$ points, then every point of S_i can participate in at most $\sqrt{n} - 1$ squares. Consequently, there are at most $n\sqrt{n}$ squares in total with at least one vertex in any subset of size at most $\sqrt{n}$. Next, consider the subsets with more than $\sqrt{n}$ points. There are less than $\sqrt{n}$ such subsets. If the points in these subsets are partitioned in new subsets of points with equal second coordinate, then these new subsets contain less than $\sqrt{n}$ points, and hence these subsets contain less than $n\sqrt{n}$ squares. The d-dimensional case follows in a similar way. $\square$

Let S be a planar set of n points. The main idea to solve the two-dimensional square problem is to partition S in a number of subsets, such that all points with equal first coordinate are in the same subset. Each pair of points in such a subset determines two possible squares, such that the pair forms the vertices of one vertical edge of the square, and it remains to determine if S contains the two vertices of an opposite edge, either to the left or to the right of the first edge. One could simply search for these four points. Unfortunately, there may be as many as $\Omega(n)$ points with equal first coordinate, hence, there can be $\Omega(n^2)$ pairs in one subset, resulting in a $O(n^2 \log n)$ time algorithm ($O(\log n)$ time for searching). To improve upon this we will show that we can avoid having to test all pairs in large subsets by observing that we could have made subsets of points with equal second coordinate equally well.

Lemma 1 *Given a sets S of n (distinct) points in the plane, then there exists an axis-parallel line that contains at least one and at most $\sqrt{n}$ points of S.*

Proof: Partition S in non-empty subsets $S_1, \ldots, S_k$ of points with equal first coordinate. If any subset contains at most $\sqrt{n}$ points, then we are done, for we can take a vertical line containing this subset. If all subsets contain more than $\sqrt{n}$ points, then obviously there are at most $\sqrt{n}$ subsets. Since two points in the same subset cannot have equal second coordinate (the points are distinct), every horizontal line will contain at most one point of each subset. Thus in this case, any horizontal line containing a point of S will do. $\square$

According to the proof of the above lemma, we can partition the set S of points in a number of subsets of all points of S with equal first coordinate, and a number of subsets of the remaining points of S with equal second coordinate, such that no subset contains more than $\sqrt{n}$ points of S.

Square (S)

1. Build a search tree T on the set S of points, using lexicographic order on the points.

2. Partition S into subsets $S_1^{(1)}, \ldots, S_{k_1}^{(1)}, S_1^{(2)}, \ldots, S_{k_2}^{(2)}$ in the following way. Two points of S are in the same subset $S_i^{(1)}$ ($1 \leq i \leq k_1$) if and only if they have equal first coordinate, and $S_i^{(1)}$ does not contain more than $\sqrt{n}$ points. Two points of S are in the same subset $S_j^{(2)}$ ($1 \leq j \leq k_2$) if and only if they have equal second coordinate, and neither of them is in some subset $S_i^{(1)}$.

3. For every subset $S_i^{(j)}$ $(1 \leq j \leq 2, 1 \leq i \leq k_j)$, and for every pair p, q in $S_i^{(j)}$, search in T whether the other two vertices of any of the two squares defined by p and q are also in S. If so, answer yes, otherwise, answer no.

This leads to the following result.

Theorem 2 *Given a set S of n points in the plane, it can be decided in time $O(n\sqrt{n}\log n)$ and $O(n)$ space whether S contains a subset of four points that form the vertices of an axis-parallel square.*

Proof: The correctness of the algorithm given above is clear, since any square will have at least one pair of vertices of an edge in one subset, and this pair will certainly be tested.

Step 1 and 2 of the algorithm clearly take $O(n \log n)$ time.

According to lemma 1, the partitioning of S in subsets as given by the algorithm results in subsets with at most $\sqrt{n}$ points. Thus the total time spent by step 3 is

$$O(\sum_{i=1}^{k_1} |S_i^{(1)}|^2 \cdot \log n + \sum_{i=1}^{k_2} |S_i^{(2)}|^2 \cdot \log n) = O(\sqrt{n} \log n (\sum_{i=1}^{k_1} |S_i^{(1)}| + \sum_{i=1}^{k_2} |S_i^{(2)}|)) =$$

$$O(n\sqrt{n}\log n).$$

$\square$

Remark: In fact, we can solve the above problem in time $O(n\sqrt{n \log n})$ in the following way. Partition S on equal first coordinate, resulting in small subsets with at most $\sqrt{n/\log n}$ points, and large subsets with more points. Treat the small subsets as above, and test every two large subsets by a simultaneous walk at two places in both (ordered) large subsets. Unfortunately this approach does not generalize to higher dimensions.

Next we consider the more-dimensional case of the problem: Given a set S of n points in d-dimensional space, does S contain a subset of 2^d points that are the vertices of an axis-parallel (hyper-) cube. Note that two points with all but the last coordinate equal-valued determine 2^{d-1} possible cubes, and every candidate cube thus found can be tested by searching in S for the $2^d - 2$ points that would be the other vertices of the cube.

Lemma 2 *Given a set S of n (distinct) points in d-dimensional space, then there exists an axis-parallel line that contains at least one and at most $n^{1/d}$ points of S.*

Proof: With induction on d. $\square$

The following algorithm solves the cube problem by partitioning the set of points S in subsets of points with all but one coordinate equal, and which contain at most $n^{1/d}$ points.

Cube (S, d)

1. Build a search tree T on the set S of points, using lexicographic order on the points.

2. Partition S into subsets $S_1^{(1)}, \ldots, S_{k_1}^{(1)}, S_1^{(2)}, \ldots, S_{k_2}^{(2)}, \ldots \ldots, S_1^{(d)}, \ldots, S_{k_d}^{(d)}$ in the following way. Two points of S are in the same subset $S_i^{(j)}$ $(1 \leq j \leq d, 1 \leq i \leq k_j)$ if and only if they have all coordinates but the jth equal, $S_i^{(j)}$ does not contain more than $n^{1/d}$ points, and neither of the points is in a subset $S_{i'}^{(j')}$ with $j' < j$.

3. For every subset $S_i^{(j)}$, and for every pair p, q in $S_i^{(j)}$, search in T whether the other $2^d - 2$ vertices of one of the 2^{d-1} cubes defined by p and q are also in S. If so, answer yes, otherwise, answer no.

This leads to the following result.

Theorem 3 *Given a set S of n points in d-dimensional space, it can be decided in time $O(n^{1+1/d} \log n)$ and $O(n)$ space whether S contains a subset of 2^d points that form the vertices of an axis-parallel cube.*

Proof: Follows in the same way as theorem 2. $\square$

3 Axis-parallel rectangles

In this section we first give upper and lower bounds on the maximal number of axis-parallel rectangles which all have their vertices in a set of n points. Then we study the problem of determining whether a planar set of points contains four points that are the vertices of an axis-parallel rectangle. Finally we consider the more-dimensional case.

Theorem 4 *For a set S of n points in d-dimensional space, the maximal number of subsets of 2^d points that form the vertices of an axis-parallel hyperrectangle is $\Theta(n^2)$.*

Proof: The lower bound is achieved by placing the points on an $n^{\frac{1}{d}} \times \cdots \times n^{\frac{1}{d}}$ grid. For the upper bound, we observe that any axis-parallel rectangle is determined uniquely by two opposite points that have no coordinates equal. Also, for every rectangle there must be such a pair. The bound follows, since there are $n(n-1)/2$ pairs of points. $\square$

The proof of the above theorem leads to a straightforward $O(n^2 \log n)$ time algorithm for arbitrary dimensional space. Just take any pair of points of which all coordinates are different, and test if the other points are also in the set. We next give a subquadratic time solution for the problem.

Let S be a planar set of n points. Like for squares, the idea is to partition S in a number of subsets of points with equal-valued first coordinate. Notice that S contains the vertices of an axis-parallel rectangle if and only if there are two different subsets that each contain two points, such that the two points in one subset have second coordinates equal to the second coordinates of the two points in the other subset. We distinguish two types of subsets: the subsets with at most $\sqrt{n}$ points are called *small*, and the subsets with more than $\sqrt{n}$ points are called *large*. Now we can distinguish three cases. Firstly, two small subsets can contain the vertices of a rectangle. Secondly, a small subset and a large subset can each contain two of the vertices of a rectangle, and thirdly, two large subsets can contain the vertices of a rectangle. Again, using lemma 1, we can show that the last case can be avoided.

To solve the first case we need a pointer structure, which is described next. Let $S_1, \ldots, S_k$ be the small subsets, and let S' be their union.

- Let Y be the list containing all different second coordinates of points in S', ordered by increasing value. With each element in Y (value of the second coordinate), an integer *count* is stored, initially zero, and a reference *first* that points to a list of all points with this second coordinate, sorted by first coordinate.

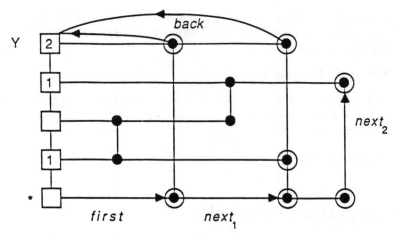

Figure 1: Pointer structure for testing the small subsets.

- Each point p in S' corresponds to a node in the structure with three references, $next_1$, $next_2$ and $back$. $next_1$ leads to the node corresponding to the first point one reaches if one moves in the direction of greater first coordinate (thus a point with equal second coordinate). More formally, the $next_1$ reference of a point $p = (p_1, p_2)$ refers to a point $q = (q_1, q_2)$ if and only if $p_2 = q_2$, $p_1 < q_1$ and for all $r = (r_1, r_2)$, $r_2 = p_2$ implies that $r_1 \leq p_1$ or $r_1 \geq q_1$. Analogously, the reference $next_2$ refers to the point with equal first coordinate and larger second coordinate. The reference $back$ of a point $p = (p_1, p_2)$ leads to the element corresponding to p_2 in the list Y. The reference $first$ in the list Y corresponding to some value y of the second coordinate, refers to the point with y as second coordinate, and smallest first coordinate among these points (see figure 1).

Solving the rectangle problem for the small subsets requires building the structure and traversing it in a special way, as given by the algorithm below.

To solve the second case, we count for every small subset S_i and every large subset L_j the number of times a point in S_i has its second coordinate equal to a point in L_j. If this number is more than one, then a rectangle is found. This is done for one small subset and all large subsets simultaneously.

As we have already noticed we will avoid the third case. Let L' be the set of all points in large subsets. Switch the roles of their first and second coordinates and repeat the algorithm. The new subsets will have at most $\sqrt{n}$ points, according to lemma 1. Thus this case is treated as the first case.

In the algorithm, we use the following notational convention. If S is a set of points, then $\tilde{S}$ denotes the set containing all points of S restricted to their second coordinate.

Rectangle (S)

1. Partition S into subsets $S_1, \ldots, S_k, L_1, \ldots, L_m$ such that two points are in the same subset if and only if they have equal first coordinate. Furthermore, let S_i $(1 \leq i \leq k)$ be the subsets with at most $\sqrt{n}$ points (the small subsets), and let L_j $(1 \leq j \leq m)$ be the subsets with more than $\sqrt{n}$ points (the large subsets).

2. Search for a rectangle in small subsets in the following way. Build the pointer structure as described above.

The structure is traversed as follows. Traverse the list Y beginning at the element with the smallest value, and for each element (which defines the bottom line of some candidate rectangle), do the following. Follow the reference $first$ and then follow the reference $next_1$ of every point p we reach, until we can go no further. For every such point p (which is a point on the bottom line), follow references $next_2$ to points q until we can go no further. For all points q we reach (which are points that lie above a point on the bottom line), follow the $back$ reference to an element in the list Y (which defines the top line of a candidate rectangle) and increase $count$ with one. If it becomes two, stop and report that S contains the vertices of a rectangle (because two top vertices on the same top line are found, which both must lie above a bottom vertex). (In figure 1, all points that are visited when the algorithm is at the marked element in the list Y, are encircled. Also, the counter values are shown.)

Before we go to the next element in the list Y, traverse the structure again in the same way and initialize all counters to zero again.

3. Search for a rectangle with two vertices in a small subset and two vertices in a large subset.

   - Let V be the set of all different second coordinates of points in $\bigcup L_j$. Store them ordered in an array A. With each index, corresponding to some value y of the second coordinate, store all names j of large subsets L_j that contain a point with y as second coordinate.

   - Make an array B of counters, with one entry for each large subset. Initialize them to zero.

   - For every small subset S_i ($1 \leq i \leq k$), compute $\tilde{S}_i \cap \tilde{L}_j$ for all j ($1 \leq j \leq m$) simultaneously in the following way. For each point $p = (p_1, p_2)$ in S_i, search with p_2 in the array A to find all large subsets L_j that have a point with p_2 as second coordinate. For every L_j thus found, increment the counter in the array B at index j. If any counter becomes 2, then stop and report the existence of the vertices of a rectangle in S. Otherwise, reinitialize the counters in array B to zero, before starting on the next small subset.

4. Search for a rectangle with four vertices in two large subsets in the following way. Let L' be the set of all points in large subsets. Switch the roles of first and second coordinates of these points, and use the same method as above to solve the problem. As stated above, only small subsets will appear.

Theorem 5 *Given a set S of n points in the plane, it can be decided in $O(n\sqrt{n})$ time and $O(n)$ space whether S contains a subset of four points that form the vertices of an axis-parallel rectangle.*

Proof: The correctness of the algorithm follows from the above discussion, and the $O(n)$ space bound is straightforward. It remains to analyse the time complexity of the steps taken by the algorithm.

The first step, partitioning the set S of points, can be performed in time $O(n \log n)$ with a straightforward algorithm.

For the second step (the construction and traversal of the pointer structure), we observe that construction can easily be done in $O(n \log n)$ time. All time spent on traversal of the structure can be charged to the number of times every node (point) is visited during the traversal. A point will only be visited once via a *first* reference or a $next_1$ reference, namely only when the second coordinate of the point is equal to the second coordinate of the element in list Y that is being considered. A point will be visited via a $next_2$ reference once for each point with equal first coordinate and smaller second coordinate. As all points connected via $next_2$ references form one small subset, any point will be visited at most $\sqrt{n}$ times via a $next_2$ reference. Thus all points together are visited $O(n\sqrt{n})$ times. As reinitialization is done in the same way as the traversal, this will at most double the time spent.

For the third step we consider how much time the algorithm takes for each of the three substeps. The first substep can be performed in $O(n \log n)$ with a straightforward algorithm. The second substep takes time $O(m)$, where m is the number of large subsets. For the third substep we consider a small subset S_i. With each point in S_i we search in array A, and possibly increment a number of counters in array B. The searching takes $O(|S_i| \log n)$ time and incrementing counters in B at most $O(m)$ time, because we can increment only m counters before a rectangle is found and we can terminate. Hence, the total time for all small subsets together is

$$O(\sum_{i=1}^{k}(|S_i| \log n + m)) = O(\sum_{i=1}^{k}(|S_i| \log n + \sqrt{n})) = O(n\sqrt{n}).$$

The time spent on the fourth step follows directly from the analysis of the time spent on the second step (as the new subsets contain at most $\sqrt{n}$ points). $\square$

Before continuing with the more-dimensional case, we show that the problem of deciding whether a planar set of points contains the vertices of an axis-parallel rectangle, is equivalent to the problem of deciding whether a bipartite graph contains a $K_{2,2}$ subgraph. (That is, a complete bipartite graph with $2+2$ nodes and thus four edges. In other words, a cycle of length 4.) The correspondence is as follows. Let S be a planar set of points. Let A be the set of all different first coordinates of the points in S, and let B be the set of all different second coordinates. A and B are the sets of nodes in the bipartite graph. Node $a \in A$ and node $b \in B$ give rise to an edge (a, b) if and only if (a, b) is a point in the set S. The correspondence goes in both directions. With this correspondence, it can easily be seen that S contains the vertices of an axis-parallel rectangle if and only if the bipartite graph consisting of the sets of nodes A, B and edges as defined above contains a $K_{2,2}$ subgraph. With our algorithm, one can obtain an $O(|E|\sqrt{|E|})$ time algorithm for deciding whether a bipartite graph (V, E) (where V is the set of nodes and E is the set of edges) contains a $K_{2,2}$ subgraph.

In [3], Clarkson et al. prove that there exists a constant c such that any bipartite graph with n nodes and more than $c \cdot n\sqrt{n}$ edges, contains a $K_{2,2}$ subgraph. Their result is a variant of a graph theoretic extremal result by Kővári, Sós and Turán [6]. This leads to:

Theorem 6 *Given a bipartite graph $G = (V, E)$, it can be decided in $O(min(|E|^{3/2}, |V|^{9/4}))$ time and $O(|V| + |E|)$ space whether G contains a $K_{2,2}$ subgraph.*

The bound is not new. In [1], Chiba and Nishizeki give an $O(|E|\sqrt{|E|})$ time algorithm for cycles of length four in arbitrary graphs.

Next we consider the more-dimensional axis-parallel rectangle problem: Given a set S of n points in d-dimensional space, does S contain a subset of 2^d points that are the vertices of an axis-parallel hyperrectangle. Contrary to the more-dimensional cube problem, we will not partition S in subsets with only the last coordinate different, but in subsets with only the first coordinate equal. Thus we obtain a partition of S in a number of hyperplanes of which the normal is parallel to the first coordinate axis. Again we distinguish small subsets, which contain at most $n^{1-1/d}$ points, and large subsets, which contain more than $n^{1-1/d}$ points. The d-dimensional problem now reduces to finding hyperfaces (which are $(d-1)$-dimensional hyperrectangles) in the subsets, and then finding two different subsets which both contain the same hyperface, only taking into consideration the first $d-1$ coordinates. Thus we search for opposite $(d-1)$-dimensional hyperrectangles. As before, we first prove a lemma which shows that we need not solve the problem of finding a hyperrectangle with both hyperfaces in large subsets.

Lemma 3 *Given a set S of n (distinct) points in d-dimensional space, then there exists a hyperplane with its normal parallel to one of the coordinate axes and which contains at least one and no more than $n^{1-1/d}$ points of S.*

Proof: Omitted. $\square$

To solve the more-dimensional rectangle problem, partition S in subsets $S_1, \ldots, S_k, L_1, \ldots, L_m$, such that two points are in the same subset if and only if they have equal first coordinate. Let $S_1, \ldots, S_k$ be the subsets with at most $n^{1-1/d}$ points, and $L_1, \ldots, L_m$ are the subsets with more than $n^{1-1/d}$ points.

To determine if there is a hyperrectangle with both hyperfaces in small subsets we need a generalization of the pointer structure used to solve the two-dimensional rectangle problem. Let S' be the union of all points in small subsets, $S' = \bigcup S_i$.

- Let $\tilde{S}'$ be the set of points of S' restricted to their last $d-1$ coordinates. Let Y be the list containing all different points in $\tilde{S}'$, ordered lexicographically increasing. With each element in Y, an integer *count* is stored, initially zero, and a reference *first*.

- Each point p in S' corresponds to a node in the structure with $d+1$ references, $next_1, \ldots, next_d$ and *back*, and each node has a counter. $next_1$ leads to the node corresponding to the first point one reaches if one moves in the direction of greater first coordinate (thus a point with the last $d-1$ coordinates equal). More formally, the $next_1$ reference of a point $p = (p_1, \ldots, p_d)$ refers to a point $q = (q_1, \ldots, q_d)$ if and only if $p_2 = q_2, \ldots, p_d = q_d$, $p_1 < q_1$ and for all $r = (r_1, \ldots, r_d)$, $r_2 = p_2, \ldots, r_d = p_d$ implies that $r_1 \leq p_1$ or $r_1 \geq q_1$. Analogously, the references $next_2, \ldots, next_d$ are defined. The reference *back* of a point $p = (p_1, \ldots, p_d)$ leads to the element corresponding to the point $(p_2, \ldots, p_d)$ in the list Y. The reference *first* in the list Y corresponding to some point $(r_2, \ldots, r_d)$, refers to the point with $(r_2, \ldots, r_d)$ as last $d-1$ coordinates, and smallest first coordinate among these points.

We need a few observations before we give the algorithm.

Observation 1 *Let H be an axis-parallel hyperrectangle in d-dimensional space, where p is the vertex of H with all coordinates smallest and q is the vertex of H with all coordinates largest. Then there are exactly $d!$ different ways of moving from p to q along the edges of H, when we are only allowed to move to a vertex with exactly one coordinate larger than the previous vertex.*

Observation 2 *Given the pointer structure for a set S of points in d-dimensional space as described above, a point p in S, and a point q in S for which all coordinates are (strictly) greater than all coordinates of p. Let $\pi = (\pi_1, \ldots, \pi_d)$ be a permutation of the integers $1, \ldots, d$. S contains the vertices of an axis-parallel hyperrectangle with p and q as opposite vertices if and only if one can move from p to q in the pointer structure for every permutation π, by following first arbitrarily many but at least one $next_{\pi_1}$ reference, then at least one $next_{\pi_2}$ reference, $\ldots$, and finally at least one $next_{\pi_d}$ reference.*

We will use the latter observation to find $(d-1)$-dimensional hyperrectangles (hyperfaces) in the small subsets when the pointer structure is traversed. Thus we consider all permutations of the coordinates $(2, \ldots, d)$, and we find a hyperface if a point q can be reached in $(d-1)!$ ways from a point p.

To solve the problem of finding a hyperrectangle with one hyperface in a small subset and one in a large subset, we compute the intersections of every small subset with every large subset, when the points are restricted to their last $d-1$ coordinates. Then we repeat the algorithm recursively in $(d-1)$-dimensional space.

In the algorithm, the set of points $\tilde{S}$ is obtained by leaving out the first coordinate of the points of S.

Hyperrectangle (S, d)

1. Let $N = n^{1-1/d}$.

2. Partition S into subsets $S_1, \ldots, S_k, L_1, \ldots, L_m$ such that two points are in the same subset if and only if they have equal first coordinate. Furthermore, let S_i $(1 \leq i \leq k)$ be the subsets with at most N points (the small subsets), and let L_j $(1 \leq j \leq m)$ be the subsets with more than N points (the large subsets).

3. Search for a hyperrectangle with both hyperfaces in small subsets in the following way. Build the pointer structure as described above.

   The structure is traversed as follows. Traverse the list Y in order, and for each element, do the following. Follow the reference $first$ and then follow the reference $next_1$ of every point p we reach, until we can go no further. For every such point p (which lies in a small subset, where we will search for hyperfaces) and for every permutation $\pi = (\pi_2, \ldots, \pi_d)$ of the numbers $(2, \ldots, d)$, follow every possible number but at least one $next_{\pi_2}$ reference, then at least one $next_{\pi_3}$ reference, $\ldots$, and finally at least one $next_{\pi_d}$ reference. For every point q we reach, given a permutation, increment the counter of that point. If it becomes $(d-1)!$, there are $(d-1)!$ ways of reaching it from p and, according to observation 2, p and q must form the opposite vertices of a $(d-1)$-dimensional hyperface. Hence, we follow the $back$ reference of q and increment the corresponding counter in the list Y. If this counter in Y becomes two, two matching hyperfaces are found and we can stop and report that S contains the vertices of a hyperrectangle.

   Before we go to the next element in the list Y, traverse the structure again in the same way and reinitialize all counters to zero.

4. Search for a hyperrectangle with one hyperface in a small subset and one hyperface in a large subset in the following way.

   - Let V be the set of all different points in $\bigcup L_j$, restricted to their last $d-1$ coordinates. Store them lexicographically ordered in an array A. With each index corresponding to some point $p = (p_2, \ldots, p_d)$, store all names j of large subsets L_j that contain a point with $(p_2, \ldots, p_d)$ as last $d - 1$ coordinates.

- Make an array B of lists, with one entry for each large subset. Initialize each entry to the empty list.

- For every small subset S_i ($1 \leq i \leq k$), compute $\tilde{S}_i \cap \tilde{L}_j$ for all j ($1 \leq j \leq m$) simultaneously in the following way. For each point $p = (p_1, \ldots, p_d)$ in S_i, search with $(p_2, \ldots, p_d)$ in the array A to find all large subsets L_j that have a point with $(p_2, \ldots, p_d)$ as last $d - 1$ coordinates. For every L_j thus found, store in the array B at index j the point $(p_2, \ldots, p_d)$. When all points in S_i are treated this way, then solve the m $(d-1)$-dimensional problems of finding a $(d-1)$-dimensional hyperrectangle for each set B_j, which are the points in the intersections of the large subset L_j with S_i. This is done recursively. If a $(d-1)$-dimensional hyperrectangle is found, it exists in both S_i and L_j and, hence, a hyperrectangle is found.

Reinitialize all lists in the array B before continuing with the next small subset.

5. Search for a hyperrectangle with both hyperfaces in two large subsets in the following way. Let L' be the set of all points in large subsets. Switch the roles of the coordinates (that is, let another coordinate take the role of the first coordinate) and repeat the algorithm at step 2. (From lemma 3, there will only be small subsets left after $d - 1$ times of switching the roles of coordinates.)

Theorem 7 *Given a set S of n points in E^d, it can be decided in $O(n^{2-1/d})$ time and $O(n)$ space whether S contains a subset of 2^d points that form the vertices of an axis-parallel hyperrectangle.*

Proof: It is not difficult to see that the algorithm uses $O(n)$ space. For the time bound, we prove by induction on d that the solution given above takes $O(n^{2-1/d})$ time. Let $N = n^{1-1/d}$.

If $d = 2$, then the time bound follows from theorem 5.

If $d > 2$, then distinguish the steps of the algorithm above.

The first two steps can easily be done in $O(n \log n)$ time.

For the third step, notice that every small set S_i is visited at most $|S_i|$ times, and at each visit, any point in S_i is visited at most a constant number of times (dependent on d, namely $(d - 1)!$). Thus traversing the structure takes time bounded by

$$c \cdot \sum_{1 \leq i \leq k} |S_i|^2 \leq c \cdot \sum_{1 \leq i \leq k} N \cdot |S_i| \leq c \cdot N \cdot n = O(n^{2-1/d})$$

(where c is some constant).

The first and second substeps of step 4 clearly take $O(n \log n)$ time. Computing the intersection of one small subset S_i with all large subsets $L_1, \ldots, L_m$ takes time bounded by $c \cdot |S_i| \cdot (m + \log n)$, thus for all small subsets we spend time bounded by

$$c \cdot \sum_{1 \leq i \leq k} |S_i| \cdot (m + \log n) \leq c \cdot \sum_{1 \leq i \leq k} |S_i| \cdot (n/N + \log n) \leq$$

$$c \cdot (n^2/N + n \log n) = O(n^{1+1/d}).$$

Solving the $(d-1)$-dimensional problem for $\tilde{S}_i \cap \tilde{L}_j$ takes time bounded by $c \cdot |\tilde{S}_i \cap \tilde{L}_j|^{2-1/(d-1)}$ by induction. Thus the total time spent to solve all $(d-1)$-dimensional problems is bounded by

$$c \cdot \sum_{1 \le i \le k} \sum_{1 \le j \le m} |\tilde{S}_i \cap \tilde{L}_j|^{2-1/(d-1)} \le c \cdot \sum_{1 \le i \le k} m \cdot |S_i|^{2-1/(d-1)} \le$$

$$c \cdot n/N \cdot \sum_{1 \le i \le k} |S_i|^{2-1/(d-1)} \le c \cdot n \cdot N^{-1/(d-1)} \cdot \sum_{1 \le i \le k} |S_i| \le$$

$$c \cdot n^2 \cdot N^{-1/(d-1)} = O(n^{2-1/d}).$$

The fourth step (and hence also the other steps) can only occur d times by lemma 3, and this step takes $O(n)$ time. $\square$

Observe that the problem of deciding whether a d-dimensional set S of points contains the vertices of an axis-parallel hyperrectangle, is equivalent to the graph problem of deciding whether a d-partite hypergraph, in which every hyperedge is incident upon exactly d nodes (one of each set), contains a $K_{2,\ldots,2}$ subgraph (d two's).

4 Arbitrarily oriented squares and rectangles

In this section we study the problems dealt with in the previous sections, but this time for not necesarily axis-parallel squares and rectangles. Again we begin with giving upper and lower bounds on the maximal number of squares and rectangles that can occur in a set of points in the plane. Then we give algorithms for recognizing (hyper-)squares and (hyper-)rectangles in sets of points in d-dimensional space. Our algorithms take time $O(n^d \log n)$ for rectangles, and time $O(n^2 \log n)$, $O(n^3)$ and $O(n^{d-1/2}\beta(n))$ for $d = 2$, $d = 3$ and $d \ge 4$, respectively, for squares. ($\beta(n)$ is related to the inverse of Ackermann's function.)

Theorem 8 *For a set S of n points in the plane, the maximal number of subsets of four points that form the vertices of a square is $\Theta(n^2)$.*

Proof: Omitted. $\square$

Theorem 9 *For a set S of n points in the plane, the maximal number of subsets of four points that form the vertices of a rectangle is $\Omega(n^2 \log n)$ and $O(n^2 \sqrt{n})$.*

Proof: The lower bound follows by placing the points on a square grid and counting carefully. We next prove the upper bound.

For two points p and q, let $C_{p,q}$ be the circle containing p and q with diameter $|\overline{pq}|$. It is easy to see that four points p_1, p_2, p_3, p_4 are the vertices of a rectangle (with diagonals $\overline{p_1 p_3}$ and $\overline{p_2 p_4}$) if and only if $C_{p_1,p_3} = C_{p_2,p_4}$.

Let $p_1, \ldots, p_n$ be the points of S. Let them define the set of different circles $\{C_1, \ldots, C_m\}$ and let d_i be the number of opposite pairs of points on C_i, i.e., $d_i = |\{(p_j, p_k) : 1 \le j < k \le n$ and $C_i = C_{p_j,p_k}\}|$. The number of rectangles of S is given by $\sum_{i=1}^m d_i(d_i - 1)/2$. Furthermore, we have $\sum_{i=1}^m d_i = n(n-1)/2$, because every pair (p_j, p_k) contributes to only one d_i, and $d_i \le n/2$ $(1 \le i \le m)$.

Assume without loss of generality that $d_1 \geq d_2 \geq \cdots \geq d_m$, and let k be such that $d_k > 2\sqrt{n}$, $d_{k+1} \leq 2\sqrt{n}$. As two circles intersect in at most two points, we have $k \leq \sqrt{n}$. (If $k > \sqrt{n}$, then the number of points in S would be at least $\sum_{i=1}^{k}(2d_i - 2(i-1)) > 4k\sqrt{n} - k(k-1) > 4n - (n + \sqrt{n}) > n$, a contradiction.) Now we have

$$\sum_{i=1}^{m} d_i(d_i - 1)/2 \leq \sum_{i=1}^{m} d_i^2 = \sum_{i=1}^{k} d_i^2 + \sum_{i=k+1}^{m} d_i^2 \leq$$

$$\sum_{i=1}^{\sqrt{n}} d_i^2 + \sum_{i=k+1}^{m} (2\sqrt{n})d_i \leq \sum_{i=1}^{\sqrt{n}} (n/2)^2 + 2\sqrt{n} \cdot n(n-1)/2 \leq 2n^2\sqrt{n}.$$

$\square$

Next we address the problem of finding the vertices of hyperrectangles and hypersquares in a set of points in d-dimensional space. For brevity we will speak of rectangles and squares in d-space. First some simple properties and observations on rectangles, squares and sets of points in d-space are given.

Observation 3 *A rectangle or square has 2^d vertices.*

Observation 4 *A rectangle or square has 2^{d-1} (parallel) edges for which the bisecting hyperplanes coincide. In this case, the 2^d endpoints of the edges form all vertices of the rectangle or square.*

Observation 5 *If there are 2^{d-1} pairs of points for which the bisecting hyperplanes coincide, and the distances between the two points in a pair are equal for every pair, then the 2^d points are the vertices of a rectangle in d-space if and only if the projections of the points on the bisecting hyperplane h are the vertices of a rectangle on h ((d − 1)-space). The 2^d points are the vertices of a square in d-space if and only if the projections of the points on the bisecting hyperplane h are the vertices of a square on h with edge length equal to the distances between the two points in the pairs.*

The above observations suggest the following algorithms, which both take every pair of points and bucket (classify) the pairs, giving rise to a simpler subproblem for every bucket.

Rectangle (S, d)

1. If $d = 1$, then report the existence of a rectangle if $|S| > 1$ and terminate.

2. If $d > 1$, then bucket every pair of points of S, such that two pairs (p_1, p_2) and (q_1, q_2) are in the same bucket if and only if the bisecting hyperplane of p_1 and p_2 is equal to the bisecting hyperplane of q_1 and q_2, and the distance between p_1 and p_2 is equal to the distance between q_1 and q_2. Let the buckets be $B_1, \ldots, B_m$.

3. For every bucket B_i corresponding to some hyperplane h, project one point of every pair in B_i on h, and solve the $(d-1)$-dimensional problem of finding a rectangle in the projected points on h recursively.

Square (S, d, e) (e is the edge length, which is not specified initially)

1. If $d = 1$ and the edge length is not specified, then report the existence of a square if $|S| > 1$ and terminate. If $d = 1$ and the edge length is specified, then report the existence of a square if there is a pair of points in S with the specified edge length, and terminate.

2. If $d > 1$ and the edge length is specified, then only consider pairs of points with distance e.

   Bucket every pair of points of S, such that two pairs (p_1, p_2) and (q_1, q_2) are in the same bucket if and only if the bisecting hyperplane of p_1 and p_2 is equal to the bisecting hyperplane of q_1 and q_2, and the distance between p_1 and p_2 is equal to the distance between q_1 and q_2. Let the buckets be $B_1, \ldots, B_m$.

3. For every bucket B_i corresponding to some hyperplane h and some distance x, project one point of every pair in B_i on h, and solve the $(d-1)$-dimensional problem of finding a square with edge length x in the projected points on h recursively.

Theorem 10 *Given a set S of n points in d-dimensional space $(d \geq 2)$, it can be decided in $O(n^d \log n)$ time and $O(n^2)$ space whether S contains a subset of 2^d points that form the vertices of a rectangle. It can be decided in $O(n^2 \log n)$ time $(d = 2)$, $O(n^3)$ time $(d = 3)$, or $O(n^{d-1/2}\beta(n))$ time $(d \geq 4)$, and $O(n^2)$ space whether S contains a subset of 2^d points that form the vertices of a square $(\beta(n)$ is related to the extremely slowly growing inverse of Ackermann's function).*

Proof: Let $T(d, n)$ denote the time taken by the above algorithm for rectangles in a set of n points in d-space, and let $|B_i|$ be the number of pairs of points in bucket B_i. Then

$$T(d, n) \leq c \cdot n^2 \log n + \sum_{i=1}^{m} T(d-1, |B_i|),$$

$$T(1, n) = O(n).$$

Using the fact that $|B_i| \leq n/2$ and $\sum_{i=1}^{m} |B_i| \leq n(n-1)/2$, one can prove that $T(d, n) = O(n^d \log n)$ for $d \geq 2$.

Next we consider the algorithm for squares. Let $T(d, n)$ denote the time taken by the square algorithm when the edge length is not specified, and $T^*(d, n)$ the time taken when the edge length is specified. Furthermore, let $M(d, n)$ be the maximal number of pairs of points in a set of n points in d-space which lie some specified distance from each other. Then

$$T(d, n) \leq c \cdot n^2 \log n + \sum_{i=1}^{m} T^*(d-1, |B_i|),$$

$$T^*(d, n) \leq c \cdot n^2 + c \cdot M(d, n) \log n + \sum_{i=1}^{m} T^*(d-1, |B_i|),$$

$$T^*(1, n) = O(n \log n),$$

and we have $|B_i| \leq n/2$, $\sum |B_i| \leq n(n-1)/2$ when the edge length is not specified, and $\sum |B_i| \leq M(d, n)$ when the edge length is specified.

The problem of determining $M(d, n)$ is called the unit-distance problem and was posed by Erdös [4, 5]. The best known bounds are $M(2, n) = O(n^{4/3})$ and $M(3, n) = O(n^{3/2}\beta(n))$ $(\beta(n)$ as in the theorem), see [3, 7]. It is known that $M(d, n) = \Theta(n^2)$ for $d \geq 4$, see [2]. Now we can prove $T^*(2, n) = O(n^2)$ and $T^*(d, n) = O(n^{d-1/2}\beta(n))$ for $d \geq 3$. The bounds of the theorem follow. $\square$

5 Concluding remarks

In this paper we have studied combinatorial and algorithmical aspects of recognizing rectangles and squares in sets of points. Both the cases of axis-parallel shapes and arbitrarily oriented shapes have been studied, in the plane as well as in more-dimensional space.

A natural question is whether it is possible to improve the $O(n^{2-1/d})$ time bound for the rectangle problem and $O(n^{1+1/d} \log n)$ for the cube problem. Furthermore, a tight bound on the maximal number of arbitrarily oriented rectangles any set of points can contain is still unknown.

The problems we have studied asked for the presence of some shape (polyhedron) in a set of points. There are many other shapes that may be present in a set of points, such as (hyper-)tetrahedrons and (hyper-)diamonds. One could ask if there exists a general method for solving this type of recognition problem.

Acknowledgements

The authors thank Mark Overmars for his quick but careful reading of this paper. Also, Mark Overmars posed most of the problems studied in this paper. Furthermore, we thank Gerhard Wöginger for the lower bound proof of theorem 9.

References

[1] Chiba, N. and T. Nishizeki, Arboricity and subgraph listing algorithms, *SIAM J. on Computing 14* (1985), pp. 210-223.

[2] Chung, F.R.K., Sphere-and-point incidence relations in high dimensions with applications to unit distances and furthest-neighbor pairs, *Discr. and Comp. Geometry 4* (1989), pp. 183-190.

[3] Clarkson, K.L., H. Edelsbrunner, L.J. Guibas, M. Sharir and E. Welzl, Combinatorial complexity bounds for arrangements of curves and surfaces, *29th Ann. Symp. on Found. of Comp. Sci.* (1988), pp. 568-579.

[4] Erdös, P., On sets of distances of n points, *Amer. Math. Monthly 53* (1946), pp. 248-250.

[5] Erdös, P., On sets of distances of n points in Euclidian space, *Magyar Tud. Akad. Mat. Kutaló Int. Kozl. 5* (1960), pp. 165-169.

[6] Kövári, T., V.T. Sós and P. Turán, On a problem of K. Zarankiewicz, *Colloquium Math. 3* (1954), pp. 50-57.

[7] Spencer, J., E. Szemerédi and W.T. Trotter, Jr., Unit distances in the Euclidean plane, *Graph Theory and Combinatorics*, Academic Press, London, 1984, pp. 293-303.

Combinatorial Properties of Abstract Voronoi Diagrams

Rolf Klein*

Abstract

Abstract Voronoi diagrams are defined by a system of bisecting curves in the plane, rather than by the concept of distance. So far, their investigation was based on the assumption that *each subfamily* of the given family of curves yields a *partition of the plane* into *connected Voronoi regions* [4,5,9]. Here we prove these conditions to be equivalent with some simple combinatorial properties of the curves that need only be verified for subfamilies of size 3. Using the simpler characterization, we are able to show that all singularities resulting from bisecting curves that share a curve segment or touch one another can be resolved by deforming the curves in a suitable way. In the new system, any two curves properly cross whereever they intersect, but each subfamily still yields the same Voronoi diagram as before, after possibly contracting some edges.

Key words: Abstract Voronoi diagram, bisector, Voronoi diagram

1 Introduction

The Voronoi diagram of a set of planar sites, S, is a partition of the plane into regions, one to each site, such that the region of site p contains all those points of the plane that are closer to p than to any other site in S.

Such partitions play an important role in different areas of science. In computer science, the Voronoi diagram belongs to the most useful data structures. Its structural properties depend on the underlying distance measure and on the type of sites considered. A survey on the variety of Voronoi diagrams that have so far been investigated in the literature is given in [1].

In order to provide a unifying concept for both the study and the computation of Voronoi diagrams, *abstract Voronoi diagrams* were introduced [4]. They are not based on the notion of distance but on systems of bisecting curves as primary objects. For any two sites, p and q, in S, let $J(p,q)$ denote a curve that is homeomorphic to the line and divides the plane into a region containing p and a region containing q. Then the Voronoi region of p with respect to S is the intersection of all p-regions, as q varies in $S - \{p\}$. The family of curves $J(p,q)$, $p \neq q$, $p,q \in S$, is called *admissible* if for each subset of sites $S' \subseteq S$ the Voronoi regions w.r.t. S' are connected and form a partition of the plane. The boundaries of the Voronoi regions form a planar graph. Thus, the theory of abstract Voronoi diagrams can be considered part of graph theory.

Abstract Voronoi diagrams based on admissible curve systems can be computed efficiently. It was shown in [4,5] how to merge two subdiagrams, $V(L)$ and $V(R)$, where S is split into L and R, in a number of steps proportional to $n = |S|$, if the bisector of the sets L and R contains no edge cycle. This result gives rise to a divide–and–conquer algorithm that needs $O(n \log n)$ steps in the

---
*Institut für Informatik, Universität Freiburg, Rheinstr. 10–12, 7800 Freiburg, West Germany

worst case, provided that S can be recursively split into subsets of about equal size whose bisector is acyclic. As to concrete metrics, sufficient conditions for the existence of such acyclic partitions were given in [6] and in [7].

An alternative approach to the computation of abstract Voronoi diagrams has been proposed by Mehlhorn, Ó'Dúnlaing and Meiser [9]. Their algorithm uses the randomized incremental construction technique introduced in [2]. If the n sites are added to the current site set one by one in random order then the total cost of updating the Voronoi diagram does not exceed $O(n \log n)$ on the average. This algorithm needs no assumptions about the existence of acyclic set bisectors, but it has to assume that the curves are in "general position" (the latest version [8], however, applies to all admissible curve systems where bisectors do not touch one another in a specific way).

Our interest in such algorithms is due to the fact that they provide us with universal tools for the computation of concrete Voronoi diagrams: if concrete sites are given together with a distance measure that yields admissible bisector systems then it only remains to implement a fixed, finite set of elementary bisector operations (such as computing the first intersection in a given direction, for example), and an efficient algorithm will result.

The first goal of this paper is to give a more tractable characterization of what an admissible curve system is.

Given a family of curves, testing for admissibility is highly inefficient if the above definition is applied, because all possible subsets of S are involved. We give an equivalent characterization that is based on simple combinatorial properties of the curves and involves subsystems of size only 3 (as a corollary, one obtains that a family of curves is admissible iff this holds true for each subfamily of size 3). First, we show that the Voronoi regions with respect to a site set S form a partition of the plane iff the transitivity relation

$$R(p,q) \cap R(q,r) \subseteq R(p,r)$$

holds for the region of p with respect to q (the region of q w.r.t. r and the region of p w.r.t. r, correspondingly), for any three sites p, q, and r in S.

Next, we prove that the Voronoi region of site p with respect to the set S' is connected iff the following condition holds. Let p, q, r be in S', and assume that the curves $J(p,q)$ and $J(p,r)$ are such oriented that the p-region lies on their left hand side. Then the two curves cross at most twice and do not give rise to a bounded clockwise oriented cycle (see Figure 5).

The above properties comprise what is combinatorially essential in most of the planar Voronoi diagram types so far studied. Besides, they enable us to decide whether a family of curves is admissible by inspecting only $O(|S|^3)$ many subproblems of size 3.

Our second goal is to prove that each admissible curve system can be brougth into general position. In concrete metrics, this can be achieved by applying an ϵ–distortion to the sites. With abstract Voronoi diagrams, the problem is non–trivial. Using the above result on admissibility we are able to show that the curves of an arbitrary admissible system can be deformed in such a way that only proper crosspoints remain and that the Voronoi diagram built from any subfamily of the new curves still resembles the corresponding old diagram in that, after shrinking some edges of the new diagram into vertices, the two diagrams differ only by a deformation. Points crossed by more than 3 curves can be resolved in a similar way.

2 Abstract Voronoi diagrams

Throughout this paper, a subset J of the plane is called a *bisecting curve* iff J is homeomorphic to the line and dissects the plane into two unbounded domains each of which has J as its complete boundary.

Now let $n \in \mathcal{N}$, and for each pair of integers p, q such that $1 \leq p \neq q \leq n$ let $J(p,q) = J(q,p)$ be a bisecting curve. We shall assume that whith each integer $p \leq n$ a site p is associated in the plane such that $J(p,q)$ separates p and q. Let $D(p,q)$ and $D(q,p)$ denote the domains containing p and q, respectively.

We define the *abstract Voronoi diagram* as follows.

Definition 2.1 Let

$$R(p,q) : = \begin{cases} D(p,q) \cup J(p,q), & \text{if } p < q \\ D(p,q), & \text{if } p > q \end{cases}$$

$$R(p,S) : = \bigcap_{\substack{q \in S \\ q \neq p}} R(p,q) \qquad \text{(Voronoi region of } p \text{ w.r.t. } S\text{)}.$$

$$V(S) : = \bigcup_{p \in S} \partial R(p,S) \qquad \text{(Voronoi diagram of } S\text{)}.$$

It is immediate that two different Voronoi regions are disjoint because $R(p,q)$ and $R(q,p)$ are complementary.

Definition 2.2 A family of bisecting curves $(S, \{J(p,q); p,q \in S, p \neq q\})$ is called *admissible* iff the following conditions are fulfilled.

1. The intersection of any two bisecting curves consists of finitely many connected components.

2. For each non-empty subset S' of S

   A) $R(p,S')$ is path-connected, for each $p \in S'$.

   B) $\Re^2 = \bigcup_{p \in S'} R(p,S')$ (disjoint).

Examples. Figure 1 displays a system of bisecting curves that is not admissible because the points of the bounded domain belong to none of the regions, thereby violating condition *B)*. The system shown in Figure 2 is admissible iff $p = \min\{p,q,r\}$, because only then does the segment T belong to $R(p, \{p,q,r\})$ and provide the required connecting arc between the two shaded subsets of this region.

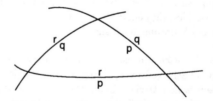

Figure 1: A system that is not admissible.

Abstract Voronoi diagrams induced by admissible curve systems resemble concrete Voronoi diagrams in many aspects. Each Voronoi region is a simply–connected set, bordered by a closed curve. This curve, however, may have multiple points as they would result from sqeezing together a simple curve (cf. Figure 2).

Locally, to each point $v \in V(S)$ there exist arbitrarily small neighborhoods where $V(S)$ looks as one might expect intuitively: a pie consisting of two pieces or more, each piece belonging to one of the regions $R(p,S)$, and some of them possibly thin as a curve. Let $p_1, \ldots, p_k$ denote the sites whose regions meet at v. The point v itself belongs to the region of $\min\{p_1, \ldots, p_k\}$, and only this region can contain more than one piece of pie. Figure 3 gives an example of a possible neighborhood

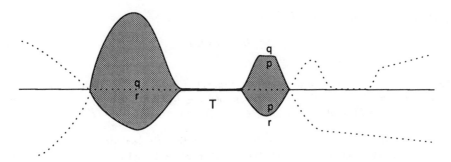

Figure 2: An admissible system iff $p < q, r$.

of a point $v \in V(S)$. Here the pieces marked with an encircled p_i belong to $R(p_i, S)$, and we have $p_0 = \min\{p_1, \ldots, p_8\}$.

Globally, the diagram $V(S)$ can be represented by a planar graph each of whose faces corresponds to a Voronoi region. In order to define vertices and edges in an appropriate way, one has to conceptually thicken the thin parts of the regions.

It is useful to encircle the interesting "finite" part of the Voronoi diagram by a simple closed curve, Γ, large enough to contain all endpoints of connected components of curve intersections. Moreover, Γ can be chosen such that precisely two (infinite) segments of each bisecting curve are in the outer domain. A detailed investigation of the local and global properties of abstract Voronoi diagrams can be found in [5].

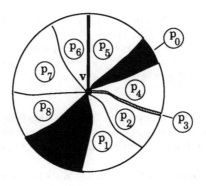

Figure 3: A neighborhood of a point

3 Admissible curve systems

In this section, we give an equivalent characterization of admissibility that involves only the subfamilies of curves associated with the subsets $S' \subseteq S$ of size 3.

First, we show that condition 2 B) of the definition of an admissible curve system (2.2) can actually be replaced by a simpler property.

Lemma 3.1 *Let* $(S, \mathcal{J})$ *be a system of bisecting curves. Then the following assertions are equivalent.*

1. If p, q and r are pairwise different sites of S then $R(p, q) \cap R(q, r) \subseteq R(p, r)$ holds. (Transitivity)

2. *For each non–empty subset $S' \subseteq S$ we have*

$$\Re^2 = \bigcup_{p \in S'} R(p, S').$$

Proof: 2) $\Longrightarrow$ 1) : Let z be contained in $R(p,q) \cap R(q,r)$. By 2), there must be a site t in $S' := \{p, q, r\}$ such that $z \in R(t, S')$. If $t = p$ then $z \in R(p, S') \subseteq R(p, r)$, as stated. Otherwise, $z \in R(q, S') \subseteq R(q, p)$ would contradict $z \in R(p, q)$, and $z \in R(r, S') \subseteq R(r, q)$ would contradict $z \in R(q, r)$.

1) $\Longrightarrow$ 2) : By induction on $| S' |$. If S' is of size ≤ 2, then the assertion is immediate. The case where $|S'| = 3$ follows directly from 1). Assume that $|S'| \geq 4$, and let z be a point in the plane. By induction hypothesis, to each $p \in S'$ there exists a site $c(p) \neq p$ such that $z \in R(c(p), S' - \{p\})$ holds.

Case 1: There exist $v \neq w$ such that $c(v) = c(w)$. Then

$$
\begin{aligned}
z \ &\in \ R(c(v), S' - \{v\}) \cap R(c(v), S' - \{w\}) \\
&\subset \ R(c(v), S' - \{v\}) \cap R(c(v), v) = R(c(v), S').
\end{aligned}
$$

Case 2: The mapping c is injective. Let p, v, w be such that $|\{p, c(p), v, w\}| = 4$. Since $c(v) \neq c(w)$, one of $c(v), c(w)$—say $c(v)$—is different from p. Because of $c(v) \neq c(p)$ we obtain the contradiction

$$
\begin{aligned}
z \ &\in \ R(c(p), S' - \{p\}) \subseteq R(c(p), c(v)) \\
z \ &\in \ R(c(v), S' - \{v\}) \subseteq R(c(v), c(p)).
\end{aligned}
$$

$\square$

Note that the first property in the lemma corresponds to the transitivity of $<$ for real metrics.

Corollary 3.2 *For each point v, by $p \prec_v q :\Leftrightarrow v \in R(p, q)$ a total order on S is induced that agrees with the given global order $<$ for those pairs (p, q) where $v \in J(p, q)$. The point v belongs to the region of the site that is minimal with respect to $\prec_v$.*

In what follows we shall confine ourselves to the situation where *a point common to two curves is a proper crossing between the curves.* Consequently, Voronoi regions cannot contain thin parts. Moreover, if $J(p, q)$ intersects $J(q, r)$ at a point v then also $J(p, r)$ passes through v, and v must be a vertex in $V(\{p, q, r\})$.

The above lemma can then be used for defining a "distance function" $d_p(v)$ to each site p, such that $V(S)$ equals the concrete Voronoi diagram obtained by means of the $d_p(v)$. However, the definition of $d_p(v)$ also depends on the other sites in S.

In the next section it will be shown how to eliminate all curve intersections that are not of cross type. The following discussion can then easily be generalized to arbitrary admissible curve systems if a reasonable definition of a crosspoint is used; see [5].

Lemma 3.3 *Let C be a simple closed curve in the plane, counterclockwise oriented, and let J be an oriented bisecting curve. Then the intersection of the domains on the left of C and J is connected iff C and J do not give rise to a bounded clockwise oriented cycle.*

See Figure 4. A proof of this result, which is intuitively clear, is given in [5].

Theorem 3.4 *A family $(S, \mathcal{J})$ of bisecting curves is admissible iff the following conditions are fulfilled.*

1. $R(p, q) \cap R(q, r) \subseteq R(p, r)$ *holds for any three sites p, q, r in S.*

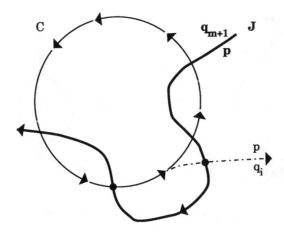

Figure 4: A forbidden clockwise cycle

2. *Any two curves $J(p,q)$ and $J(p,r)$ cross at most twice and do not constitute a clockwise cycle in the plane.*

Proof: Due to Lemma 3.1, we may concentrate on the connectedness of the Voronoi regions and assume that the regions form a partition of the plane. If the system is admissible then $V(p,q,r)$, encircled by Γ, is a planar graph with exactly 4 faces each of whose vertices is of degree at least 3. By the Euler Formula, there can be at most 4 vertices. Two of them must be situated on Γ since at least two edges of the original diagram tend to infinity. Thus, $J(p,q)$ and $J(p,r)$ cross at most twice. A simple case analysis shows that no clockwise cycle can arise from the two curves; see Figure 5.

Conversely, assume that the conditions of the theorem are fulfilled. From the case analysis we know that the regions with respect to a site set of size 3 are connected. Now we show by induction on m the following: If $R := R(p, \{p, q_1, \ldots, q_m\})$ is connected then $R \cap R(p, q_{m+1}) = R(p, \{p, q_1, \ldots, q_m, q_{m+1}\})$ is connected, too. Let the curves $J(p,q)$ be such oriented that the p-region is on the left. Suppose $R \cap R(p, q_{m+1})$ were not connected. Assume first that R is bounded. Then $C := \partial R$ and $J(p, q_{m+1})$ would form a clockwise cycle, due to Lemma 3.3. By investigating the prolongations to infinity of the curves that contribute to the contour of R one sees that $J(p, q_{m+1})$ together with some curve $J(p, q_i), i \leq m$, must form a clockwise cycle—a contradiction (cf. Figure 4). If R is unbounded we intersect it with the inner domain of the encircling curve Γ. Let R' denote the resulting domain, and let C' be its contour. It follows from the properties of Γ that $R' \cap R(p, q_{m+1})$ is still disconnected and that the clockwise cycle that is therefore generated by C' and by $J(p, q_{m+1})$ does not involve pieces of Γ. Now the same reasoning applies. $\qquad\Box$

Note, however, that two bisecting curves $J(p,q)$ and $J(r,t)$ may cross arbitrarily often if the sites are pairwise different.

Corollary 3.5 *A family $(S, \mathcal{J})$ of bisecting curves is admissible iff each subfamily associated with a subset $S' \subseteq S$ of size 3 is admissible.*

4 General position of bisecting curves

In this section we show how to remove from an admissible curve system all those intersections that are not of cross–point type (i.e., all intersections where two curves share a segment or touch), without changing the structure of the abstract Voronoi diagram that results from *any of the subfamilies* of the

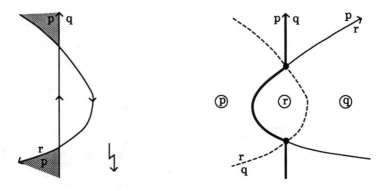

Figure 5: Two crosspoints of $J(p,q)$ and $J(p,r)$; the first picture shows a forbidden clockwise oriented cycle.

system. Note that this is much harder a task than just separating the coinciding edges and vertices of a single Voronoi diagram. However, constructing a Voronoi diagram usually does involve diagrams of subsets of S.

We start with separating the curves that share a segment.

Lemma 4.1 *Let $(S,\mathcal{J})$ be an admissible curve system, and let v be a point in the plane. In a suitable neighborhood U of v the curves of $\mathcal{J}$ passing v can be deformed in such a way that the resulting system $(\tilde{S},\tilde{\mathcal{J}})$ is admissible and has the following properties.*

1. *If $v \in J(p,q) \cap J(r,s)$ then v is the only point of intersection of the deformed curves, $\tilde{\mathcal{J}}(p,q)$ and $\tilde{\mathcal{J}}(r,s)$, inside U.*

2. *For each non-empty $T \subseteq S$ the cyclical sequence of edges radiating from v is the same in $V(T)$ as in $V(\tilde{T})$.*

Here S and $\tilde{S}$ are equal as sets of indices; ~refers to the modified curves.

Proof: Let U be a neighborhood of v. Assume that there are two or more curves of $\mathcal{J}$ that have an arc α in common which extends from v to ∂U (endpoints not included). We have to separate these curves by spreading them out like the wires of a cable, without touching the neighbor curves of α; see Figure 6. To prove the theorem, it is sufficient to show how to separate the wires of *one* cable incident with v.

By D_l (resp. D_r) we denote the area of U on the left (resp. on the right) of and close to α. A bisecting curve $J(p,q)$ that runs through α is denoted by $p \mid q$ iff $D_l \subset D(p,q)$ (and consequently, $D_r \subset D(q,p)$) holds, and by $q \mid p$, otherwise; note that no ambiguity can arise since bisecting curves are simple.

Now assume that $J(p,q)$ and $J(p,r)$ are both passing through α. In order to separate them, we have to decide which one comes first in left-to-right order, as seen from v. Due the property 2), this order must agree with the situation at v in $V(\{p,q,r\})$. Here α can induce one or two edges, depending on $J(q,r)$ and on the order relations between p,q, and r. A complete case analysis is displayed in Figure 7. On the right hand side the correct order of curves is shown which turns out to be uniquely determined in each case; edges are drawn bold, and Voronoi regions are marked by encircling site labels.

Note that in case 1.3 the relations $r < p < q$ are in fact impossible. For, $p < q$ would imply

$$\alpha \subset J(p,q) \subset R(p,q),$$

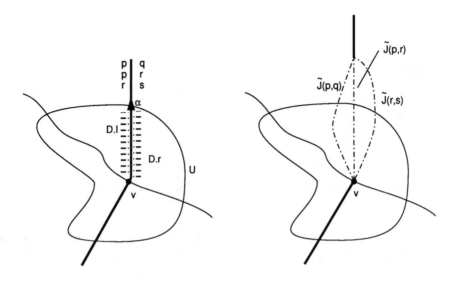

Figure 6: Separation of curve segments.

hence

$$\alpha \subset R(p, \{p, q, r\}) \subset R(p, r)$$

which contradicts $\alpha \subset J(r, p) \subset R(r, p)$ if $r < p$.

In case 3, D_l must belong to r-land since it can't belong to p-land nor to q-land. Similary, D_r is contained in q-land. Hence, $r|q$ must be in α. Now the intended order of curves is defined as follows.

Definition 4.2 Let $p\,|\,q$, $p\,|\,r$, and $q\,|\,p$, $r\,|\,p$, be in α.

$$\left.\begin{array}{lll} p\,|\,q \longrightarrow p\,|\,r & : & \Longleftrightarrow \\ r\,|\,p \longrightarrow q\,|\,p & : & \Longleftrightarrow \end{array}\right\} \quad \begin{array}{l} \alpha \text{ induces an edge separating the regions of} \\ p \text{ and } q \text{ in } V(\{p, q, r\}) \end{array}$$

Obviously, the second line follows from the first by applying reflection. An investigation of Figure 7 immediately shows that this definition captures the correct order.

Claim 1 If $J(p, q)$ and $J(p, r)$ contain α then they are comparable in the transititve closure of $\longrightarrow$, and the correct order of separation is given by the direction of arrows.

Here at most two arrows are necessary to relate $J(p, q)$ to $J(p, r)$; this happens in case 3 and case 4. Claim 1 assures us that the above definition is necessary in order to fulfill property 2) of the theorem. The problem is in proving that relation $\longrightarrow$ can in fact be extended to an order relation; there could be conflicts arising from other site sets $\{p, q, s\}, \{q, r, s\}, \ldots$ which cause cycles in $\longrightarrow$.

Claim 2 If $\quad p\,|\,q \longrightarrow p\,|\,r \longrightarrow p\,|\,s \quad$ then $\quad p\,|\,q \longrightarrow p\,|\,s.$
If $\quad s\,|\,p \longrightarrow r\,|\,p \longrightarrow q\,|\,p \quad$ then $\quad s\,|\,p \longrightarrow q\,|\,p.$

Proof: By straightforward case analysis. $\qquad\qquad\square$

Now the key observation is the following.

Claim 3

$$\begin{array}{lll} \text{If} \quad p\,|\,q \longrightarrow p\,|\,r & \text{then} & (p < q \Longrightarrow p < r). \\ \text{If} \quad r\,|\,p \longrightarrow q\,|\,p & \text{then} & (r < p \Longrightarrow q < p). \end{array}$$

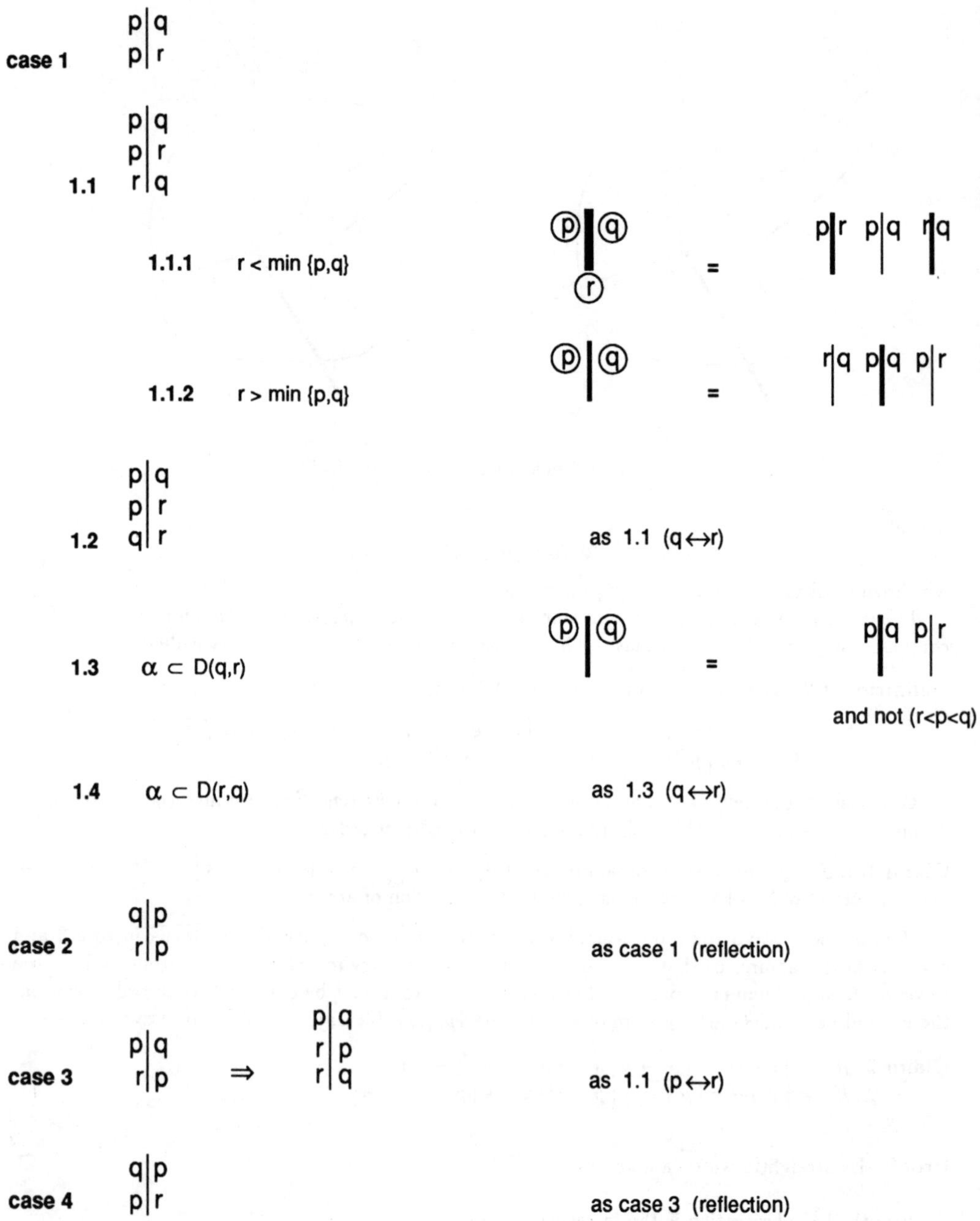

Figure 7: Situations possible at α in $V(\{p,q,r\})$ if $\alpha \subset J(p,q) \cap J(p,r)$.

The proof is by case analysis, guided by Figure 7. Note that again the second line follows from the first by reflection, because the right hand side implications are equivalent.

Claim 4 *Relation* $\longrightarrow$ *is acyclic.*

Proof: Otherwise, let C be a cycle of minimum length. Due to Claim 3, we may assume that $p < q$ holds for each term $p \mid q$ that occurs in C (should always $q < p$ be true we could apply reflection). Because of Claim 2 and the supposed minimality of C, transitions of type

$$w \mid q \longrightarrow p \mid q, \qquad p \mid q \longrightarrow p \mid r$$

must interchange in C; in particular, C is of even length. For each transition of type $p \mid q \longrightarrow p \mid r$ in C the arc α either contains $r \mid q$, or is contained in $D(q,r)$. This follows by checking up on the remaining cases shown in Figure 7; note that only those situations need be examined where α induces *one* borderline of $V(\{p,q,r\})$, since $p < q$ and $p < r$ hold! Hence, we obtain

$$p \mid q \longrightarrow p \mid r \text{ in } C \implies D_r \subset D(q,r).$$

Now suppose that C is the sequence of transitions

$$p_i \mid q_i \longrightarrow p_{i+1} \mid q_i \longrightarrow p_{i+1} \mid q_{i+1}, \qquad 1 \le i \le m-1,$$

where $p_m = p_1$ and $q_m = q_1$. Then

$$D_r \subset \bigcap_{i=1}^{m-2} D(q_i, q_{i+1}) \cap D(q_{m-1}, q_1),$$

which is impossible because the points of D_r would then not be contained in any region of the Voronoi diagram $V(\{q_1, q_2, \ldots, q_{m-1}\})$. □

Since relation $\longrightarrow$ is acyclic it can be embedded in a linear order which is afterwards extended in an arbitrary way to all those curves passing through α that do not belong to the domain of $\longrightarrow$. Let $(\widetilde{S}, \widetilde{\mathcal{J}})$ denote the curve system where the curves of $\mathcal{J}$ formerly running through α have been bent in an appropriate way, according to the order just defined.

Claim 5 *For each non-empty set $T \subset S$, the regions $R(p, \widetilde{T})$ form a partition of the plane.*

Proof: Due to Lemma 3.1 we need only prove this if $T = \{p, q, r\}$ is of size 3. If none, or only one, of the associated bisecting curves of $\mathcal{J}$ is passing α, the assertion follows immediately (in the latter case the bent curve in $\widetilde{\mathcal{J}}$ doesn't interfere with its two companions.

Assume that two or three curves of $\mathcal{J}$ associated with T contain α. Then Claim 1 guarantees that the neighborhood of v looks the same in $V(T)$ as in $V(\widetilde{T})$, so that each point is contained in a Voronoi region. □

It is also clear that we have not caused a region $R(p, \widetilde{T})$ to be disconnected, if $|T| = 3$. Hence, the curve system of T in $\widetilde{\mathcal{J}}$ is admissible and fulfills property 2) of the theorem. Now let $T \subseteq S$ be of arbitrary size ≥ 4. Let $p_1, p_2, \ldots, p_m$ denote the sites of the regions of $V(\widetilde{T})$ as they now appear where α has been before, see Figure 8.

Assume that $m \ge 4$. Let $Q_1 = \{p_1, p_2, p_3\}$ and $Q_2 = \{p_2, p_3, p_4\}$. Then in $V(\widetilde{Q_1})$ (resp. $V(\widetilde{Q_2})$) the regions of p_1, p_2, p_3 (resp. p_2, p_3, p_4) appear in the same order. Since this must be the original order in $V(Q_1)$ and $V(Q_2)$, we conclude $p_2 < p_1, p_3$, and $p_3 < p_2, p_4$, which is a contradiction. Hence, $m \le 3$. This implies that the regions in $V(\widetilde{T})$ are still pathwise connected, and that property 2) holds for T. Thus, the system $(\widetilde{S}, \widetilde{\mathcal{J}})$ is admissible, fulfills 2), and the wires that have formerly belonged to cable α are separated now.

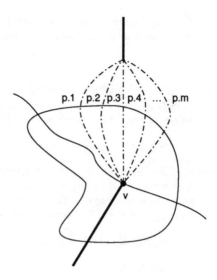

Figure 8: Not a possible neighborhood of v in $V(\tilde{T})$.

This completes the proof of Lemma 4.1. □

Clearly, the way two wires get separated is the same at each point v of a cable α because it depends on the global order $<$ on the set of sites, S. If we apply Lemma 4.1 to each point of $V(S)$ that is not an endpoint of a connected component of an intersection of curves we arrive at a new family of admissible curves that do no longer share segments. So far, the Voronoi diagrams obtained from the subfamilies have not been changed (up to deformations).

Next, we remove those singularities that arise from two curves, J and J', that touch (but do not cross) at a point v, as depicted in Figure 9. This is done by simply pulling off one of them, say J. The point v is thereby streched and becomes a cable, α, connecting two points, w and w', of J and J', respectively. Exactly the curves that were formerly crossing J are now running through α.

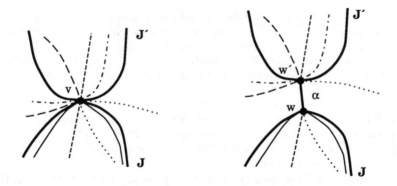

Figure 9: Separating curves that touch at a point.

To show that the resulting curve system is still admissible we apply Theorem 3.4. First, the above deformation has not introduced additional crosspoints. Moreover, each point z is still contained in a Voronoi region. This is obvious if z is not contained in a bisecting curve, because a point of the same relative position with respect to all the curves has existed before. The same holds for all other points that are not contained in the closed segment α. Now the following lemma is helpful.

Lemma 4.3 *Let $(S, \mathcal{J})$ be a system of bisecting curves, where $|S| = 3$, and assume that each point not contained in a bisecting curve belongs to a Voronoi region. Then the full plane is covered by the regions, unless the situation looks as in the first picture of Figure 10.*

Proof: Let $S = \{p, q, r\}$. If $z \in J(p, q)$ belongs to the other two curves, too, then it is contained in a region, due to Definition 2.1. If z belongs to no other curve two cases arise: A piece of $J(p, q)$ containing z is common border of p–land and q–land, or it is contained in r–land. In either case, this piece belongs to a Voronoi region. In the remaining case, z lies in—say—$J(p, r) \cap D(q, r)$. If $p < q \& p < r$ then $z \in R(p, S)$. If $q < p$ then $z \in R(q, S)$. Assume $r < p < q$. In a neighborhood of z only points of $R(p, S) \cup R(q, S)$ can occur (since $z \in D(q, r)$). By assumption, the points outside of $J(p, q)$ do belong to one of the regions. Since $J(p, q)$ passes through z we are in the situation of Figure 10. Note that $J(p, r)$ cannot run through $D(p, q)$ or there would be points of $R(r, S)$ at z. In fact, the segment β of $J(p, q) \cap J(p, r)$ that contains z does not belong to any Voronoi region. $\square$

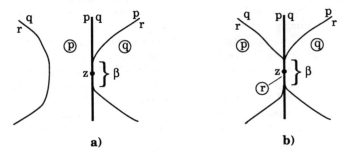

a)   b)

Figure 10: Assume $r < p < q$. In a), no region contains β. In b), β belongs to $R(r, \{p, q, r\})$.

The exceptional case of Lemma 4.3 can in fact result from separating touching curves, see the second picture of Figure 10. But here β is a connected component of $R(r, \{p, q, r\})$, a situation impossible for an admissible curve system (a region is connected and contains a site in its interior).

Thus, we have shown that the "pull" deformation described above leads to an admissible curve system. Clearly, the Voronoi diagrams resulting from the new system may contain α as an additional edge. If we shrink α into v we obtain diagrams that are homeomorphic to the original ones.

Theorem 4.4 *Each admissible family $(S, \mathcal{J})$ of bisecting curves can be deformed in such a way that the resulting system $(\widetilde{S}, \widetilde{\mathcal{J}})$ has the following properties.*

1. *$(\widetilde{S}, \widetilde{\mathcal{J}})$ is admissible.*

2. *No other intersections than crosspoints exist between the curves in $\widetilde{\mathcal{J}}$. At most 3 curves cross at a point (general position).*

3. *For each non–empty subset $S' \subseteq S$: After contracting some edges, $V(\widetilde{S'})$ differs from $V(S')$ only by a deformation.*

Proof: If we apply Lemma 4.1 first and then perform a pull operation on each point where two or more curves touch, the total number of touchpoints may increase. But each time a point passed by m curves is pulled apart, the two resulting points are passed by at most $m - 1$ bisecting curves; cf. Figure 9. Therefore, iterating this process leads to a system without touchpoints, after finitely many steps. If the pull operation is also applied to points crossed by more than 3 curves, we arrive at a curve system that is in general position. □

It is very instructive (and not too time consuming) to apply this algorithm to the Euclidean bisector system of 4 cocircular points.

5   Concluding remarks

We have shown that the admissibility of a system of bisecting curves separating n sites can be verified by inspecting only $O(n^3)$ subproblems of size 3. Up to now, it was necessary to look at exponentially many subproblems of size between 1 and n. Our result also reveals essential combinatorial properties common to the majority of Voronoi diagrams studied in the literature.

Furthermore, we have shown that each admissible curve system can be brought into general position. Algorithms for the computation of abstract Voronoi diagrams greatly benefit from this property. Though our proof is constructive, it does not immediately provide us with an efficient preprocessing algorithm. For, the procedure described in Theorem 4.4 can take an exponentially many steps and must be applied to an unknown number of points; cf. the remark following the proof of Theorem 3.4. On the other hand, deforming curves needs a primitive that allows to draw a cross–cut through a given domain. This operation (which is also needed if one wants to apply Kirkpatrick's triangulation technique [3] to abstract Voronoi diagrams) is *not* as elementary as determining intersections.

However, future algorithms may profit from the techniques described above by applying them *lazily*. The computation of one Voronoi diagram, $V(S)$ usually involves constructing the diagram $V(S')$ for certain subsets S' of S but not for all possible subsets. Intersections between curves that belong to disjoint pairs of sites may be disregarded. Finally, only a linear number of edges and vertices actually appears in a Voronoi diagram, so that it may not be necessary to desingularize the full curve system.

References

[1] F. Aurenhammer. Voronoi diagrams—a survey. Graz Technical University, 1988.

[2] K. L. Clarkson and P. W. Shor. Algorithms for diametral pairs and convex hulls that are optimal, randomized, and incremental. In *Proc. 4th ACM Symp. on Computational Geometry*, pages 12–17, Urbana–Champaign, 1988.

[3] D. G. Kirkpatrick. Optimal search in planar subdivisions. *SIAM J. Comput.*, 12(1):28–35, 1983.

[4] R. Klein. Abstract Voronoi diagrams and their applications (extended abstract). In H. Noltemeier, editor, *Computational Geometry and its Applications (CG '88)*, pages 148–157, Würzburg, 1988. LNCS 333.

[5] R. Klein. *On a generalization of planar Voronoi diagrams*. University of Freiburg, Habilitationsschrift, 1988. To appear in LNCS.

[6] R. Klein. Voronoi diagrams in the Moscow metric (extended abstract). In J. van Leeuwen, editor, *Graphtheoretic Concepts in Computer Science (WG '88)*, pages 434–441, Amsterdam, 1988. LNCS 344.

[7] R. Klein and D. Wood. Voronoi diagrams based on general metrics in the plane. In R. Cori and M. Wirsing, editors, *Proc. 5th Ann. Symp. on Theoretical Aspects of Computer Science (STACS)*, pages 281–291, Bordeaux, 1988. LNCS 294.

[8] R. Klein, K. Mehlhorn, and St. Meiser. Five sites are enough. Manuscript, University of Saarbrücken, 1989.

[9] K. Mehlhorn, C. Ó'Dúnlaing, and St. Meiser. On the construction of abstract Voronoi diagrams. Presented at *Computational Geometry and its Applications (CG '89)*, Freiburg, 1989. Submitted for publication.

List of Participants of WG'89

Dr. Balcazar, Jose L., Departament de Llenguates i Sistemes Informatics, UPC, Edifici FIB, Pau Gargallo 5, E-08028 Barcelona, Spain.

Dr. Bein, Wolfgang W., Dept. of Computer Science, The University of New Mexico, Farris Engineering Center, Rm. 312, Albuquerque, NM 87131, USA.

de Berg, Mark, Dept. of Computer Science, University of Utrecht, P.O. Box 80.089, NL-3508 TB Utrecht, the Netherlands.

Dr. Bodlaender, Hans, Dept. of Computer Science, University of Utrecht, P.O. Box 80.089, NL-3508 TB Utrecht, the Netherlands.

Prof. Dr. Brandenburg, Franz J., Lehrstuhl f. Informatik , Universität Passau, Postfach 2540, D-8390 Passau.

Prof. Dr. Bunke, Horst, Institut f. Informatik u. Angew. Math., Universität Bern, Länggassstr. 51, CH-3012 Bern, Switzerland.

Das, A., H-915, Dept. of Electr. Engg., Concordia University, 1455 de Maisonneuve Blvd. W.w., Montreal, Canada, H3G 1M8.

Prof. Dr. Dehne, Frank, School of Computer Science, Herzberg Bldg., Carleton University, Ottawa, Canada, K1S 5B6.

Gargano, Luisa, Dipartimento di Informatica ed Applicazioni, Universita di Salerno, I-84081 Baronissi (SA), Italy.

Guan, Yong Gang, Lehrstuhl f. Informatik, F.B. Informatik, Universität des Saarlandes, Am Stadtwald, D-6600 Saarbrücken.

Dr. Habel, Annegret, Fachbereich Mathematik u. Informatik, Universität Bremen, Bibliothekstraße, D-2800 Bremen.

Henrich, Andreas, Fachbereich Mathematik u. Informatik, Prakt. Informatik III, Fernuniversität Gesamthochschule, Feithstr. 140, D 5800-Hagen.

Prof. Dr. Indermark, Klaus, Lehrstuhl f. Informatik II, RWTH Aachen, Ahornstr. 55, D-5100 Aachen.

Dr. Icking, Christian, Institut f. Informatik, Universität Freiburg, Rheinstr. 10 - 12, D-7800 Freiburg.

Kahan, Simon, Computer Science Dept., FR-35 University of Washington, Seattle, WA 98195, USA.

PD Dr. Klein, Rolf, Institut f. Informatik, Universität Freiburg, Rheinstr. 10 - 12, D-7800 Freiburg.

Kloks, A.J.J., Dept. of Computer Science, University of Utrecht, P.O. Box 80.089, NL–3508 TB Utrecht, the Netherlands.

Prof. Dr. Knödel, Walter, Institut f. Informatik, Universität Stuttgart, Azenbergstr. 12, D–7000 Stuttgart.

Prof. Dr. Kreowski, Hans–Jörg, FB Mathematik und Informatik, Universität Bremen, Bibliothekstr., D–2800 Bremen.

van Kreveld, Marc, Dept. of Computer Science, University of Utrecht, P.O. Box 80.089, NL–3508 TB Utrecht, the Netherlands.

Lawry, Joan, Computer Science Dept., FR–35, University of Washington, Seattle, WA 98195, USA.

La Poutre, J.A., Dept. of Computer Science, University of Utrecht, P.O. Box 80.089, NL–3508 TB Utrecht, the Netherlands.

Prof. Dr. van Leeuwen, Jan, Dept. of Computer Science, University of Utrecht, P.O. Box 80.089, NL–3508 TB Utrecht, the Netherlands.

Dr. Lefmann, Hanno, IBM Deutschland GmbH., Wissenschaftl. Zentrum, Tiergartenstr. 15, D–6900 Heidelberg.

Lozano, Antonio, Department de Llenguates i Sistemes Informatics, UPC, Edifici FIB, Pau Gargallo 5, E–008028 Barcelona, Spain.

Madhavapeddy, Seshu, Computer Science Dept., The University of Texas at Dallas, MS MP 3.1, Box 830688, Richardson, Texas 75083–0688, USA.

Menzel, Knut, Fachbereich 17, Universität Paderborn, D–4790 Paderborn.

Metivier, Y., Department d'Informatique, Universite de Bordeaux I, 351 Cours de la Liberation, F–33405 Talence Cedex, France.

Prof. Dr. Möhring, Rolf H., FB 3 Mathematik Sekr. MA 6- 1, Technische Universität Berlin, Straße des 17. Juni 136, D–1000 Berlin 12.

Prof. Dr. Nagl, Manfred, Lehrstuhl f. Informatik III, RWTH Aachen, Ahornstr. 55, D–5100 Aachen.

Prof. Dr. Noltemeier, Hartmut, Lehrstuhl f. Informatik, Universität Würzburg, Am Hubland, D–8700 Würzburg.

Overmars, Mark H., Dept. of Computer Science, University of Utrecht, P.O. Box 80.089, NL–3508 TB Utrecht, the Netherlands.

Prof. Dr. Paz, Azaria, Technion, Israel Institute of Technology, Computer Science Dept., Technion City, Haifa 32 000, Israel.

Reich, Gabriele, Institut f. Informatik, Universität Freiburg, Rheinstr. 10 - 12, D–7800 Freiburg.

Roos, Thomas, Lehrstuhl f. Informatik I, Universität Würzburg, Am Hubland, D–8700 Würzburg.

Dr. Ruland, Detlev, Lehrstuhl f. Informatik I, Universität Würzburg, Am Hubland, D–8700 Würzburg.

Dr. Schiermeyer, Ingo, Lehrstuhl C für Mathematik, RWTH Aachen, Templergraben 55, D–5100 Aachen.

Prof. Dr. Schneider, Hans–J., Lehrstuhl f. Programmiersprachen, Universität Erlangen–Nürnberg, Martensstr. 3, D–5100 Aachen.

Prof. Dr. Schmidt, Gunther, Fakultät f. Informatik, Universität der Bundeswehr, Werner–Heisenberg–Weg 39, D–8014 Neuiberg.

Schürr, Andy, Lehrstuhl f. Informatik III, RWTH Aachen, Ahornstr. 55, D–5100 Aachen.

Schulhoff, Robert, Technion, Israel Institute of Technology, Computer Science Dept., Technion City, Haifa 32 000, Israel.

Dr. Simon, Klaus, Institut f. Theoretische Informatik, ETH Zürich, ETH–Zentrum, CH–8092 Zürich, Switzerland.

Prof. Dr. Six, H.–W., FB Mathematik u. Informatik, Lehrgebiet Prakt. Informatik III, Fernuniversität, Gesamthochscule, Postfach 940, D–5800 Hagen 1.

Sopena, Eric, I.U.T. "A" Department Informatique, F–33405 Talence Cedex, France.

Dr. Speckenmeyer, Ewald, FB Informatik, Lehrstuhl Vi, Universität Dortmund, Postfach 500 500, D–4600 Dortmund 50.

Dr. Syslo, Maciej M., Institute of Computer Science, University of Wroclaw, Przesmyciego 20, PL–51–151 Wroclaw, Poland.

Prof. Dr. Thomas, Wolfgang, Lehrstuhl f. Informatik II, RWTH Aachen, Ahornstr. 55, D–5100 Aachen.

Prof. Dr. Vaccaro, Ugo, Dipartimento di Informatica ed Applicazioni, Universita di Salerno, I–84081 Baronissi (SA), Italy.

Dr. Vogler, Walter, Institut f. Informatik, Technische Universität Arcisstr. 21, D–8000 München 2.

Prof. Dr. Widmayer, Peter, Institut f. Informatik, Universität Freiburg, Rheinstr. 10 – 12, D–7800 Freiburg.

Authors' Index

Proceedings of the Workshops on
Graph–Theoretic Concepts in Computer Science
WG'76 – WG'89

The Proceedings of these annual workshops have always been published. Please check whether your library (research group library, institute library, department library, university library, company library) possesses a complete series of volumes. To facilitate ordering of missing volumes please use the following list.

U. Pape (Ed.): *Graph Languages and Algorithms (Proc. WG'75)*, 236 p., Munich: Hanser-Verlag (1976).

H. Noltemeier (Ed.): *Graphs, Algorithms, Data Structures (Proc. WG'76)*, 336 p., Munich: Hanser-Verlag (1977). ISBN 3–446–12330–X

J. Mühlbacher (Ed.): *Data Structures, Graphs, Algorithms, (Proc. WG' 77)*, 368 p., Munich: Hanser-Verlag (1978). ISBN 3–446–12526–4

M Nagl, H. J. Schneider (Eds.): *Graphs, Data Structures, Algorithms, (Proc. WG'78)*, 320 p., Munich: Hanser-Verlag (1979). ISBN 3–446–12748–8

U. Pape (Ed.): *Discrete Structures and Algorithms (Proc. WG'79)*, 270 p., Munich: Hanser-Verlag (1980). ISBN 3–446–13135–3

H. Noltemeier (Ed.): *Graphtheoretic Concepts in Computer Science (Proc. WG'80)*, Lecture Notes in Computer Science 100, 403 p., Berlin/New York: Springer Verlag (1981). ISBN 0–387–10291–4

J. Mühlbacher (Ed.): *Proceedings WG'81*, 355 p., Munich: Hanser-Verlag (1982). ISBN 3–446–13538–3

H. J. Schneider, H. Göttler (Eds.): *Proceedings WG'82*, 280 p., Munich: Hanser-Verlag (1982). ISBN 3–446–13778–5

M. Nagl, J. Perl (Eds.): *Proceedings WG'83*, 397 p., Linz: Trauner-Verlag (1983). ISBN 3–85320–311–6

U. Pape (Ed.): *Proceedings WG'84*, 381 p., Linz: Trauner-Verlag (1984). ISBN 3–85320–334–5

H. Noltemeier (Ed.): *Proceedings WG'85*, 443 p., Linz: Trauner-Verlag (1985). ISBN 3–85320–357–4

G. Tinhofer/G.Schmidt (Eds.): *Proceedings WG'86*, 305 p., Lecture Notes in Computer Science 246, Berlin/New York: Springer-Verlag (1987). ISBN 0–387–17218–1

H. Göttler/H. J. Schneider (Eds.): *Proceedings WG'87*, 254 p., Lecture Notes in Computer Science 314, Berlin/New York: Springer-Verlag (1988). ISBN 0–387–19422–3

J. van Leeuwen (Ed.): *Proceedings WG'89*, 457 p., Lecture Notes in Computer Science 344, Berlin/New York: Springer-Verlag (1989). ISBN 0–387–50728–0

M. Nagl (Ed.): *Proceedings WG'89*, Lecture Notes in Computer Science, Berlin/New York: Springer-Verlag (1989).